Lecture Notes in Physics

Springer
Berlin
Heidelberg
New York
Hong Kong
London
Milan
Paris
Tokyo

The Editorial Policy for Edited Volumes

The series *Lecture Notes in Physics* (LNP), founded in 1969, reports new developments in physics research and teaching - quickly, informally but with a high degree of quality. Manuscripts to be considered for publication are topical volumes consisting of a limited number of contributions, carefully edited and closely related to each other. Each contribution should contain at least partly original and previously unpublished material, be written in a clear, pedagogical style and aimed at a broader readership, especially graduate students and nonspecialist researchers wishing to familiarize themselves with the topic concerned. For this reason, traditional proceedings cannot be considered for this series though volumes to appear in this series are often based on material presented at conferences, workshops and schools.

Acceptance

A project can only be accepted tentatively for publication, by both the editorial board and the publisher, following thorough examination of the material submitted. The book proposal sent to the publisher should consist at least of a preliminary table of contents outlining the structure of the book together with abstracts of all contributions to be included. Final acceptance is issued by the series editor in charge, in consultation with the publisher, only after receiving the complete manuscript. Final acceptance, possibly requiring minor corrections, usually follows the tentative acceptance unless the final manuscript differs significantly from expectations (project outline). In particular, the series editors are entitled to reject individual contributions if they do not meet the high quality standards of this series. The final manuscript must be ready to print, and should include both an informative introduction and a sufficiently detailed subject index.

Contractual Aspects

Publication in LNP is free of charge. There is no formal contract, no royalties are paid, and no bulk orders are required, although special discounts are offered in this case. The volume editors receive jointly 30 free copies for their personal use and are entitled, as are the contributing authors, to purchase Springer books at a reduced rate. The publisher secures the copyright for each volume. As a rule, no reprints of individual contributions can be supplied.

Manuscript Submission

The manuscript in its final and approved version must be submitted in ready to print form. The corresponding electronic source files are also required for the production process, in particular the online version. Technical assistance in compiling the final manuscript can be provided by the publisher's production editor(s), especially with regard to the publisher's own LaTeX macro package which has been specially designed for this series.

LNP Homepage (springerlink.com)

On the LNP homepage you will find:
—The LNP online archive. It contains the full texts (PDF) of all volumes published since 2000. Abstracts, table of contents and prefaces are accessible free of charge to everyone. Information about the availability of printed volumes can be obtained.
—The subscription information. The online archive is free of charge to all subscribers of the printed volumes.
—The editorial contacts, with respect to both scientific and technical matters.
—The author's / editor's instructions.

W. Hergert A. Ernst M. Däne (Eds.)

Computational Materials Science

From Basic Principles to Material Properties

 Springer

Editors

Prof. W. Hergert
M. Däne
Martin-Luther-Universität Halle-Wittenberg
Fachbereich Physik, Fachgruppe Theoretische Physik
Von-Seckendorff-Platz 1
06120 Halle, Germany

Dr. A. Ernst
Max Planck Institute for Microstructure Physics
Weinberg 2
06120 Halle, Germany

W. Hergert A. Ernst M. Däne (Eds.), *Computational Materials Science*, Lect. Notes Phys.
642 (Springer, Berlin Heidelberg 2004), DOI 10.1007/b11279

Library of Congress Control Number: 2004102310

Bibliographic information published by Die Deutsche Bibliothek Die Deutsche Bibliothek
lists this publication in the Deutsche Nationalbibliografie; detailed bibliographic data is
available in the Internet at <http://dnb.ddb.de>

ISSN 0075-8450
ISBN 3-540-21051-2 Springer-Verlag Berlin Heidelberg New York

Springer-Verlag is a part of Springer Science+Business Media

springeronline.com

© Springer-Verlag Berlin Heidelberg 2004
Printed in Germany

The use of general descriptive names, registered names, trademarks, etc. in this publication
does not imply, even in the absence of a specific statement, that such names are exempt
from the relevant protective laws and regulations and therefore free for general use.

Typesetting: Camera-ready by the authors/editor
Data conversion: PTP-Berlin Protago-TeX-Production GmbH
Cover design: *design & production*, Heidelberg

Printed on acid-free paper
54/3141/ts - 5 4 3 2 1 0

Preface

Computational modelling of novel materials is an increasingly powerful tool being used in the development of advanced materials and their device applications. Computational materials science is a relatively new scientific field, in which known concepts and recent advancements in physics, chemistry, mathematics and computer science are combined and applied numerically. The unique advantage of such modelling lies in the possibility to predict macroscopic properties of materials based on calculations of microscopic quantities, i.e., at the atomic level. This has been made possible by the spectacular increase in computational power over recent decades, allowing us to solve numerically and with unprecedented accuracy, fundamental equations at the atomic level. Today, based only on our knowledge of a single atom, we can predict how the material formed by that atom type will look, what properties that material will have and how it will behave under certain conditions. By simply changing the arrangement of constituent atoms, or by adding atoms of a different type, the macroscopic properties of all materials can be modified. It is in this way that one can learn how to improve mechanical, optical and/or electronic properties of known materials, or one can predict properties of new materials, those which are not found in nature but are designed and synthesized in the laboratory. Supercomputers (both vector and parallel) and modern visualization techniques are utilized to generate direct comparisons with experimental conditions, and in some cases experiments may become redundant.

The authors of this book have endeavoured to give an overview of the techniques, which operate at various levels of sophistication to describe microscopic and macroscopic properties of wide range of materials. The most important methods used today in computational physics are addressed and, in general, each topic is illustrated by a number of applications. The book starts with basic aspects of density functional theory and the discussion of modern methods to calculate the electronic structure of materials. A rapidly developing field of scientific interest over the last years is nanophotonics. Two articles discuss how properties of photonic nanostructures can be computed. The main part of the book contains contributions dealing with different aspects of simulation methods. Ab initio calculations of free and supported molecules and clusters are discussed. The application of molecular-dynamics in biology, chemistry and physics is studied. The articles give a representative cross section of different simulation methods on the one hand and of their application to different materials on the other hand. Essential for the field of the Computational Material Science is the availability of effective algorithms and numerical methods. Therefore multigrid methods and strategies for the implementation of sparse and irregular algorithms are discussed as well.

The editors are grateful to the authors for their valuable contributions to the book. The chapters are based, to some extent, on lectures given at the WE-Heraeus course of the same name as the present book, held from 16th to 27th September 2002 in Halle. We gratefully acknowledge the support of the Wilhelm und Else Heraeus Stiftung.

Halle/Saale,
January 2004

Wolfram Hergert
Arthur Ernst
Markus Däne

Table of Contents

1 Introduction 1

Part I Basic Description of Electrons and Photons in Crystals

2 The Essentials of Density Functional Theory and the Full-Potential Local-Orbital Approach
H. Eschrig 7
2.1 Density Functional Theory in a Nutshell 7
2.2 Full-Potential Local-Orbital Band Structure Scheme (FPLO) ... 11
    2.2.1 The Local Orbital Representation 12
    2.2.2 Partitioning of Unity 14
    2.2.3 Density and Potential Representation 14
    2.2.4 Basis Optimization 15
    2.2.5 Examples 15
    2.2.6 Comparison of Results from FPLO and WIEN97 19

3 Methods for Band Structure Calculations in Solids
A. Ernst, M. Lüders 23
3.1 The Green's Function and the Many-Body Method 23
    3.1.1 General Considerations............................... 23
    3.1.2 Quasi-Particles...................................... 29
    3.1.3 Self-Energy 31
    3.1.4 Kohn-Sham Approximation for the Self-Energy.......... 34
3.2 Methods of Solving the Kohn-Sham Equation 39
3.3 GW Approximation 47

4 A Solid-State Theoretical Approach to the Optical Properties of Photonic Crystals
K. Busch, F. Hagmann, D. Hermann, S.F. Mingaleev, M. Schillinger . 55
4.1 Introduction 55
4.2 Photonic Bandstructure Computation........................ 57
4.3 Defect Structures in Photonic Crystals 59
    4.3.1 Maximally Localized Photonic Wannier Functions 60
    4.3.2 Defect Structures via Wannier Functions 61

4.3.3   Localized Cavity Modes 62
4.3.4   Dispersion Relations of Waveguides 64
4.3.5   Photonic Crystal Circuits 66
4.4   Finite Photonic Crystals 68
4.5   Conclusions and Outlook 71

5 Simulation of Active
and Nonlinear Photonic Nano-Materials
in the Finite-Difference Time-Domain (FDTD) Framework
A. Klaedtke, J. Hamm, O. Hess................................. 75
5.1   Introduction 75
5.2   Finite-Difference in Time-Domain 76
5.3   Uniaxial Perfectly Matching Layers (UPML) Boundary Conditions   80
5.4   Time-Domain Full Vectorial Maxwell-Bloch Equations 87
5.5   Computational Costs 90
5.6   Test Runs 92
5.7   Microdisk Laser Dynamics................................ 95
5.8   Conclusion 98
        Appendix A: Relations 99

6 Symmetry Properties
of Electronic and Photonic Band Structures
W. Hergert, M. Däne, D. Ködderitzsch 103
6.1   Introduction 103
6.2   Group Theory Packages for Computer Algebra Systems 104
6.3   Basic Concepts in Group Theory 105
6.4   Representation Theory 106
        6.4.1   Matrix Representations of Groups 106
        6.4.2   Basis Functions of Irreducible Representations 109
6.5   Symmetry Properties
        of Schrödinger's Equation and Maxwell's Equations............. 110
6.6   Consequences of Lattice Periodicity 111
6.7   Electronic Band Structure 115
        6.7.1   Compatibility Relations 116
        6.7.2   Symmetry-Adapted Basis Functions 118
6.8   Discussion of Photonic Band Structures 120
        6.8.1   Assignment of the IRs to the Photonic Band Structure ... 120
6.9   Conclusions.. 123

Part II Simulation from Nanoscopic Systems to Macroscopic Materials

7 From the Cluster to the Liquid: Ab-Initio Calculations on Realistic Systems Based on First-Principles Molecular Dynamics

C. Massobrio, M. Celino, Y. Pouillon, I.M.L. Billas 129
7.1 Introduction 129
7.2 Theoretical Methods 133
    7.2.1 First-Principles Molecular Dynamics 133
    7.2.2 Details of Calculations 134
7.3 Selected Applications to Clusters and Disordered Systems 135
    7.3.1 CuO_n Clusters 135
    7.3.2 Si-Doped Heterofullerenes $C_{59}Si$ and $C_{58}Si_2$ 140
    7.3.3 Disordered Network-Forming Materials: Liquid $SiSe_2$ 146
7.4 Concluding Remarks 154

8 Magnetism, Structure and Interactions at the Atomic Scale

V.S. Stepanyuk, W. Hergert 159
8.1 Introduction 159
8.2 Theoretical Methods 160
    8.2.1 Calculation of Electronic Structure 160
    8.2.2 Molecular Dynamics Simulations 161
8.3 Magnetic Properties of Nanostructures on Metallic Surfaces..... 163
    8.3.1 Metamagnetic States of 3d Nanostructures on the Cu(001)Surface 164
    8.3.2 Mixed Co-Cu Clusters on Cu(001) 166
    8.3.3 Effect of Atomic Relaxations on Magnetic Properties of Adatoms and Small Clusters........................ 166
8.4 Quantum Interference and Interatomic Interactions 169
8.5 Strain and Stress on the Mesoscale 171
    8.5.1 The Concept of Mesoscopic Misfit 171
    8.5.2 Strain and Adatom Motion on Mesoscopic Islands 172
    8.5.3 Mesoscopic Relaxation in Homoepitaxial Growth 172

9 Molecular Dynamics Simulations in Biology, Chemistry and Physics

P. Entel, W.A. Adeagbo, M. Sugihara, G. Rollmann, A.T. Zayak, M. Kreth, K. Kadau... 177
9.1 Molecular Dynamics as a Multidisciplinary Numerical Tool 177
9.2 Simulation of Biochemical Systems 180
    9.2.1 Molecular Dynamics Simulation of Liquid Water 181
    9.2.2 Simulation of β-Cyclodextrin-Binaphtyl and Water 187
    9.2.3 Simulation of Bovine Rhodopsin...................... 189

9.3   Simulation of Chemical Reactions in the Gas Phase   193
9.4   Simulation of Structural Transformations in Solids and Particles.   196
      9.4.1   Simulation of the Phase Diagram
              of Fe-Ni and Ni-Mn-Ga Alloys   196
      9.4.2   Simulation of the Structural Transformation
              in Fe-Ni Particles....................................   198
      9.4.3   Simulation of the Melting of Al Clusters...............   201

10 Computational Materials Science with Materials Studio®:
Applications in Catalysis
M.E. Grillo, J.W. Andzelm, N. Govind, G. Fitzgerald, K.B. Stark....   207
10.1   Introduction   208
10.2   Geometry Optimization in Delocalised Internal Coordinates   208
10.3   Transition State Searching.................................   213
10.4   Transition State Confirmation Algorithm   214
10.5   Chemical Bonding and Elastic Properties
       of Corundum-Type Oxides: The Rhodium Oxide Case   217
10.6   Summary...   220

11 Integration of Modelling
at Various Length and Time Scales
S. McGrother, G. Goldbeck-Wood, Y.M. Lam   223
11.1   Introduction   223
11.2   Structure-Activity and Structure-Property Approaches   225
11.3   Atomistic and Mesoscale Simulations and Their Parameterisation   225
       11.3.1 Atomistic Simulation   226
       11.3.2 Mesoscale Methods   227
       11.3.3 Applications of Mesoscale Modeling...................   229
11.4   Multiscale Modeling   230
       11.4.1 From the Molecular to the Mesoscale   230
       11.4.2 From Mesoscale to Finite Element Simulation   231
11.5   Conclusion   232

12 Simulation of the Material Behavior
from the Engineering Point of View – Classical Approaches
and New Trends
H. Altenbach   235
12.1   Introduction   235
12.2   Principles of Modelling....................................   238
12.3   Phenomenological Models   240
12.4   Classical and Nonclassical Material Behavior Models...........   244
12.5   Analysis of Thin-Walled Structures   248

Part III Modern Methods of Scientific Computing

13 Parallel Implementation Strategies for Algorithms from Scientific Computing

T. Rauber, G. Rünger 261

13.1 Introduction 261

13.2 A Short Introduction to MPI 263

13.3 Modeling the Execution Time of MPI Operations............. 266

13.4 Example: Solving Systems of Linear Equations 267

    13.4.1 Standard Iterative Methods........................... 268

    13.4.2 Sparse Iteration Matrices............................ 269

    13.4.3 Red-Black Ordering 272

13.5 Task and Data Parallel Execution 274

    13.5.1 Overview of the Tlib Library 275

    13.5.2 Example: Strassen Matrix Multiplication 278

14 Multi-Grid Methods – An Introduction

G. Wittum 283

14.1 Introduction 283

    14.1.1 Historical Overview and Introduction
    to Multi-Grid Methods............................... 283

    14.1.2 Additive Multigrid 290

14.2 Convergence Theory 291

    14.2.1 General Setting 291

    14.2.2 The Smoothing Property 293

    14.2.3 Approximation Property 295

14.3 Robustness 297

    14.3.1 Robustness for Anisotropic Problems 297

    14.3.2 Robustness for Convection-Diffusion Problems 302

14.4 Treatment of Systems of PDE............................... 306

14.5 Adaptive Multigrid 307

Index 313

Part III. Modern Methods of Bayesian Computation

18. Parallel Implementation Strategies for Algorithms and
Reductive Computing

List of Contributors

W.A. Adeagbo
Universität Duisburg-Essen
Institut für Physik,
Theoretische Tieftemperaturphysik
Duisburg Campus, Lotharstraße 1
47048 Duisburg, Germany
adeagbo@thp.uni-duisburg.de

H. Altenbach
Martin-Luther-Universität
Halle-Wittenberg
Fachbereich Ingenieurwissenschaften
06099 Halle, Germany
silvia.runkel@iw.uni-halle.de

J. W. Andzelm
Accelrys Inc.
9685 Scranton Rd
San Diego, CA 92121, USA
jwa@accelrys.com

I.M.L. Billas
Département de Biologie
et de Génomique Structurales,
Institut de Génetique et de Biologie
Moléculaire et Cellulaire,
1 rue Laurent Fries, BP 10142
67404 Illkirch, France
billas@igbmc.u-strasbg.fr

K. Busch
Department of Physics and School
of Optics: CREOL & FPCE,
University of Central Florida,
Orlando, FL 32816, USA
kbusch@physics.ucf.edu

M. Celino
Ente per le Nuove Tecnologie,
l'Energia e l'Ambiente,
C.R. Casaccia, CP 2400
00100 Roma, Italy
and
Istituto Nazionale per la Fisica
della Materia,
Unità di Ricerca Roma1, Italy
Massimo.Celino@casaccia.enea.it

M. Däne
Martin-Luther-Universität
Halle-Wittenberg
Fachbereich Physik,
Fachgruppe Theoretische Physik
Von-Seckendorff-Platz 1
06120 Halle, Germany
daene@physik.uni-halle.de

P. Entel
Universität Duisburg-Essen
Institut für Physik,
Theoretische Tieftemperaturphysik
Duisburg Campus, Lotharstraße 1
47048 Duisburg, Germany
entel@thp.uni-duisburg.de

A. Ernst
Max Planck Institute
of Microstructure Physics
Weinberg 2
06120 Halle, Germany
aernst@mpi-halle.mpg.de

H. Eschrig
Leibniz Institute for Solide State
and Materials Research Dresden
PF 27 01 16
01171 Dresden, Germany
h.eschrig@ifw-dresden.de

G. Fitzgerald
Accelrys Inc.
9685 Scranton Rd
San Diego, CA 92121, USA
gxf@accelrys.com

G. Goldbeck-Wood
Accelrys Inc
334 Science Park
Cambridge CB4 0WN, UK
Goldbeck-Wood@accelrys.com

N. Govind
Accelrys Inc.
9685 Scranton Rd
San Diego, CA 92121, USA
nxg@accelrys.com

M.E. Grillo
Accelrys, GmbH
Inselkammerstr. 1
82008 Unterhaching, Germany
mariag@accelrys.com

F. Hagmann
Institut für Theorie der Kondensierten Materie,
Universität Karlsruhe
76128 Karlsruhe, Germany
hagmann@
tkm.physik.uni-karlsruhe.de

J. Hamm
Advanced Technology Institute
School of Electronics
and Physical Sciences
University of Surrey
GU2 7XH, UK
j.hamm@surrey.ac.uk

W. Hergert
Martin-Luther-Universität
Halle-Wittenberg
Fachbereich Physik, Fachgruppe
Theoretische Physik
Von-Seckendorff-Platz 1
06120 Halle, Germany
hergert@physik.uni-halle.de

D. Hermann
Institut für Theorie
der Kondensierten Materie,
Universität Karlsruhe
76128 Karlsruhe, Germany
hermann@
tkm.physik.uni-karlsruhe.de

O. Hess
Advanced Technology Institute
School of Electronics and Physical
Sciences
University of Surrey
GU2 7XH, UK
o.hess@surrey.ac.uk

K. Kadau
Universität Duisburg-Essen
Institut für Physik,
Theoretische Tieftemperaturphysik
Duisburg Campus, Lotharstraße 1
47048 Duisburg, Germany
kadau@comphys.uni-duisburg.de

A. Klaedtke
Advanced Technology Institute
School of Electronics
and Physical Sciences
University of Surrey
GU2 7XH, UK
a.klaedtke@surrey.ac.uk

D. Ködderitzsch
Martin-Luther-Universität
Halle-Wittenberg
Fachbereich Physik, Fachgruppe
Theoretische Physik
Von-Seckendorff-Platz 1
06120 Halle, Germany
d.koedderitzsch@
physik.uni-halle.de

M. Kreth
Universität Duisburg-Essen
Institut für Physik,
Theoretische Tieftemperaturphysik
Duisburg Campus, Lotharstraße 1
47048 Duisburg, Germany
magnus@thp.uni-duisburg.de

M. Lüders
Daresbury Laboratory,
Daresbury, Warrington,
Cheshire, WA4 4AD, UK
m.lueders@dl.ac.uk

C. Massobrio
Institut de Physique et Chimie
des Matériaux de Strasbourg
23 Rue du Loess BP 43
67034 Strasbourg Cedex 2, France
Carlo.Massobrio@
ipcms.u-strasbg.fr

S. McGrother
Accelrys Inc.
9685 Scranton Rd
San Diego, CA 92121, USA
smcgrother@accelrys.com

S.F. Mingaleev
Institut für Theorie
der Kondensierten Materie,
Universität Karlsruhe
76128 Karlsruhe, Germany
smino@
tkm.physik.uni-karlsruhe.de

Y. Pouillon
Ente per le Nuove Tecnologie,
l'Energia e l'Ambiente,
C.R. Casaccia, CP 2400
00100 Roma, Italy
and
Istituto Nazionale per la Fisica
della Materia,
Unità di Ricerca Roma1, Italy
pouillon@pcpm.ucl.ac.be

T. Rauber
Universität Bayreuth
Fakultät für Mathematik und Physik
95440 Bayreuth, Germany
rauber@uni-bayreuth.de

G. Rollmann
Universität Duisburg-Essen
Institut für Physik,
Theoretische Tieftemperaturphysik
Duisburg Campus, Lotharstraße 1
47048 Duisburg, Germany
georg@thp.uni-duisburg.de

G. Rünger
Technische Universität Chemnitz
Fakultät für Informatik
Professur Praktische Informatik
Straße der Nationen 62
09107 Chemnitz, Germany
ruenger@
informatik.tu-chemnitz.de

M. Schillinger
Institut für Theorie
der Kondensierten Materie,
Universität Karlsruhe
76128 Karlsruhe, Germany
matthias@
tkm.physik.uni-karlsruhe.de

K.B. Stark
Accelrys Inc.
9685 Scranton Rd
San Diego, CA 92121, USA
kstark@accelrys.com

V. Stepanyuk
Max Planck Institute
of Microstructure Physics
Weinberg 2
06120 Halle, Germany
stepanyu@mpi-halle.de

M. Sugihara
Universität Duisburg-Essen
Institut für Physik,
Theoretische Tieftemperaturphysik
Duisburg Campus, Lotharstraße 1
47048 Duisburg, Germany
minoru@thp.uni-duisburg.de

G. Wittum
Interdisziplinäres Zentrum für
Wissenschaftliches Rechnen (IWR)
Simulation in Technology Center
Im Neuenheimer Feld 368
69120 Heidelberg, Germany
wittum@iwr.uni-Heidelberg.de

A.T. Zayak
Universität Duisburg-Essen
Institut für Physik,
Theoretische Tieftemperaturphysik
Duisburg Campus, Lotharstraße 1
47048 Duisburg, Germany
alexei@thp.uni-duisburg.de

1 Introduction

The longstanding goal of computer simulation of materials is gaining accurate knowledge on structural and electronic properties of realistic condensed-matter systems. Since materials are complex in nature, past theoretical investigations were restricted mostly to simple models. Recent progress in computer technology made a realistic description of a wide range of materials possible. Since then the computational material science has rapidly developed and expanded into new fields in science and technologies, and computer simulations have become an important tool in many areas of academic and industrial research, such as physics, chemistry, biology and nanotechnology. This book focuses on the foundations and practical aspects of computational physics.

Most of the methods in computational physics are based on the density functional theory (DFT). The density functional theory deals with inhomogeneous systems of electrons. This approach is based on the theorem of Hohenberg and Kohn which states that the ground state properties of a many-particle system can be exactly represented in terms of the ground state density. This allows to replace the many-particle wave function by the particle density or the current density of the system. The desired ground state quantities can be obtained by minimization of an unique energy functional, which is decomposed into one-electron contributions and the so-called exchange-correlation energy functional, which contains all many-body effects. In practical applications it is usually approximated by some model density functionals. The variational problem can be reduced to an effective one-electron Kohn-Sham equation describing non-interacting electrons in an effective potential and in principle reproducing exact ground-state density. One of the most popular functionals is the local density approximation (LDA), in which all many-body effects are included on the level of the homogeneous electron gas. Such approach enables one to carry out the so-called *"first-principles"* or *"ab-initio"* calculations (direct calculations of material properties from fundamental quantum mechanical theory). Many fundamental properties, for example bond strength and reaction energies, can be estimated from first principles.

By construction the stationary density functional theory is designed for the ground state and can not be expected to describe excited state properties. A classical tool for this problem is the Green's function formalism which is an important technique for studying correlations in many-body systems. The Green's function provides spectral densities for occupied and unoccupied states and its poles can be interpreted as single-particle excitations. To calculate the Green's function is in general a non-trivial task and it requires to solve the Dyson equation which involves a non-interacting Green's function and a self-energy operator. In practice, it is necessary to make approximations for the self-energy which can be handled. In most first-principles methods the self-energy is approximated by a local exchange-correlation potential. In this case the Kohn-Sham eigenvalues are interpreted as excitation energies. This simple approach works surprisingly well in many applications and is widely used in a variety of methods on the first-principles level. However, the electron correlations and many-body effects are not adequately represented by the local Kohn-Sham approximation which fails to describe, for example, band gaps in semiconductors or lifetime in spectroscopies. At an *ab-initio* level the self-energy can be implemented e.g. within the random-phase approximation (RPA). This approximation treats the electron correlations on the basis of many-particle theory and is much more accurate but also much more time-consuming then the Kohn-Sham approach.

The development of first-principles methods is a difficult and challenging task. Firstly, an *"ab-inito"* method should be constructed as general as possible to be applicable to a wide class of problems. Usually, a first principles code consists of many thousand lines. Secondly, a development of such programs requires a deep knowledge of numerical methods and programming tools. Since the majority of physical systems exhibit intrinsic symmetries, one should use a symmetry analysis and the group theory to optimize and to speed up computational process. The group theory as a mathematical tool plays an important role to classify the solutions within the context of the underlying symmetries. Extensive use of group theory has been made to simplify the study of electronic structure or vibrational modes of solids or molecules. The group theory is especially useful in understanding the degeneracy of electronic energy levels and also photonic energy bands.

First-principles methods enable to describe fundamental processes in biology, chemistry and physics as accurately as possible with moderate computational effort. Nevertheless, the application of first-principles methods to study the real-time evolution of complex systems is strongly limited. This can be done by using first-principles molecular-dynamics (MD) methods, which can simulate the evolution of atomic and electronic motions without assuming empirical parameters. Molecular dynamics has proved to be an optimal numerical recipe applicable to problems with many degrees of freedom from quite different fields of science. The knowledge of the energy or potential landscape of interacting particles, like electrons and atoms, enables one to calculate the forces acting on the particles and to study the evolution of the system with time. Overall, first-principles molecular dynamics appears as a convincing method to corroborate experimental work and make reliable predictions based on well-established electronic structure techniques.

Due to of similarities between the Schrödinger equation and the Maxwell's equations, the concept of electronic structure methods can be adopted for the study of photonic nano-materials, which have important applications in micro- and nano-electronics. Such computer simulations help to understand quantum dots or photonic crystals that act as new source of a coherent radiation, cages or guided pathways, operating on smaller and smaller scales at steadily increasing speed.

In this book the above-mentioned aspects of computational physics will be discussed from frameworks to practical applications. The general organization of this book is as follows (see the table of Contents). Part I gives a basic description of electrons and photons in crystals. Part II reviews the various techniques of simulations of nanoscopic and macroscopic materials. Some practical aspects of parallel computing and multi-grid methods will be presented in Part III. The interested reader can find references of several other works in the bibliography sections at the end of each chapter.

Part I

Basic Description of Electrons and Photons in Crystals

2 The Essentials of Density Functional Theory and the Full-Potential Local-Orbital Approach

H. Eschrig

IFW Dresden, P.O.Box 27 00 16, 01171 Dresden, Germany

Abstract. Density functional theory for the ground state energy in its modern understanding which is free of representability problems or other logical uncertainties is reported. Emphasis is on the logical structure, while the problem of modeling the unknown universal density functional is only very briefly mentioned. Then, a very accurate and numerical effective solver for the self-consistent Kohn-Sham equations is presented and its power is illustrated. Comparison is made to results obtained with the WIEN code.

2.1 Density Functional Theory in a Nutshell

Density functional theory deals with inhomogeneous systems of identical particles. Its general aim is to eliminate the monstrous many-particle wave function from the formulation of the theory and instead to express chosen quantities of the system directly in terms of the particle density or the particle current density. There are basically three tasks: (i) to prove that chosen quantities are unique functionals of the density and to indicate how in principle they can be obtained, (ii) to find constructive expressions of model density functionals which are practically tractable and approximate the unique functionals in a way to provide predictive power, and (iii) to develop tools for an effective solution of the resulting problems.

As regards task (i), final answers have been given for the ground state energy [2.2, and citations therein]. In the following these results are summarized. Task (iii) is dealt with in the next section as well as in Chap. 3.

Central quantities of the *density functional theory for the ground state energy* are:

- the external potential $v(\boldsymbol{r})$ or its spin-dependent version

$$\check{v} = v_{ss'}(\boldsymbol{r}) = v(\boldsymbol{r})\delta_{ss'} - \mu_{\mathrm{B}}\boldsymbol{B}(\boldsymbol{r}) \cdot \boldsymbol{\sigma}_{ss'}, \tag{2.1}$$

- the ground state density $n(\boldsymbol{r})$ or spin-matrix density

$$\check{n} = n_{ss'}(\boldsymbol{r}) \hat{=} \left\{ \begin{array}{l} n(\boldsymbol{r}) = \sum_s n_{ss}(\boldsymbol{r}), \\ \boldsymbol{m}(\boldsymbol{r}) = \mu_{\mathrm{B}} \sum_{ss'} n_{ss'}(\boldsymbol{r})\boldsymbol{\sigma}_{s's} \end{array} \right\}, \tag{2.2}$$

- the ground state energy

$$E[\check{v}, N] = \min_{\Gamma} \left\{ H_{\check{v}}[\Gamma] \mid N[\Gamma] = N \right\}. \tag{2.3}$$

H. Eschrig, The Essentials of Density Functional Theory and the Full-Potential Local-Orbital
Approach, Lect. Notes Phys. **642**, 7–21 (2004)
http://www.springerlink.com/

Here, Γ means a general (possibly mixed) quantum state,

$$\Gamma = \sum_{M\alpha} |\Psi_{M\alpha}\rangle p_{M\alpha} \langle\Psi_{M\alpha}| , \quad p_{M\alpha} \geq 0 , \quad \sum_{M\alpha} p_{M\alpha} = 1 \qquad (2.4)$$

where $\Psi_{M\alpha}$ is the many-body wave function of M particles in the quantum state α. In (2.3), $H_{\check{v}}[\Gamma]$ and $N[\Gamma]$ are the expectation values of the Hamiltonian with external potential $\check{v}$ and of the particle number operator, resp., in the state Γ. In the (admitted) case of non-integer N, non-pure (mixed) quantum states are unavoidable.

The variational principle by Hohenberg and Kohn states that there exists a density functional $H[\check{n}]$ so that

$$E[\check{v}, N] = \min_{\check{n}} \left\{ H[\check{n}] + (\check{v} \,|\, \check{n}) \,\,\Big|\, (\check{1} \,|\, \check{n}) = N \right\} , \qquad (2.5)$$

$$(\check{v} \,|\, \check{n}) = \sum_{ss'} \int d^3r v_{ss'} n_{s's} = \int d^3r (vn - \boldsymbol{B} \cdot \boldsymbol{m}), \quad (\check{1} \,|\, \check{n}) = \sum_{s} \int d^3r n_{ss} .$$
$$(2.6)$$

Given an external potential $\check{v}$ and a (possibly non-integer) particle number N, the variational solution yields $E[\check{v}, N]$ and the minimizing spin-matrix density $\check{n}(\boldsymbol{r})$, the ground state density.

There is a unique solution for energy since $H[\check{n}]$ is convex by construction. The solution for $\check{n}$ is in general non-unique since $H[\check{n}]$ need not be strictly convex. The ground state (minimum of $H[\check{n}] + (\check{v} \,|\, \check{n})$) may be degenerate with respect to $\check{n}$ for some $\check{v}$ and N (cf. Fig. 2.1).

In what follows, only the much more relevant spin dependent case is considered and the checks above v and n are dropped.

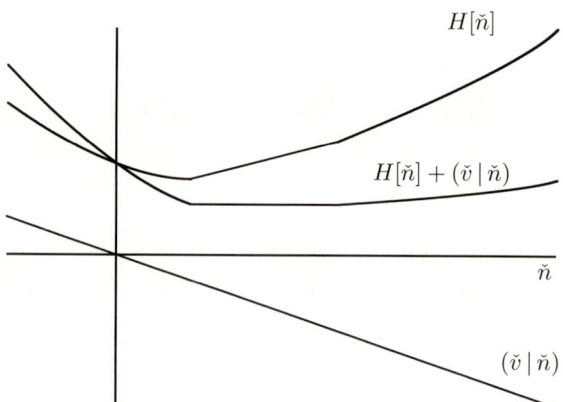

Fig. 2.1. The functionals $H[\check{n}]$, $(\check{v} \,|\, \check{n})$ and $H[\check{n}] + (\check{v} \,|\, \check{n})$ for a certain direction in the functional $\check{n}$-space and a certain potential $\check{v}$.

The mathematical basis of the variational principle is (for a finite total volume, for instance provided by periodic boundary conditions, to avoid formal difficulties with a continuous energy spectrum) that $E[v, N]$ *is convex in* N *for fixed* v *and concave in* v *for fixed* N, *and*

$$E[v + \text{const.}, N] = E[v, N] + \text{const.} \cdot N. \tag{2.7}$$

Because of these simple properties of the ground state energy (which are not even mentioned by Hohenberg and Kohn in their seminal paper [2.5]) it can be represented as a double Legendre transform,

$$E[v, N] = \inf_{n} \sup_{\mu} \left\{ H[n] + (v|n) + [N - (1|n)]\mu \right\}, \tag{2.8}$$

which is equivalent to (2.5) because the μ-supremum is $+\infty$ unless $(1|n) = N$. The inverse double Legendre transformation yields the universal density functional:

$$H[n] = \inf_{N} \sup_{v} \left\{ E[v, N] - (n|v) \right\}. \tag{2.9}$$

Universality means that given a particle-particle interaction (Coulomb interaction between electrons say) a single functional $H[n]$ yields the ground state energies and densities for all (admissible) external potentials.

The expression (2.9) need not be the only density functional which provides (2.5). Generally two functions which have the same convex hull have the same Legendre transform. (Here the situation is more involved because of the intertwined double transformation.) Nevertheless, the outlined analysis can be put to full mathematical rigor, and the domain of admissible potentials is very broad and contains for instance the Coulomb potentials of arbitrary arrangements of nuclei. There are no representability problems. For details see [2.2].

This solves task (i) for the ground state energy as chosen quantity, which, for given v and N, is uniquely obtained via (2.5) from the functional $H[n] + (v|n)$. There are attempts to consider other quantities as excitation spectra or time-dependent quantities which are so far on a much lower level of rigor. Of course, $H[n]$ is unknown and of the same complexity as $E[v, N]$. It can only be modeled by guesses. This turns out to be uncomparably more effective than a direct modeling of $E[v, N]$.

Modeling of $H[n]$ starts with the Kohn-Sham (KS) parameterization [2.7] of the density by KS orbitals $\phi_k(rs)$ and orbital occupation numbers n_k:

$$n(\mathbf{r}) = n_{ss'}(\mathbf{r}) = \sum_{k} \phi_k(\mathbf{r}s) \, n_k \, \phi_k^*(\mathbf{r}s'), \tag{2.10}$$

$$0 \leq n_k \leq 1, \quad \langle \phi_k | \phi_{k'} \rangle = \delta_{kk'}, \quad (1 \,|\, n) = \sum_{k} n_k = N . \tag{2.11}$$

Model functionals consist of an orbital variation part K and a local density expression L:

$$H[n] = K[n] + L[n] \ ,$$

$$K[n] = \min_{\{\phi_k, n_k\}} \left\{ k[\phi_k, n_k] \Big| \sum_k \phi_k n_k \phi_k^* = n \right\} \ ,$$

$$L[n] = \int d^3 r n(\mathbf{r}) l\left(n_{ss'}(\mathbf{r}), \boldsymbol{\nabla} n, \dots\right) \ , \tag{2.12}$$

which cast the variational principle (2.5) into the KS form

$$E[v, N] = \min_{\{\phi_k, n_k\}} \left\{ k[\phi_k, n_k] + L[\Sigma \phi n \phi^*] + (\Sigma \phi n \phi^* \,|\, v) \Big| \right.$$

$$\left. \Big| \langle \phi_k | \phi_{k'} \rangle = \delta_{kk'}, 0 \le n_k \le 1, \sum_k n_k = N \right\} \ . \tag{2.13}$$

ϕ_i^*, ϕ_i and n_i must be varied independently. The uniqueness of solution now depends on the convexity of $k[\phi_k, n_k]$ and $L[n]$.

Variation of ϕ_k^* yields the (generalized) KS equations:

$$\frac{1}{n_k} \frac{\delta k}{\delta \phi_k^*} + \left(\frac{\delta L}{\delta n} + v \right) \phi_k = \phi_k \epsilon_k \ . \tag{2.14}$$

Since n_k and ϕ_k^* enter in the combination $n_k \phi_k^*$ only, the relation

$$n_k \frac{\partial}{\partial n_k} = \left\langle \phi_k \Big| \frac{\delta}{\delta \phi_k^*} \right. \tag{2.15}$$

is valid which yields Janak's theorem:

$$\frac{\partial}{\partial n_k} \left(k + L + (v \,|\, n) \right) = \epsilon_k \ . \tag{2.16}$$

Variation of n_k, in view of the side conditions, yields the Aufbau principle: Let $n_{k'} < n_k$, then (cf. Fig. 2.2)

$$\delta n \left(\frac{\partial}{\partial n_{k'}} - \frac{\partial}{\partial n_k} \right) \left(k + L + (v \,|\, n) \right) \begin{cases} \ge 0 \text{ for } n_{k'} = 0 \text{ or } n_k = 1 \ , \\ = 0 \text{ for } 0 < n_{k'}, n_k < 1 \ . \end{cases} \tag{2.17}$$

Hence,

$$\begin{aligned} n_k &= 1 & \text{for } \epsilon_k < \epsilon_N \ , \\ 0 \le n_k &\le 1 & \text{for } \epsilon_k = \epsilon_N \ , \\ n_k &= 0 & \text{for } \epsilon_k > \epsilon_N \ . \end{aligned} \tag{2.18}$$

Finding suitable expressions for k and L mainly by physical intuition is the way task (ii) is treated. The standard L(S)DA or GGA is obtained by putting

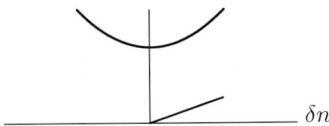

Fig. 2.2. A free minimum of a function of δn and a minimum under the constraint $\delta n > 0$.

$$k[\phi_k, n_k] = \sum_k n_k \langle \phi_k | - \frac{\nabla^2}{2} | \phi_k \rangle +$$

$$+ \sum_{kk'} \frac{n_k n_{k'}}{2} \sum_{ss'} \int d^3r d^3r' \frac{|\phi_k(\boldsymbol{r}s)|^2 |\phi_{k'}(\boldsymbol{r}'s')|^2}{|\boldsymbol{r} - \boldsymbol{r}'|} . \tag{2.19}$$

This completes the brief introduction to the state of the art of density functional theory of the ground state energy.

Just to mention one other realm of possible density functionals, quasiparticle excitations are obtained from the coherent part (pole term) of the single particle Green's function ([2.3–2.5])

$$G_{ss'}(\boldsymbol{r}, \boldsymbol{r}'; \omega) = \sum_k \frac{\chi_s(\boldsymbol{r}) \eta_{s'}^*(\boldsymbol{r}')}{\omega - \varepsilon_k} + G_{ss'}^{\text{incoh}}(\boldsymbol{r}, \boldsymbol{r}'; \omega) , \tag{2.20}$$

$$\sum_{s'} \int d^3r' \left[\delta(\boldsymbol{r} - \boldsymbol{r}') \left(-\frac{\nabla^2}{2} + u(\boldsymbol{r}) + u_{\text{H}}(\boldsymbol{r}) \right) \right.$$

$$\left. + \Sigma_{ss'}(\boldsymbol{r}, \boldsymbol{r}'; \epsilon_k) \right] \chi_{s'}(\boldsymbol{r}') = \chi_s(\boldsymbol{r}) \varepsilon_k . \tag{2.21}$$

Here, in the inhomogeneous situation of a solid, the self-energy Σ is among other dependencies a functional of the density. This forms the shaky ground (with rather solid boulders placed here and there on it, see for instance also [2.4]) for interpreting a KS band structure as a quasi-particle spectrum. In principle from the full $\Sigma_{ss'}(\boldsymbol{r}, \boldsymbol{r}'; \omega)$ the total energy might also be obtained.

2.2 Full-Potential Local-Orbital Band Structure Scheme (FPLO)

This chapter deals with task (iii) mentioned in the introduction to Chap. 1. A highly accurate and very effective tool to solve the KS equations self-consistently is sketched. The basic ideas are described in [2.6, see http://www.ifw-dresden.de/agtheo/FPLO/ for actual details of the implementation].

The KS (2.14) represents a highly non-linear set of functional-differential equations of the form

$$\hat{H}\phi_i = \left[-\frac{\nabla^2}{2} + v_{\text{eff}}\right]\phi_i = \phi_i \epsilon_i \tag{2.22}$$

since the effective potential parts contained in $\delta k/\delta\phi_i^*$ and in $\delta L/\delta n$ depend on the solutions ϕ_i. The general iterative solving procedure is as follows:

Guess a density $n_{ss'}^{(\text{in})}(\boldsymbol{r})$.

– Determine the potentials $v_{\text{H}}(\boldsymbol{r})$ (part of $\delta k/\delta\phi_i^*$ from the second line of (2.19)) and $v_{\text{xc},ss'}(\boldsymbol{r}) = \delta L/\delta n_{s's}$.
– Solve the KS equation for $\phi_i(\boldsymbol{r}s), \epsilon_i$.
– Determine the density $n_{ss'}^{(\text{out})}(\boldsymbol{r}) = \sum_i \phi_i(\boldsymbol{r}s)\theta(\mu - \epsilon_i)\phi_i^*(\boldsymbol{r}s')$ with $\mu = \mu(N)$ from $\sum_i \theta(\mu - \epsilon_i) = N$.
– Determine a new input density $n_{ss'}^{(\text{in})}(\boldsymbol{r}) = f\big(n_{ss'}^{(\text{out})}(\boldsymbol{r}), n_{ss'}^{(\text{in},j)}(\boldsymbol{r})\big)$ from $n^{(\text{out})}$ of the previous step and $n^{(\text{in},j)}$ of a number of previous cycles; f has to be chosen by demands of convergence.

Iterate until $n^{(\text{out})} = n^{(\text{in})} = n^{(\text{SCF})}$.
SCF density: $n_{ss'}(\boldsymbol{r}) \hat{=} (n(\boldsymbol{r}), \boldsymbol{m}(\boldsymbol{r}))$,
Total energy: $E[v, N] = H[n] + (v \,|\, n)$.

In the following the most demanding second step is sketched.

2.2.1 The Local Orbital Representation

The KS orbitals ϕ_{kn} of a crystalline solid, indexed by a wave number $\boldsymbol{k}$ and a band index n, are expanded into a nonorthogonal local orbital minimum basis (one basis orbital per band or per core state):

$$\phi_{kn}(\boldsymbol{r}) = \sum_{\boldsymbol{R}sL} \varphi_{sL}(\boldsymbol{r} - \boldsymbol{R} - s)C_{Ls,kn}e^{i\boldsymbol{k}(\boldsymbol{R}+s)}. \tag{2.23}$$

This leads to a secular equation of the form

$$HC = SC\epsilon , \tag{2.24}$$

$$H_{s'L',sL} = \sum_{\boldsymbol{R}} \langle 0s'L'|\hat{H}|\boldsymbol{R}sL\rangle e^{i\boldsymbol{k}(\boldsymbol{R}+s-s')} , \tag{2.25}$$

$$S_{s'L',sL} = \sum_{\boldsymbol{R}} \langle 0s'L'|\boldsymbol{R}sL\rangle e^{i\boldsymbol{k}(\boldsymbol{R}+s-s')} . \tag{2.26}$$

By definition, core states are local eigenstates of the effective crystal potential which have no overlap to neighboring core states and are mutually orthogonal. This gives the overlap matrix (2.26) a block structure (indices c and v denote core and valence blocks) allowing for a simplified Cholesky decomposition into left and right triangular factors:

$$S = \begin{pmatrix} 1 & S_{cv} \\ S_{vc} & S_{vv} \end{pmatrix} = \begin{pmatrix} 1 & 0 \\ S_{vc} & S_{vv}^L \end{pmatrix} \begin{pmatrix} 1 & S_{cv} \\ 0 & S_{vv}^R \end{pmatrix} = S^L S^R , \qquad (2.27)$$

$$S_{vv}^L S_{vv}^R = S_{vv} - S_{vc} S_{cv} . \qquad (2.28)$$

The corresponding block structure of the Hamiltonian matrix (2.25) is

$$H = \begin{pmatrix} \epsilon_c 1 & \epsilon_c S_{cv} \\ S_{vc} \epsilon_c & H_{vv} \end{pmatrix}, \quad \epsilon_c = \mathrm{diag}(\cdots, \epsilon_{sL_c}, \cdots) . \qquad (2.29)$$

With these peculiarities the secular problem for H may be converted into a much smaller secular problem of a projected Hamiltonian matrix $\tilde{H}_{vv}$ as follows:

$$HC = SC\epsilon$$

$$(S^{L-1} H S^{R-1})(S^R C) = (S^R C)\epsilon$$

$$\Downarrow$$

$$\tilde{H}_{vv} \tilde{C}_{vv} = \tilde{C}_{vv} \epsilon_v \qquad (2.30)$$

$$\tilde{H}_{vv} = S_{vv}^{L-1}(H_{vv} - S_{vc} H_{cc} S_{cv}) S_{vv}^{R-1}$$

$$C = \begin{pmatrix} 1 & -S_{cv} S_{vv}^{R-1} \tilde{C}_{vv} \\ 0 & S_{vv}^{R-1} \tilde{C}_{vv} \end{pmatrix} .$$

This exact reduction of the secular problem saves a lot of computer time in solving (2.25), by a factor of about 3 in the case of fcc Cu (with $3s, 3p$-states treated as valence states for accuracy reasons) up to a factor of about 40 in the case of fcc Au (again with $5s, 5p$-states treated as valence states). With slightly relaxed accuracy demands and treating the $3s, 3p$- and $5s, 5p$-states, resp., as core states, the gain is even by factors of 8 and 110.

2.2.2 Partitioning of Unity

The use of a local basis makes it desirable to have the density and the effective potential as lattice sums of local contributions. This is not automatically provided: the density comes out form summation over the occupied KS orbitals (2.22) as a double lattice sum, and the effective potential has anyhow a complicated connection with the KS orbitals. The decisive tool here is a partitioning of unity in r-space.

There may be chosen:

- a locally finite cover of the real space $\mathbb{R}^3$ by compact cells Ω_i, that is, $\mathbb{R}^3 = \cup_i \Omega_i$ and each point of $\mathbb{R}^3$ lies only in finitely many Ω_i,
- a set of n-fold continuously differentiable functions $f_i(r)$ with $\mathrm{supp} f_i \subset \Omega_i$, that is $f_i(r) = 0$ for $r \notin \Omega_i$,
- $0 \le f_i(r) \le 1$ and $\sum_i f_i(r) = 1$ for all r.

In the actual context, $\Omega_i = \Omega_{Rs}$ indexed by atom positions.

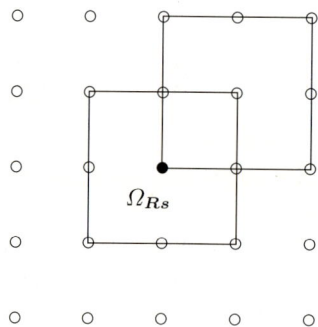

Fig. 2.3. A locally finite cover of the $\mathbb{R}^2$ by squares.

Figure 2.3 shows a locally finite cover of the plane by a lattice of overlapping squares.

2.2.3 Density and Potential Representation

The decomposition of the density

$$n(r) = \sum_{Rs} n_s(r - R - s) \tag{2.31}$$

is obtained by an even simpler one-dimensional partitioning along the line joining the two centers of a two-center contribution.

The potential is decomposed according to

$$v(\boldsymbol{r}) = \sum_{\boldsymbol{Rs}} v_s(\boldsymbol{r} - \boldsymbol{R} - \boldsymbol{s}), \quad v_s(\boldsymbol{r} - \boldsymbol{R} - \boldsymbol{s}) = v(\boldsymbol{r})f_{\boldsymbol{Rs}}(\boldsymbol{r}) \qquad (2.32)$$

with use of the functions f of the previous subsection.

Now, in the local items, radial dependencies are obtained numerically on an inhomogeneous grid (logarithmic equidistant), and angular dependencies are expanded into spherical harmonics (typically up to $l = 12$). To compute the overlap and Hamiltonian matrices, one has

- one-center terms: 1D numerical integrals,
- two-center terms: 2D numerical integrals,
- three-center terms: 3D numerical integrals.

2.2.4 Basis Optimization

The essential feature which allows for the use of a minimum basis is that the basis is not fixed in the course of iterations, instead it is adjusted to the actual effective crystal potential in each iteration step and it is even optimized in the course of iterations.

Take $\bar{v}_s$ to be the total crystal potential, spherically averaged around the site center s. Core orbitals are obtained from

$$(\hat{t} + \bar{v}_s)\varphi_{sL_c} = \varphi_{sL_c}\epsilon_{sL_c} . \qquad (2.33)$$

Valence basis orbitals, however, are obtained from a modified equation

$$\left(\hat{t} + \bar{v}_s + \left(\frac{r}{r_{sL_v}}\right)^4\right)\varphi_{sL_v} = \varphi_{sL_v}\epsilon_{sL_v} . \qquad (2.34)$$

The parameters r_{sL_v} are determined by minimizing the total energy. There are two main effects of the r_{sL_v}-potential:

- The counterproductive long tails of basis orbitals are suppressed.
- The orbital resonance energies ϵ_{sL_v} are pushed up to close to the centers of gravity of the orbital projected density of states of the Kohn-Sham band structure, providing the optimal curvature of the orbitals and avoiding insufficient completeness of the local basis.

In the package FPLO the optimization is done automatically by applying a kind of force theorem during the iterations for self-consistency.

2.2.5 Examples

In order to illustrate the accuracy of the approach, the simple case of fcc Al is considered. Figure 2.4 shows the dependence of the calculated total energy

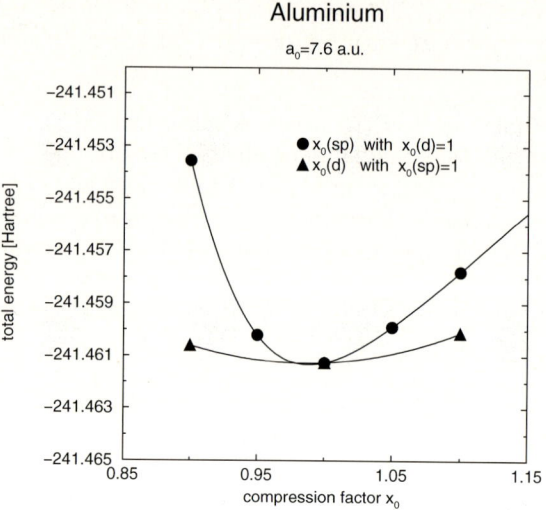

Fig. 2.4. Total energy of aluminum as a function of the parameters $x_0 = r_0/r_{\text{NN}}^{3/2}$.

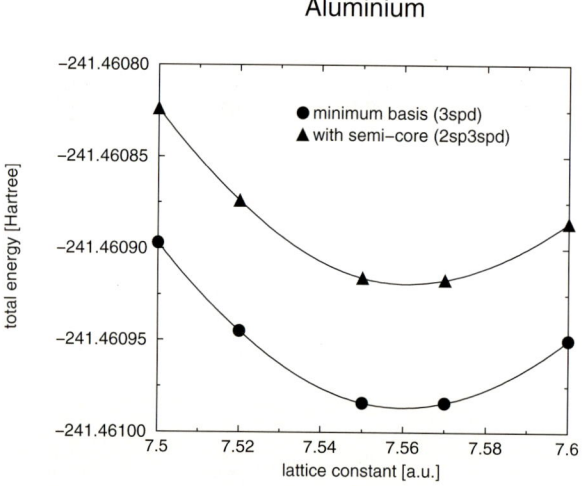

Fig. 2.5. Total energy vs. lattice constant of aluminum for two basis sets.

as a function of the basis optimization parameters $x_0(L_v) = r_{L_v}/r_{NN}^{3/2}$ while Fig. 2.5 shows the effect of treating the $2s, 2p$-states either as core states or as valence states (called semi-core states in the latter case). Note that neglecting the neighboring overlap of $2s, 2p$-states is an admitted numerical error and not a question of basis completeness; the more accurate total energy in this case is the higher one with the semi-core treatment.

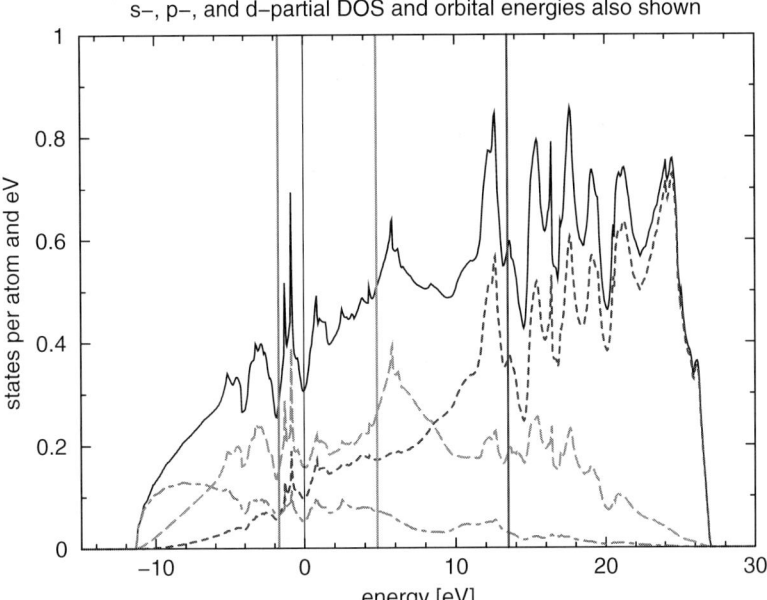

Fig. 2.6. DOS and l-projected DOS of aluminum; the vertical lines indicate the orbital resonance energies.

Next, in Fig. 2.6 the density of states (DOS) and the orbital projected DOS of Al are shown. (The energy zero here and in the following is put at the Fermi level.) Vertical lines mark the basis orbital resonance energies ϵ_{L_v} of (2.34). As a result of optimizing the parameters r_{L_v} of this equation, clearly the energies are close to the centers of gravity of the corresponding projected DOS. This proves optimization of basis completeness within the fixed number of basis orbitals. Even the down shift of ϵ_d from the corresponding center of gravity of the d density of states is correct since completeness is needed only in the lower, occupied part.

For illustration, the KS band structure of Al is shown on Fig. 2.7. Remarkably, the third band is above the Fermi level at point W: The LDA Fermi surface of Al has the right topology and is quantitatively very correct (Fig. 2.8). The frequently asserted failure of the LDA not to produce the right FS topology of Al is a muffin-tin problem.

On Fig. 2.9 it is illustrated on the example of Sr_2CuO_3 how the automate basis optimization works; x_0 steps are those self-consistency steps which adjust the x_0 parameters. The corresponding convergence of the total energy is also shown.

Al, a$_0$ = 7.56 a.u. (LDA minimum), scalar relativistic

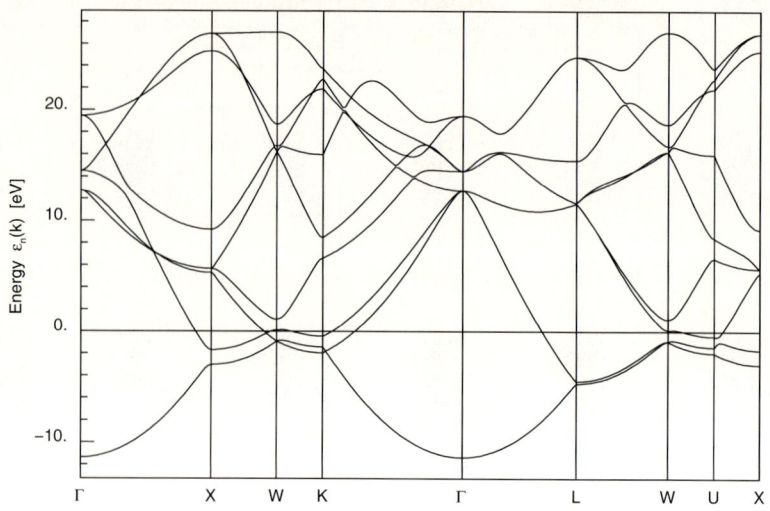

Fig. 2.7. KS band structure of aluminum.

Fig. 2.8. LDA Fermi surface of aluminum.

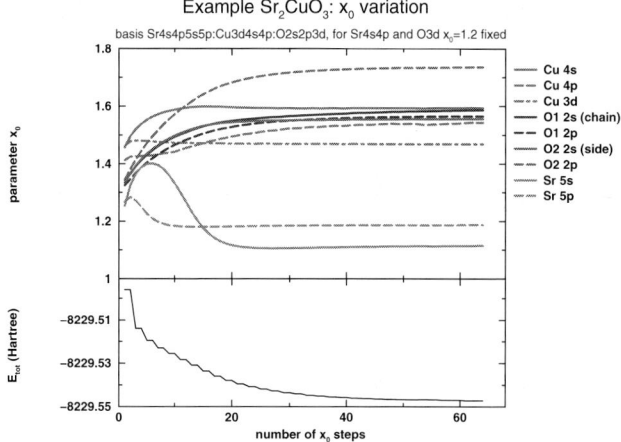

Fig. 2.9. Automatic basis optimization for Sr_2CuO_3 in the course of iterations.

For this example, the position of the basis orbital resonance energies relative to the corresponding orbital projected DOS are shown on Fig. 2.10. The same correlation as for Al is observed.

2.2.6 Comparison of Results from FPLO and WIEN97

To get some feeling on the absolute accuracy of calculated total energies, a number of comparisons is made between results obtained with exactly the same density functional (occasionally very slightly different from the previous examples) but with solvers working with totally different basis sets: augmented plane waves vs. local orbitals. The results were carefully converged within both approaches: in WIEN97 [2.1] with the number of plane waves (far beyond the default) and in FPLO with basis optimization and generally including $3d$ polarization orbitals for oxygen.

On Fig. 2.11 the obtained densities of states for $CaCuO_2$ are presented. The FPLO basis was

$\quad$ Ca: $\{1s, 2s, 2p\}_c, \{3s, 3p, 3d, 4s, 4p\}_v$

$\quad$ Cu: $\{1s, 2s, 2p\}_c, \{3s, 3p, 3d, 4s, 4p\}_v$

$\quad$ O: $\{1s\}_c, \{2s, 2p, 3d\}_v$.

The differences are mainly due to a different $\boldsymbol{k}$-integration routine used in the codes to calculate the density of states from the band energies.

Table 1 contains the absolute values of total energies obtained with both approaches for a number of elemental metals and compounds. The agreement is to our knowledge unprecedented so far which speaks of the quality of both codes.

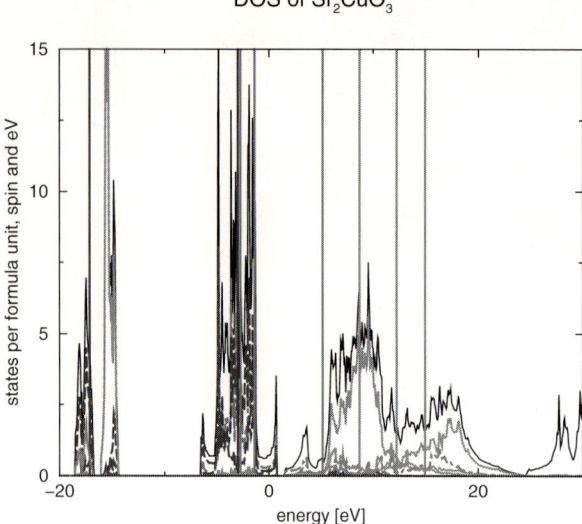

Fig. 2.10. DOS and l-projected DOS of Sr_2CuO_3; vertical lines indicate orbital resonance energies. dashed line: O $2s, 2p$; dot-dashed line: Cu $3d, 4s, 4p$; dotted line: Sr $4d, 5s, 5p$ in ascending order of energies.

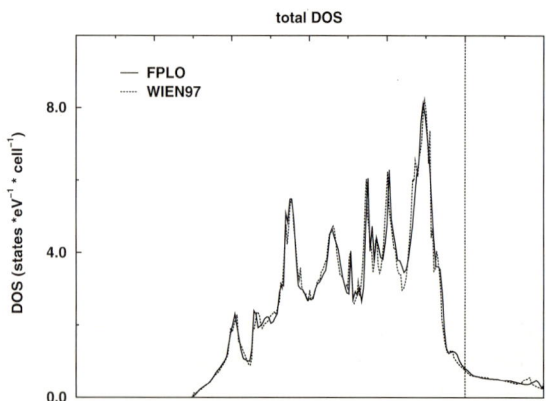

Fig. 2.11. Comparison of the DOS of $CaCuO_2$ from FPLO and from WIEN97.

It should be mentioned that due to the extremely small basis of FPLO the computing time for matrix algebra and diagonalization does not dominate the total computing time even for rather large unit cells: up to at least 100 atoms per cell the time scales roughly as $N^{1.5}$ which allows for instance for large numbers of $\boldsymbol{k}$ points where this is needed.

A fully relativistic four-component version and a flexible CPA implementation for substitutional alloys are available.

Table 2.1. Total energies: (Al, Fe, Ni non-relativistic, all others scalar-relativistic)

solid	N	E_{FPLO} [Hartree]	E_{WIEN} [Hartree]	$\Delta E/N$ [mHartree]
fcc-Al	1	-241.464 0	-241.465 5	1.5
FM-bcc-Fe	1	-1 261.456 5	-1 261.457 2	0.7
FM-fcc-Ni	1	-1 505.875 5	-1 505.877 -	1.5
fcc-Cu	1	-1 652.483 2	-1 652.484 1	0.9
$CaCuO_2$	4	-2 480.992 6	-2 480.995 5	0.7
Sr_2CuO_3	6	-8 229.534 4	-8 229.544 1	1.6
$Sr_2CuO_2Cl_2$	7	-9 075.266 6	-9 075.286 3	1.4
Cu_2GeO_4	14	-11 400.481 1	-11 400.503 6	1.6

Acknowledgements

In preparing this text, many discussions with M. Richter, K. Koepernik, H. Rosner and U. Nitzsche were of great help.

References

[2.1] P. Blaha, K. Schwarz, and J. Luitz. WIEN'97, a full potential linearized augmented plane wave package for calculating crystal properties. Technical report, Technical University of Vienna, 1999.

[2.2] H. Eschrig. *The Fundamentals of Density Functional Theory.* Edition am Gutenbergplatz, Leipzig, 2003.

[2.3] L. Hedin. New method for calculating the one-particle green's function with application to the electron-gas problem. *Phys. Rev.*, 139:A796–A823, 1965.

[2.4] L. Hedin and B.I. Lundqvist. Explicit local exchange and correlation potentials. *J. Phys. C: Solid St. Phys.*, 4:2064–2083, 1971.

[2.5] P. Hohenberg and W. Kohn. Inhomogeneous electron gas. *Phys. Rev.*, 136:B864–B871, 1964.

[2.6] K. Koepernik and H. Eschrig. Full-potential nonorthogonal local-orbital minimum-basis band-structure scheme. *Phys. Rev.*, B59:1743–1757, 1999.

[2.7] W. Kohn and L.J. Sham. Self-consistent equations including exchange and correlation effects. *Phys. Rev.*, 140:A1133–A1138, 1965.

3 Methods for Band Structure Calculations in Solids

A. Ernst[1] and M. Lüders[2]

[1] Max-Planck-Institut für Mikrostrukturphysik, Weinberg 2, 06120 Halle, Germany
[2] Daresbury Laboratory, Warrington WA4 4AD, United Kingdom

Abstract. The calculation of the ground-state and excited-state properties of materials is one of the main goals of condensed matter physics. While the most successful first-principles method, the density-functional theory (DFT), provides, in principle, the exact ground-state properties, the many-body method is the most suitable approach for studying excited-state properties of extended systems. Here we discuss general aspects of the Green's function and different approximations for the self-energy to solve the Dyson equation. Further we present some tools for solving the Dyson equation with several approximations for the self-energy: a highly precise combined basis method providing the band structure in the Kohn-Sham approximation, and some implementations for the random-phase approximation.

3.1 The Green's Function and the Many-Body Method

3.1.1 General Considerations

The Green's function is a powerful tool for studying ground state and excited state properties of condensed matter. The basic idea of the Green's function has its origin in the theory of differential equations. The solution of any inhomogeneous differential equation with a Hermitian differential operator $\hat{H}$, a complex parameter z, and a given source function $u(\boldsymbol{r})$ of the form

$$\left[z - \hat{H}(\boldsymbol{r}) \right] \psi(\boldsymbol{r}) = u(\boldsymbol{r}) \tag{3.1}$$

can be represented as an integral equation

$$\psi(\boldsymbol{r}) = \varphi(\boldsymbol{r}) + \int G(\boldsymbol{r}, \boldsymbol{r}'; z) u(\boldsymbol{r}') \mathrm{d}^3 r'. \tag{3.2}$$

Here the so-called Green's function $G(\boldsymbol{r}, \boldsymbol{r}'; z)$ is a coordinate representation of the resolvent of the differential operator $\hat{H}$, i.e. $\hat{G} = [z - \hat{H}]^{-1}$, which obeys the differential equation

$$\left[z - \hat{H}(\boldsymbol{r}) \right] G(\boldsymbol{r}, \boldsymbol{r}'; z) = \delta(\boldsymbol{r} - \boldsymbol{r}'), \tag{3.3}$$

and the function $\varphi(\boldsymbol{r})$ is a general solution of the homogeneous equation associated with (3.1), i.e. for $u(\boldsymbol{r}) = 0$. The integral equation (3.2) contains, in

A. Ernst and M. Lüders, Methods for Band Structure Calculations in Solids, Lect. Notes Phys.
642, 23–54 (2004)
`http://www.springerlink.com/`   © Springer-Verlag Berlin Heidelberg 2004

contrast to the differential equation (3.1), also information about the boundary conditions, which are built into the function $\varphi(\boldsymbol{r})$. This method is in many cases very convenient and widely used in many-body physics. In this review we shall consider the application of the Green's function method in condensed matter physics at zero temperature. The formalism can be generalised to finite temperatures but this is beyond the scope of this papers and can be found in standard textbooks [3.22, 3.42, 3.24]

The evolution of an N-body system is determined by the time-dependent Schrödinger equation (atomic units are used throughout)

$$i\frac{d\Psi(t)}{dt} = \hat{H}\Psi(t), \tag{3.4}$$

where $\Psi(t) \equiv \Psi(\boldsymbol{r}_1, \boldsymbol{r}_2, ..., \boldsymbol{r}_N; t)$ is a wave function of the system and $\hat{H}$ is the many-body Hamiltonian

$$\hat{H} = \hat{H}_0 + \hat{V}, \tag{3.5}$$

which includes the kinetic energy and the external potential

$$\hat{H}_0 = -\sum_i \frac{\nabla_i^2}{2} + \sum_i v_{\text{ext}}(\boldsymbol{r}_i), \tag{3.6}$$

and the interactions between the particles

$$\hat{V} = \frac{1}{2}\sum_{i,j}\frac{1}{|\boldsymbol{r}_i - \boldsymbol{r}_j|}. \tag{3.7}$$

Knowing the wave function $\Psi(t)$, the average value of any operator $\hat{A}$ can be obtained from the equation:

$$\mathcal{A}(t) = \langle \Psi^*(t)\hat{A}\Psi(t)\rangle. \tag{3.8}$$

The many-body wave function can be expanded in a complete set of time-independent either anti-symmetrized (for fermions) or symmetrized (for bosons) products of single particle wave-functions: $\Phi_{i_1,...,i_n}(\boldsymbol{r}_1, ..., \boldsymbol{r}_N)$

$$\Psi(\boldsymbol{r}_1, \boldsymbol{r}_2, .. \boldsymbol{r}_N; t) = \sum_{\{i_1, i_2, ..., i_N\}} C_{\{i_1, i_2, ..., i_N\}}(t)\,\Phi_{i_1, ..., i_n}(\boldsymbol{r}_1, ..., \boldsymbol{r}_N). \tag{3.9}$$

These (anti-) symmetrized products of single-particle states $\phi_i(\boldsymbol{r})$ are given by

$$\Phi_{i_1,...,i_n}(\boldsymbol{r}_1, ..., \boldsymbol{r}_N) = \sum_{\mathcal{P}}(\pm 1)^{\mathcal{P}}\phi_{\mathcal{P}(i_1)}(\boldsymbol{r_1})\phi_{\mathcal{P}(i_2)}(\boldsymbol{r_2})...\phi_{\mathcal{P}(i_N)}(\boldsymbol{r}_N) \tag{3.10}$$

where $\mathcal{P}$ denotes all permutations of the indices $i_1, ..., i_N$ and $(-1)^{\mathcal{P}}$ yields a minus sign for odd permutations. In case of Fermions this anti-symmetrized product can conveniently be written as a determinant, the so-called Slater determinant,

$$\Phi_{i_1,\ldots,i_n}(\boldsymbol{r}_1,\ldots,\boldsymbol{r}_N) = |\phi_{i_1}(\boldsymbol{r}_1),\phi_{i_2}(\boldsymbol{r_2}),\ldots\phi_{i_N}(\boldsymbol{r}_N)|, \tag{3.11}$$

which already fulfills Pauli's exclusion principle. The basis set can be arbitrary, but in practice one uses functions which are adequate for the particlar problem. For example, the plane wave basis is appropriate for the description of a system with free or nearly free electrons. Systems with localised electrons are usually better described by atomic-like functions. For systems with a large number of particles solving equation (3.4) using the basis expansion (3.9) is a quite formidable task.

To describe a many-body system one can use the so-called second quantisation: instead of giving a complete wave function one specifies the numbers of particles to be found in the one-particle states $\phi_1(\boldsymbol{r}), \phi_2(\boldsymbol{r}), .., \phi_N(\boldsymbol{r})$. As a result the many-body wave function is defined by the expansion coefficients at the occupation numbers and the Hamiltonian, as well as any other operator can be expressed in terms of the so-called creation and annihilation operators $\hat{c}_i^+$ and $\hat{c}_i$, obeying certain commutation or anticommutation relations according to the statistical properties of the particles (fermions or bosons). The creation operator $\hat{c}_i^+$ increases the number of particles by one, while the annihilation operators $\hat{c}_i$ decreases the occupation number of a state by one. Any observable can be represented as some combination of these operators. For example, a one-particle operator $\hat{A}$ can be expressed as

$$\hat{A} = \sum_{i,k} A_{ik}\hat{c}_i^+ \hat{c}_k, \tag{3.12}$$

where A_{ik} are matrix elements of $\hat{A}$. Often it is convenient to use the field operators [1]

$$\hat{\psi}_\sigma^+(\boldsymbol{r}) = \sum_i \phi_{i\sigma}(\boldsymbol{r})\hat{c}_{i\sigma}^+$$

$$\hat{\psi}_\sigma(\boldsymbol{r}) = \sum_i \phi_{i\sigma}(\boldsymbol{r})\hat{c}_{i\sigma} \ , \tag{3.13}$$

which can be interpreted as creation and annihilation operators of a particle with spin σ at a given point $\boldsymbol{r}$. In this representation a single particle operator is given as

$$\hat{A} = \sum_\sigma \int \mathrm{d}^3r\hat{\psi}_\sigma^+(\boldsymbol{r})\hat{A}\hat{\psi}_\sigma(\boldsymbol{r}), \tag{3.14}$$

---

[1] Here we have explicitly included the spin-indices. In the remainder of the paper, we include the spin in the other quantum numbers, wherever not specified explicitly

where A is of the same form as the operator $\hat{A}$ in first quantisation, but without coordinates and momenta beeing operators.

Suppose we have a system with N particles in the ground state, which is defined by the exact ground state wave function Ψ_0. If at time $t_0 = 0$ a particle with quantum number i is added into the system, the system is described by $c_i^+|\Psi_0\rangle$. The evolution of the system in time will now proceed according to $e^{-i\hat{H}(t-t_0)}c_i^+|\Psi_0\rangle$. The probability amplitude for finding the added particle in the state j, is the scalar product of $e^{-i\hat{H}(t-t_0)}c_i^+|\Psi_0\rangle$ with the function $c_j^+e^{-i\hat{H}(t-t_0)}|\Psi_0\rangle$, describing a particle in the state j, added to the ground state at time t. The resulting probability amplitude is given by $\langle\Psi_0|e^{i\hat{H}(t-t_0)}c_je^{-i\hat{H}(t-t_0)}c_i^+|\Psi_0\rangle$. Analogously, a particle removed from a state can be described with the function $\pm\langle\Psi_0|e^{-i\hat{H}(t-t_0)}c_i^+e^{i\hat{H}(t-t_0)}c_j|\Psi_0\rangle$, where plus sign applies to Bose statistics and minus sign to Fermi statistics. Both processes contribute to the definition of the one-particle causal Green's function:

$$G(j,t;i,t_0) = -i\langle\Psi_0|T\left[\hat{c}_j(t)\hat{c}_i^+(t_0)\right]|\Psi_0\rangle. \tag{3.15}$$

Here we have used Heisenberg representation of the operators $\hat{c}_i$ and $\hat{c}_i^+$:

$$\hat{c}_i(t) = e^{i\hat{H}t}\hat{c}_ie^{-i\hat{H}t}. \tag{3.16}$$

The symbol T (3.15) is Wick's time-ordering operator which rearranges a product of two time-dependent operators so that the operator referring to the later time appears always on the left:

$$T\left[\hat{c}_j(t)\hat{c}_i^+(t_0)\right] = \begin{cases} \hat{c}_j(t)\hat{c}_i^+(t_0) & (t > t_0) \\ \pm\hat{c}_i^+(t_0)\hat{c}_j(t) & (t < t_0) \end{cases}. \tag{3.17}$$

The physical meaning of the Green's function in this representation is that for $t > t_0$ $G(i,t_0;j,t)$ describes the propagation of a particle created at time t_0 in the state i and detected at time t in the state j. For $t < t_0$, the Green's function describes the propagation of a hole in the state j emitted at time t into the state i at time t_0. Analogously to the above, one can write down the Green's function in the space-time representation:

$$G(\boldsymbol{r}_0,t_0;\boldsymbol{r},t) = -i\langle\Psi_0|T\left[\hat{\psi}(\boldsymbol{r}_0,t_0)\hat{\psi}^+(\boldsymbol{r},t)\right]|\Psi_0\rangle, \tag{3.18}$$

where $\hat{\psi}(\boldsymbol{r},t)$ and $\hat{\psi}^+(\boldsymbol{r},t)$ are particle annihilation and creation operators in the Heisenberg representation.

The time-evolution of the Green's function is controlled by the equation of motion. For $V = 0$ this is reduced to

$$\left(i\frac{\partial}{\partial t} - \hat{H}(\boldsymbol{r})\right)G(\boldsymbol{r}t,\boldsymbol{r}'t') = \delta(\boldsymbol{r}-\boldsymbol{r}')\delta(t-t'), \tag{3.19}$$

which follows directly from the equation of motion of the field operators:

$$i\frac{\partial\hat{\psi}(\boldsymbol{r},t)}{\partial t} = \left[\hat{\psi}(\boldsymbol{r},t),\hat{H}\right], \tag{3.20}$$

and a similar equation for the creation operator $\hat{\psi}^+(\boldsymbol{r},t)$. Taking the Fourier transform of (3.19) into frequency space, we get

$$\left[\omega - \hat{H}(\boldsymbol{r})\right] G(\boldsymbol{r},\boldsymbol{r}';\omega) = \delta(\boldsymbol{r}-\boldsymbol{r}'), \tag{3.21}$$

which demonstrates, that $G(\boldsymbol{r},\boldsymbol{r}';\omega)$ is a Green's function in the mathematical sence, as described above in (3.3).

The one-particle Green's function has some important properties which make the use of the Green's function method in condensed matter physics attractive. The Green's function contains a great deal of information about the system: knowing the single-particle Green's function, one can calculate the ground state expectation value of any single-particle operator:

$$A(t) = \pm i \int \left[\lim_{t'\to t+0}\lim_{\boldsymbol{r}'\to\boldsymbol{r}} \hat{A}(\boldsymbol{r})\, G(\boldsymbol{r},t;\boldsymbol{r}',t')\right] \mathrm{d}^3 r. \tag{3.22}$$

As a consequence, in particular, the charge density and the total energy can be found for any system in the ground state at zero temperature. Furthermore the one-particle Green's function describes single-particle excitations. In what follows, we shall discuss the latter in more detail.

For simplicity, in the following paragraphs we consider the homogeneous electron gas. Due to the translational invariance, the momentum $\boldsymbol{k}$ is a good quantum number, and we can use the basis functions $\phi_{\boldsymbol{k}}(\boldsymbol{r}) = \frac{1}{\sqrt{N}}e^{i\boldsymbol{k}\cdot\boldsymbol{r}}$. It can easily be verified that the Green's function of the homogeneous electron gas is diagonal in momentum space and depends only on the time difference:

$$G(\boldsymbol{k}t,\boldsymbol{k}'t') = \delta_{\boldsymbol{k},\boldsymbol{k}'}G(\boldsymbol{k},t-t') \tag{3.23}$$

The time-ordering operator T can be mathematically expressed using the Heaviside function $\theta(t)$, which leads to the following equation for the Green's function

$$iG_{\boldsymbol{k}}(t-t') = \theta(t-t')\sum_n e^{-i\left[E_n^{N+1}-E_0^N\right](t-t')}|\langle N+1,n|c_{\boldsymbol{k}}^+|N,0\rangle|^2$$

$$\pm\theta(t'-t)\sum_n e^{-i\left[E_0^N-E_n^{N-1}\right](t-t')}|\langle N-1,n|c_{\boldsymbol{k}}|N,0\rangle|^2. \tag{3.24}$$

Here E_n^{N+1} and E_n^{N-1} are all the exact eigenvalues of the $N+1$ and $N-1$ particle systems respectively, n represents all quantum numbers necessary to specify the state completely, and E_0^N is the exact ground state energy for the system with N particles ($n=0$). Using the integral form of the Heaviside function,

$$\theta(t) = -\lim_{\Gamma \to 0} \frac{1}{2\pi i} \int_{-\infty}^{\infty} \frac{e^{-i\omega t}}{\omega + i\Gamma},$$

(3.25)

the Green's function can easily be Fourier transformed into the frequency representation:

$$G_{\boldsymbol{k}}(\omega) = \lim_{\Gamma \to 0} \left[\sum_n \frac{|\langle N+1, n|c_{\boldsymbol{k}}^+|N, 0\rangle|^2}{\omega - \left[E_n^{N+1} - E_0^N\right] + i\Gamma} \right.$$
$$\left. \mp \sum_n \frac{|\langle N-1, n|c_{\boldsymbol{k}}|N, 0\rangle|^2}{\omega - \left[E_0^N - E_n^{N-1}\right] - i\Gamma} \right].$$

(3.26)

Equation (3.26) provides insight into the analytical properties of the single-particle Green's function. The frequency ω appears only in the denominators of the above equation. The Green's function is a meromorphic function of the complex variable ω, and all its singularities are simple poles, which are infinitesimally shifted into the upper half-plane of ω when $\omega > 0$ and into the lower one if $\omega < 0$. Each pole corresponds to an excitation energy. If we now set

$$E_n^{N+1} - E_0^N = (E_n^{N+1} - E_0^{N+1}) + (E_0^{N+1} - E_0^N) = \omega_n - \mu$$
$$E_n^{N-1} - E_0^N = (E_n^{N-1} - E_0^{N-1}) + (E_0^{N-1} - E_0^N) = \mu' - \omega_n', \quad (3.27)$$

then ω_n and ω_n' denote excitation energies in the $(N+1)$-and $(N-1)$-particle systems respectively and μ and μ' are changes of the ground state energy when a particle is added to the N-particle system or otherwise is removed from the N-particle system, , known as the chemical potentials. In the thermodynamic limit $(N \to \infty, V \to \infty, N/V = \text{const})$ one finds within an error of the order N^{-1} that the chemical potential and the excitation energies are independent of the particle number, i.e.

$$\omega_n \approx \omega_n', \quad \mu \approx \mu'.$$

Another simple property of the Green's function which follows from (3.26) is the asymptotic behaviour for large $|\omega|$:

$$G_{\boldsymbol{k}}(\omega) \sim \frac{1}{\omega}.$$

(3.28)

It is convenient to introduce the spectral densities:

$$A_{\boldsymbol{k}}^+(\epsilon) = \sum_n |\langle N+1, n|c_{\boldsymbol{k}}^+|N, 0\rangle|^2 \delta(\epsilon - \omega_n)$$

(3.29 a)

$$A_{\boldsymbol{k}}^-(\epsilon) = \sum_n |\langle N-1, n|c_{\boldsymbol{k}}|N, 0\rangle|^2 \delta(\epsilon - \omega_n),$$

(3.29 b)

which are real and positive functions, and whose physical interpretation is simple. The spectral density function $A_k^+(\omega)$ gives the probability that the original N-particle system with a particle added into the state k will be found in an exact eigenstate of the $(N+1)$-particle system. In other words, it counts the number of states with excitation energy ω and momentum k which are connected to the ground state through the addition of an extra particle. Similarly, the function $A_k^-(\omega)$ is the probability that the original N-particle system and a hole will be found at an exact eigenstate of the $(N-1)$-particle system. Using the spectral functions (3.29), we may write the causal Green's function (3.26) in the Lehmann representation:

$$G_k(\omega) = \lim_{\Gamma \to 0} \int_0^\infty d\epsilon \left[\frac{A_k^+(\epsilon)}{\omega - (\epsilon + \mu) + i\Gamma} \mp \sum_n \frac{A_k^-(\epsilon)}{\omega + \epsilon - \mu - i\Gamma} \right]. \quad (3.30)$$

In addition, the spectral functions (3.29) may by expressed via the causal Green's function (3.26):

$$A_k^+(\omega - \mu) = -\frac{1}{\pi} \mathrm{Im}\, G_k(\omega), \quad \omega > \mu \quad (3.31\ \text{a})$$

$$A_k^-(\mu - \omega) = \pm \frac{1}{\pi} \mathrm{Im}\, G_k(\omega), \quad \omega < \mu \quad (3.31\ \text{b})$$

3.1.2 Quasi-Particles

From the Lehmann representation of the Green's function (3.30), it easy to see that the special features of the Green's function originate from the denominator whose zeros can be interpreted as single-particle excitations. If the Green's function has a pole ω_k in the momentum state k, then the spectral function $A_k^+(\omega)$ will have a strong maximum at the energy $\omega_k = \omega - \mu$. If $c_k^+|N\rangle$ was an eigenstate, the peak would be a δ-function. In the presence of an interaction, the state $c_k^+|N\rangle$ will not be, in general, an eigenstate. The system will have many other states with the same momentum. An exact eigenstate will be a linear combination of the respective Slater determinants with energies spread out by the interaction. The shape of the function $A_k^+(\omega)$ will depend strongly on the interaction: the stronger the interaction the larger the spread of energies and hence the larger the width of the function $A_k^+(\omega)$. Inserting the spectral functions back into the time-representation of the Green's function, one sees that a finite width of the spectral function gives rise to a loss of coherence with increasing time, and hence to a damping of the propagation. The behaviour for positive times will be approximately:

$$iG_k(t) \sim z_k e^{-i\omega_k t - \Gamma_k t} + iG_k^{\text{incoherent}}(t), \quad \Gamma_k > 0, \ t > 0 \quad (3.32)$$

where $\omega_{\boldsymbol{k}}$ defines the quasi-particle energy, $\Gamma_{\boldsymbol{k}}$ the quasi-particle inverse life-time. The factor $z_{\boldsymbol{k}}$ is called the quasi-particle weight and describes the amount of coherence in the quasi-particle Green's function.

The above quasi-particle Green's function reads in frequency space:

$$G_{\boldsymbol{k}}(\omega) = \frac{z_{\boldsymbol{k}}}{\omega - \omega_{\boldsymbol{k}} + i\Gamma_{\boldsymbol{k}}} + G_{\boldsymbol{k}}^{\text{incoherent}}(\omega), \quad \omega > \mu, \tag{3.33}$$

Now, in contrast to (3.26), $\Gamma_{\boldsymbol{k}}$ is finite because it is determined by the inter-actions. The incoherent part of the Green's function is a smooth and mainly structureless function of frequency. This form gives rise to the spectral function

$$A_{\boldsymbol{k}}^{+}(\omega) \sim \left| \frac{\text{Re}z_{\boldsymbol{k}} + \text{Im}z_{\boldsymbol{k}}(\omega - \omega_{\boldsymbol{k}})}{(\omega - \omega_{\boldsymbol{k}}) + \Gamma_{\boldsymbol{k}}^{2}} \right|. \tag{3.34}$$

The last equation shows that the shape of $A_{\boldsymbol{k}}^{+}(\omega)$ is determined by the pole in the complex plane, and in the special case, $\text{Im}z_{\boldsymbol{k}} = 0$, it has the symmetric Lorentzian form. In general, the spectral function has the asymmetric Breit-Wigner shape as illustrated in Fig. 3.1. The peak in $A_{\boldsymbol{k}}^{+}(\omega)$ is associated with a quasi-particle state or elementary excitation. The physical meaning of $\Gamma_{\boldsymbol{k}}$ is clearly seen from the time-representation (3.33).

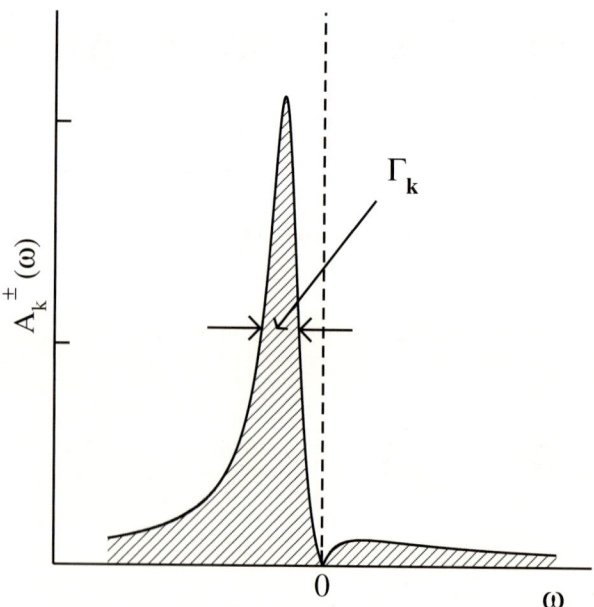

Fig. 3.1. Spectral functions $A_{\boldsymbol{k}}^{\pm}(\omega)$ with a quasi-particle peak of energy $\omega_{\boldsymbol{k}} > \mu$ with lifetime $\Gamma_{\boldsymbol{k}}^{-1}$.

3.1.3 Self-Energy

The exact explicit expression for the single-particle Green's function or its spectral function is only known for a few systems. In general one has to resort to some approximations for the Green's function. One class of approximations is derived via the equation of motion (3.19). It is useful to split the Hamiltonian into its non-interacting part $\hat{H}_0$, which now also includes the Coulomb potential from the electron charge density, and the interaction $\hat{V}$. Application of these equations to the definition of the Green's function leads to the equation of motion for the single-particle Green's function

$$\left[i\frac{\partial}{\partial t} - \hat{H}_0\right] G_{\sigma,\sigma'}(\boldsymbol{r}\,t, \boldsymbol{r}'t')$$

$$+i\int d^3 r'' v(\boldsymbol{r}, \boldsymbol{r}'')\langle \Psi_0|T\left[\hat{\Psi}^+_{\sigma''}(\boldsymbol{r}''t)\hat{\Psi}_{\sigma''}(\boldsymbol{r}''t)\hat{\Psi}_{\sigma}(\boldsymbol{r}t)\hat{\Psi}^+_{\sigma'}(\boldsymbol{r}'t')\right]|\Psi_0\rangle$$

$$= \delta(\boldsymbol{r} - \boldsymbol{r}')\delta_{\sigma,\sigma'}\delta(t - t'). \tag{3.35}$$

where

$$v(\boldsymbol{r}, \boldsymbol{r}') = \frac{1}{|\boldsymbol{r} - \boldsymbol{r}'|} \tag{3.36}$$

is the Coulomb kernel. This equation involves a two-particle Green's function. The equation of motion of the two-particle Green's function contains the three-particle Green's function, and so on. Subsequent application of 3.20 generates a hierarchy of equations, which relate an n-body Green's function to an n+1 body Green's function. Approximations can be obtained when this hierarchy is truncated at some stage by making an decoupling ansatz for higher order Green's functions in terms of lower order Green's functions.

An alternative approach is to define a generalized, non-local and energy-dependent potential, which formally includes all effects due to the interaction. The equation

$$\left[i\frac{\partial}{\partial t} - \hat{H}_0\right] G_{\sigma,\sigma'}(\boldsymbol{r}\,t, \boldsymbol{r}'t')$$

$$-\sum_{\sigma''}\int d^3 r'' \int dt'' \Sigma_{\sigma,\sigma''}(\boldsymbol{r}\,t, \boldsymbol{r}''t'')\,G_{\sigma''\sigma1'}(\boldsymbol{r}''t'', \boldsymbol{r}'t')$$

$$= \delta_{\sigma,\sigma'}\delta(\boldsymbol{r} - \boldsymbol{r}')\delta(t - t') \tag{3.37}$$

implicitly defines this potential $\Sigma_{\sigma,\sigma''}(\boldsymbol{r}t, \boldsymbol{r}''t'')$, which is called the self-energy operator, or mass operator.

Introducing the Green's function of the non-interacting part of the Hamiltonian, $\hat{H}_0$, which obeys the equation

$$\left[i\frac{\partial}{\partial t} - \hat{H}_0\right] G^0_{\sigma,\sigma'}(\boldsymbol{r}\,t, \boldsymbol{r}'t') = \delta(\boldsymbol{r} - \boldsymbol{r}')\delta_{\sigma,\sigma'}\delta(t - t') \tag{3.38}$$

one easily sees that the full and the non-interacting Green's function are related by the Dyson equation:

$$G_{\sigma,\sigma'}(\boldsymbol{r},\boldsymbol{r}';\omega) = G^0_{\sigma,\sigma'}(\boldsymbol{r},\boldsymbol{r}';\omega)$$
$$+ \int d^3x \int d^3x' \sum_{\sigma_1,\sigma_2} G^0_{\sigma,\sigma_1}(\boldsymbol{r},\boldsymbol{x};\omega)\, \Sigma_{\sigma_1,\sigma_2}(\boldsymbol{x},\boldsymbol{x}';\omega) G_{\sigma_2,\sigma'}(\boldsymbol{x}',\boldsymbol{r}';\omega). \quad (3.39)$$

The Dyson equation can also be derived using Feynman's diagrammatic technique. More details about that can be found in standard text books on the many-body problem [3.22, 3.24, 3.42].

The simplest approximation for the self-energy is the complete neglect of it, i.e.

$$\Sigma \equiv \Sigma_H = 0, \quad (3.40)$$

corresponding to the case of $\hat{V} = 0$. Since the classical electrostatic potential is already included in H_0 this reduces to the Hartree approximation to the many-body problem. Due to the structure of the Dyson equation an approximation for the self-energy of finite order corresponds to an infinite order perturbation theory. The self-energy can be evaluated by using Wick's theorem, Feynman's diagram technique or by Schwinger's functional derivative method.

Here we follow Hedin and Lundquist [3.29] and present an iterative method for generating more and more elaborate approximations to the self-energy. The self-energy is a functional of the full Green's function and can formally also be expressed as a series expansion in a dynamically screened interaction W. In turn, the screened Coulomb interaction can be expressed via the polarisation function P which is related to the dielectric function. One can show that all functions appearing in the evaluation of the full Green's function and the self-energy, form a set of coupled integral equations which are known as the Hedin equations. Here we present this set of equations, as common in literature, in the real-space/time representation

$$G(1,2) = G_0(1,2) + \int d(3,4)G_0(1,3)\Sigma(3,4)G(4,2), \quad (3.41)$$

$$\Sigma(1,2) = i \int d(3,4)W(1,3^+)G(1,4)\Gamma(4,2;3), \quad (3.42)$$

$$W(1,2) = v(1,2) + \int d(3,4)v(1,3)P(3,4)W(4,2), \quad (3.43)$$

$$P(1,2) = -i \int d(3,4)G(2,3)\Gamma(3,4;1)G(4,2^+), \quad (3.44)$$

$$\Gamma(1,2;3) = \delta(1-2)\delta(2-3) + \int d(4,5,6,7)\frac{\delta\Sigma(1,2)}{\delta G(4,5)}G(4,6)G(7,5)\Gamma(6,7;3),$$
$$(3.45)$$

where we have used an abbreviated notation $(1) = (\boldsymbol{r}_1, \sigma_1, t_1)$ (the symbol t^+ means $\lim \eta \to 0(t + \eta)$ where η is a positive real number). $v(1, 2) = v(\boldsymbol{r}, \boldsymbol{r}')\delta(t_1 - t_2)$ is the bare Coulomb potential, and $\Gamma(1, 2; 3)$ is the vertex function containing fluctuations of the charge density. The Hedin equations should be solved self-consistently: one starts with $\Sigma = 0$ in (3.45), then one calculates, with some starting Green's function, the polarisation function (3.44), the screened Coulomb function (3.43), the self-energy (3.42), and finally, with the Dyson equation (3.41), the new Green's function which should be used with the self-energy for evaluation of the vertex function (3.45). This process should be repeated until the resulting Green's function coincides with the starting one. For real materials such calculations are extremely difficult, mainly because of the complexity of the vertex function (3.45). In practice one usually truncates the self-consistency cycle and approximates one or more functions, appearing in the Hedin equations (3.41)-(3.45).

Before we turn to practical approaches for the self energy, we shall consider its characteristic features following from quite general considerations [3.39].

The formal solution of Dyson's equation (for translational invariant systems in momentum space) is:

$$G_{\boldsymbol{k}}(\omega) = \frac{1}{\omega - \omega_{\boldsymbol{k}}^0 - \Sigma_{\boldsymbol{k}}(\omega)}. \tag{3.46}$$

The singularities of the exact Green's function $G_{\boldsymbol{k}}(\omega)$, considered as a function of ω, determine both the excitation energies $\omega_{\boldsymbol{k}}$ of the system and their damping $\Gamma_{\boldsymbol{k}}$. From the Lehman representation (3.30) and the Dyson equation (3.46) it follows that the excitation energy $\omega_{\boldsymbol{k}}$ is given by

$$\omega_{\boldsymbol{k}} = \omega_{\boldsymbol{k}}^0 + \operatorname{Re}\Sigma_{\boldsymbol{k}}(\omega_{\boldsymbol{k}}). \tag{3.47}$$

Analogously, the damping $\Gamma_{\boldsymbol{k}}$ is defined by the imaginary part of the self-energy:

$$\Gamma_{\boldsymbol{k}} = [1 - \frac{\partial \operatorname{Re}\Sigma_{\boldsymbol{k}}(\omega_{\boldsymbol{k}})}{\partial \omega}]^{-1} \operatorname{Im}\Sigma_{\boldsymbol{k}}(\omega_{\boldsymbol{k}}). \tag{3.48}$$

Using the behaviour of the Green's function for large $|\omega|$, we can write an assymtotic series

$$[G_{\boldsymbol{k}}(\omega)]^{-1} = \omega + a_{\boldsymbol{k}} + b_{\boldsymbol{k}}/\omega + ..., \tag{3.49}$$

and from (3.46) it follows that the self-energy is a regular function at infinity:

$$\Sigma_{\boldsymbol{k}}(\omega) = -(\omega_{\boldsymbol{k}}^0 + a_{\boldsymbol{k}} + b_{\boldsymbol{k}}/\omega + ...) \tag{3.50}$$

Further, since the imaginary part of the Green's function never vanishes except on the real axis, the Green's function $G_{\boldsymbol{k}}(\omega)$ has no complex zeros. From the analyticity of $G_{\boldsymbol{k}}(\omega)$ it follows that $\Sigma_{\boldsymbol{k}}(\omega)$ is analytic everywhere in the complex plane with the possible exception of the real axis. Another important property of the self-energy is its behaviour in the vicinity of the chemical potential. If one finds

$$|\mathrm{Im}\Sigma_{\boldsymbol{k}}(\omega)| \sim (\omega - \mu)^2 \tag{3.51}$$

then the system is called a Fermi liquid [3.40]. This relation is not valid in general because one of its consequences is the existence of a sharp Fermi surface, which is certainly not present in some systems of fermions with attractive forces between particles.

3.1.4 Kohn-Sham Approximation for the Self-Energy

In the last section we have discussed general properties of the Green's function and how the Green's function is related to quasi-particle excitations. Now we shall consider some practical approaches for the self-energy, with which we can solve the Dyson equation (3.39) on a first-principles (or ab-initio) level.

The first method we discuss is the direct application of density functional theory (DFT) to the calculation of the Green's function.

DFT is one of the most powerful and widely used ab-initio methods [3.14, 3.19, 3.45]. It is based on the Hohenberg and Kohn theorem [3.30], which implies, that all ground state properties of an inhomogeneous electron gas can be described by a functional of the electron density, and provides a one-to-one mapping between the ground-state density and the external potential. Kohn and Sham [3.34] used the fact that this one-to-one mapping holds both for an interacting and a non-interacting system, to define an effective, non-interacting system, which yields the same ground-state density as the interacting system. The total energy can be expressed in terms of this non-interacting auxiliary system as:

$$E_0 \quad = \quad \min_{\rho}\left\{T[\rho] + \int v_{\mathrm{ext}}(\boldsymbol{r})\rho(\boldsymbol{r})\mathrm{d}^3r + \frac{1}{2}\int\int\frac{\rho(\boldsymbol{r})\rho(\boldsymbol{r}')}{|\boldsymbol{r}-\boldsymbol{r}'|}\mathrm{d}^3r\mathrm{d}^3r'+ \right.$$
$$\left. +E_{xc}[\rho]\right\}. \tag{3.52}$$

Here the first term is kinetic energy of a non-interacting system with the density ρ, the second one is the potential energy of the external field $v_{\mathrm{ext}}(\boldsymbol{r})$, the third is the Hartree energy, and the exchange-correlation energy $E_{xc}[\rho]$ entails all interactions, which are not included in the previous terms. The charge density $\rho(\boldsymbol{r})$ of the non-interacting system, which by construction equals the charge density of the full system, can be expressed through the orthogonal and normalized functions $\varphi_i(\boldsymbol{r})$ as

$$\rho(\boldsymbol{r}) = \sum_i^{\mathrm{occ.}} |\varphi_i(\boldsymbol{r})|^2. \tag{3.53}$$

Variation of the total energy functional (3.52) with respect to the function $\varphi_i(\boldsymbol{r})$ yields a set of equations, the Kohn-Sham (KS) equations, which have to be solved self-consistently, and which are of the form of a single-particle Schrödinger equation:

$$\left[-\frac{\nabla^2}{2} + v_{\text{eff}}(\boldsymbol{r}) \right] \varphi_i(\boldsymbol{r}) = \varepsilon_i \varphi_i(\boldsymbol{r}). \tag{3.54}$$

This scheme corresponds to a single-particle problem, in which electrons move in the effective potential

$$v_{\text{eff}}(\boldsymbol{r}) = v_{\text{ext}}(\boldsymbol{r}) + v_{\text{H}}(\boldsymbol{r}) + v_{\text{xc}}(\boldsymbol{r}). \tag{3.55}$$

Here $v_{\text{H}}(\boldsymbol{r})$ is the Hartree potential and $v_{\text{xc}}(\boldsymbol{r}) = \frac{\delta E_{\text{xc}}(\rho(\boldsymbol{r}))}{\delta \rho(\boldsymbol{r})}$ is the exchange-correlation potential.

The energy functional (3.52) provides in principle the exact ground state energy if the exchange-correlation energy $E_{xc}[\rho]$ is known exactly. This functional is difficult to find, since this would be equivalent to the solution of the many-body problem, and remains a topic of current research in density-functional theory. For applications the exchange-correlation energy $E_{xc}[\rho]$ is usually approximated by some known functionals obtained from some simpler model systems. One of the most popular approaches is the local-density approximation (LDA), in which the exchange-correlation energy $E_{xc}[\rho]$ of an inhomogeneous system is approximated by the exchange-correlation energy of a homogeneous electron gas, which can be evaluated accurately, e.g., by Quantum Monte Carlo techniques. Thereby all many-body effects are included on the level of the homogeneous electron gas in the local exchange-correlation potential, which depends on the electronic density and some parameters obtained from many-body calculations for a the homogeneous electron gas [3.57, 3.13, 3.58, 3.48, 3.47]. In many cases the LDA works well and is already for three decades widely used for great variety of systems (see review by R.O. Jones and O. Gunnarsson [3.31], and some text books on DFT [3.14, 3.19, 3.45]). The simplicity of the local density approximation makes it possible to solve the Kohn-Sham equation (3.54) with different basis sets and for different symmetry cases. When the on-site Coulomb interaction dominates the behaviour of electrons (strongly correlated systems), the local-density does not work well, and another approximation of $E_{xc}[\rho]$ is needed.

As was already mentioned above, standard density functional theory is designed for the study of ground state properties. In many cases it appears reasonable to interpret the eigenvalues ε_i in the one-particle equation (3.54) as excitation energies, but there is no real justification for such an interpretation [3.46, 3.51]. However, also the Green's function and thus the self-energy are, among other dependencies, functionals of the electronic density. Sham and Kohn argue that for sufficiently homogenous systems, and for energies in the close vicinity of the Fermi-surface, the approximation

$$\Sigma(\boldsymbol{r}, \boldsymbol{r}'; \omega) \approx v_{xc}[\rho](\boldsymbol{r})\delta(\boldsymbol{r} - \boldsymbol{r}') \tag{3.56}$$

should be good. The range of energies, in which this approximation can be expected to work, depends on the effective mass of a homogeneous electron gas with a density, corresponding to some average of the actual density of the system. Using this approximation in the Dyson equation, one sees that the full Green's function is approximated by the Green's function of the non-interacting Kohn-Sham system, which can be expressed in terms of the KS orbitals and KS energies:

$$G_{\mathrm{KS},\sigma,\sigma'}(\boldsymbol{r}, \boldsymbol{r}'; \omega) = \lim_{\Gamma \to 0} \left(\sum_i^{\mathrm{occ}} \frac{\varphi_{i\sigma}(\boldsymbol{r})\varphi^*_{i\sigma'}(\boldsymbol{r}')}{\omega - \mu + \epsilon_i + i\Gamma} + \sum_i^{\mathrm{unocc}} \frac{\varphi_{i\sigma}(\boldsymbol{r})\varphi^*_{i\sigma'}(\boldsymbol{r}')}{\omega - \mu + \epsilon_i - i\Gamma} \right)$$
$$\tag{3.57}$$

This argument explains why the otherwise unjustified interpretation of the Kohn-Sham Green's function, often gives surprisingly good results for spectroscopy calculations. Despite the lack of a proper justification, this fact led to the development of variety of methods on the first-principle level, which are successfully applied for many spectroscopy phenomena (see recent reviews in [3.16, 3.15]). An example, how the density functional theory within the LDA does work, is illustrated in Fig. 3.2. Here we present magneto-optical spectra for iron calculated by a self-consistent LMTO method [3.3] and the experimental results [3.37]. Because magneto-optics in the visible light belongs to the low-energy spectroscopy, one can expect, that the use of the LDA is reasonable. Indeed, the calculated polar Kerr rotation and Kerr ellipticity agree very well with the experimental curve. The theoretical curve reproduces all main features of the experimental result. With increasing energy the agreement with experiment is getting worse, as expected from the deterioration of the LDA approximation for higher energies. The theory also could not represent the magnitude of the experimental curve for the whole energy range; this is related to the damping of the quasi-particle states. The main failure of the LDA in the description of spectroscopic phenomena is the inability to reproduce the damping of single-particle excitation, which is given by the imaginary part of the self-energy (3.48) and which is not present in the approximation (3.81). The calculated spectrum is usually artificially smeared by a Lorentzian broadening with some constant width Γ, but this is not a satisfactory approximation, because in reality the damping has a more complicated structure. An evident case when the LDA does not work is shown in Fig. 3.3. Here we present photoemission spectra for silver at a photon energy of 26 eV. The solid line shows a theoretical spectrum calculated by these authors using a self-consistent Green's function method [3.18,3.38] within the LDA. The dashed line reproduces the experiment [3.43]. The low energy part of the spectra (up to 5.5 eV below the Fermi level) is adequately represented by theory. At the energy 5.1 eV below the Fermi level the experiment shows a peak corresponding to a 4d-state, which is predicted by theory at 3.6 eV

Fe

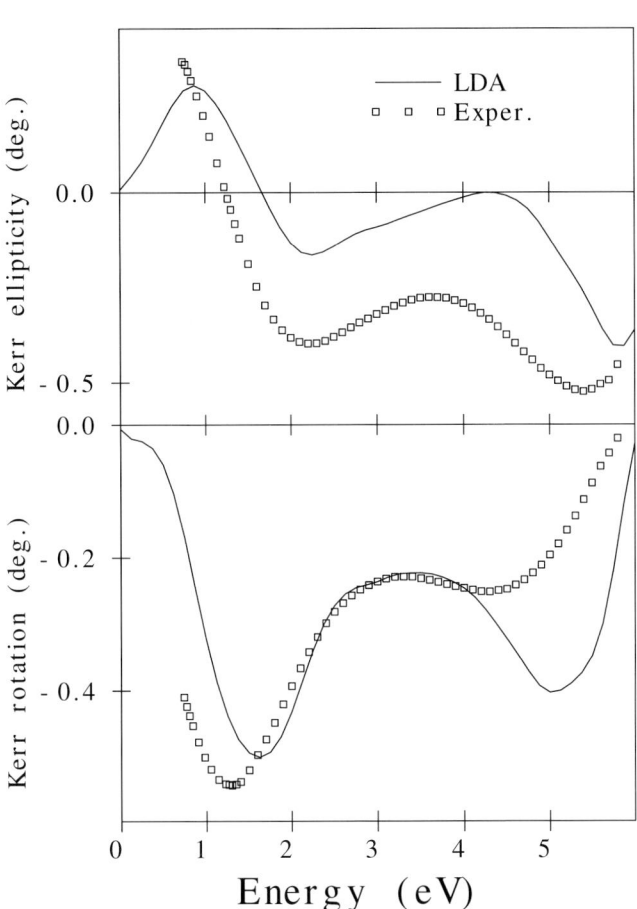

Fig. 3.2. Calculated [3.3] and experimental [3.37] magneto-optical spectra of Fe

below the Fermi level. The discrepancies are related to the inadequacy of the local approximation of $v_{xc}(r)$ and to the failure of correlating Kohn-Sham eigenvalues with excitation energies in the photoemission experiment. Below we point some serious faults of the DFT and the LDA in the description of ground state and quasi-particle state properties:

- The approximation of the exchange-correlation energy is a crucial point of DFT calculations. Existing approximations are usually not applicable for systems with partially filled inner shells. In presence of strong-correlated electrons the LDA does not provide reasonable results.
- The LDA is not completely self-interaction free. The unphysical interaction of an electron with itself can approximately be subtracted if the electron is sufficiently localized [3.48]. This remarkably improves the total energy and

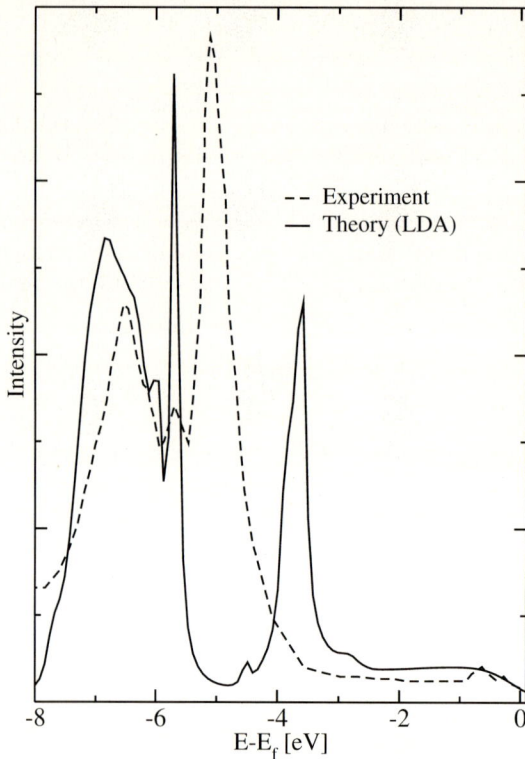

Fig. 3.3. Photoemission spectra of silver at a photon energy of 26 eV: theory (solid line) and experiment (dashed line) [3.43]

other ground state properties, but quasi-particle excitations are still badly described because the Kohn-Sham approach (3.81) is even less appropriate for strongly localized electrons. Application the SIC method for extended systems is difficult since the correction term, as proposed by Perdew and Zunger [3.48] disappears for delocalized electrons. The method can be generalized by applying the correction to localized Wannier states [3.54, 3.53].

– The band gaps in *sp*-semiconductors like Si, GaAs, Ge, etc. are by 70-100% systematically underestimated.

– The damping of excitation states can not be conceptually described within a direct interpretation of the Kohn-Sham system.

The main reason for all those problems is the fact, that (standard) density functional theory is not designed for excited states. The approximation (3.81) is empirically right only for specific materials and for a limited range of energies. Optical excitations (two-particle excitations) can, in principle, be obtained from time-dependent DFT [3.49, 3.44].

3.2 Methods of Solving the Kohn-Sham Equation

In this chapter we describe two methods of calculating the electronic band structures of crystals, which is equivalent to calculating the KS Green's function. These methods represent two general approaches of solving the Kohn-Sham equation.

In the first approach the Kohn-Sham equation can be solved in some basis set. The wave function $\varphi_i(\mathbf{r})$ in the Kohn-Sham equation (3.54) can be represented as linear combination of N appropriate basis functions $\phi_i(\mathbf{r})$:

$$\varphi(\mathbf{r}) = \sum_i^N C_i \phi_i(\mathbf{r}) \tag{3.58}$$

The basis set $\{\phi_i(\mathbf{r})\}$ should correspond to the specifics of the solving problem such as the crystal symmetry, the accuracy or special features of the electronic structures. According to the variational principle the differential equation (3.54) represented in the basis (3.58) can be transformed to a set of linear equations:

$$\sum_j^N (H_{ij} - \varepsilon S_{ij})C_j = 0, \quad j = 1, 2, .., N. \tag{3.59}$$

The matrices

$$H_{ij} = \int d^3 r \phi_i^*(\mathbf{r}) \hat{H}(\mathbf{r}) \phi_i(\mathbf{r}), \tag{3.60}$$

$$S_{ij} = \int d^3 r \phi_i^*(\mathbf{r}) \phi_i(\mathbf{r}) \tag{3.61}$$

are the Hamiltionan and overlap matrices respectively. Using identity

$$S = \left(S^{1/2}\right)^T S^{1/2} \tag{3.62}$$

the system of (3.59) can be easily transformed to the ordinary eigenvalue problem

$$\sum_j^N (\tilde{H}_{ij} - \varepsilon \delta_{ij}) \tilde{C}_j = 0 \tag{3.63}$$

with $\tilde{H} = \left(S^{1/2}\right)^T H S^{1/2}$ and $\tilde{C} = S^{1/2}C$. This ordinary eigenvalue problem (3.63) can by solved by the diagonalization of the matrix $\tilde{H}$.

All variational methods differ from each other only by the choice of the basis functions $\{\phi_i(\mathbf{r})\}$ and by the construction of the crystal potential. Several efficient basis methods have been developed in last four decades and are widely used for band structure calculations of solids. A designated choice of basis functions serves for specific purposes. For example the *linearized muffin-tin orbital* (LMTO) method [3.1] or *augmented spherical wave* (ASW) [3.60] method provide very fast band structure calculations, with an accuracy which is sufficient for many applications in solids. A *tight-binding* representation of the basis [3.21,3.2] is very useful for different models with parameters determined from band structure calculations. Usually simple basis methods are very fast but not very accurate. Methods with more complicated basis and potential constructions are appropriate for high precision electronic structure calculations (see *augmented plane wave* (APW) method [3.52], *full-potential linearized augmented plane wave* (FPLAPW) method [3.10], *projected augmented wave* (PAW) method [3.12], *full-potential local-orbital minimum-basis* method [3.32], several norm-conserving pseudopotential methods [3.26,3.56]). Such methods are slower than the fast simple basis methods mentioned above, but on modern computers they are successfully applied even to extended systems like large super-cells, surfaces and interfaces.

Another efficient way to solve the the Kohn-Sham equation (3.54) is the Green's function method. This approach is based on a corresponding mathematical scheme of solving differential equations. Basically the method uses Green's function technique to transform the Schrödinger equation into an equivalent integral equation. In the crystal one can expand the crystal states in a complete set of functions which are solutions of the Schrödinger equation within the unit cell, and then determine the coefficients of the expansion by requiring that the crystal states satisfy appropriate boundary conditions. This method was proposed originally by Korringa [3.35], Kohn and Rostoker [3.33], in different though equivalent form. More details about the *Korringa-Kohn-Rostoker* (KKR) method can be found in [3.23,3.59].

As an example of ab-initio band structure methods we shall discuss below a high precision full-potential combined basis method [3.20,3.17], which is a flexible generalisation of the LCAO computational scheme. Before we start the discussion about this approach, we consider some general features which are typical for any basis method.

In a crystal the effective potential $v_{\text{eff}}(\boldsymbol{r})$ in the Kohn-Sham equation (3.54) is a periodic function of direct lattice vectors $\boldsymbol{R}_i$: $v_{\text{eff}}(\boldsymbol{r}+\boldsymbol{R}_i) = v_{\text{eff}}(\boldsymbol{r})$. A wave function $\varphi_n(\boldsymbol{k};\boldsymbol{r})$ satisfies the Bloch theorem:

$$\varphi_n(\boldsymbol{k};\boldsymbol{r}+\boldsymbol{R}_i) = \exp\left(i\boldsymbol{k}\cdot\boldsymbol{R}_i\right)\varphi_n(\boldsymbol{k};\boldsymbol{r}) \qquad (3.64)$$

Here $\boldsymbol{k}$ is a wave vector of an electron. According to the Bloch theorem solutions of the Kohn-Sham equation (3.54) depend on the wave vector $\boldsymbol{k}$ and can generally be represented as

$$\varphi_n(\boldsymbol{k};\boldsymbol{r}) = \exp\left(i\boldsymbol{k}\cdot\boldsymbol{r}\right)u_n(\boldsymbol{k};\boldsymbol{r}), \qquad (3.65)$$

where $u_n(\boldsymbol{k};\boldsymbol{r})$ is a lattice periodic function. The index n is known as the band index and occurs because for a given $\boldsymbol{k}$ there will be many independent eigenstates. In any basis method the crystal states are expanded in a complete set of Bloch type functions. The trial wave function expanded in the basis set should be close to the true wave function in the crystal. In crystals with almost free valence electrons an appropriate basis would be plane waves while the atomic-like functions are a proper choice for systems with localized valence electrons. Different methods of band calculations can be classified depending on which of the two above approaches is followed. Pseudopotential methods or *orthogonalized plane wave* (OPW) method use plane waves or modified plane waves as the basis set. The tight-binding methods like the LMTO or LCAO are based on the second concept. There are also approaches which combine both delocalized and localized functions. The aim of these schemes is to find the best fit for true wave functions in systems which contain different types of electron states. Moreover, the true wave function changes its behaviour throughout the crystal: close to the nuclear the wave function is usually strongly localized and in the interstitial region it has more free electron character. To this class of methods belongs the very popular FPLAPW method [3.10] in which the wave functions are represented by localized functions in the *muffin-tin* sphere which are smoothly matched to plane waves in the interstitial region. Here we shall discuss another combined basis approach which is based on the LCAO scheme. In the LCAO method, originally suggested by Bloch [3.11], the atomic orbitals of the atoms (or ions) inside the unit cell are used as basic expansion set for the Bloch functions. This procedure is convenient only for low energy states because the atomic orbitals are very localized and poorly describe the wave functions in the region where the crystal potential is flat. The atomic orbitals can be optimized as suggested in [3.21]. In this approach the atomic-like functions are squeezed by an additional attractive potential. The extention of the basis functions is tuned by a parameter that can be found self-consistently [3.32]. One of the main difficulties of the LCAO scheme is an abundance of multicenter integrals, which must be performed to arrive at a reasonable accuracy of band structure calculations. This difficulty is evidently most critical for solids which have a close packed structure and therefore a great number of neighbours within a given distance. To avoid these difficulties, one usually imposes some restrictions on the basis set and, as a rule, this leads to its incompleteness, which, on the other hand, is most pressing for open structures. Completeness can be regained by adding plane waves to the LCAO's. By increasing the number of plane waves in the basis we can decrease the spatial extent of localised valence orbitals and thereby reduce the number of multi-center integrals. This provides a flexibility which goes far beyond usual pseudo potentials. Retaining some overlap between the local valence orbitals, we improve our plane-wave basis set and can receive a good converged Bloch function for both valence electrons and excited states using a relatively small number of plane waves. Orthogonalization of both LCAO's and plane waves to core states can be done before forming of Hamiltonian and overlap matrices. Thus, the application of combined basis sets in the combination of LCAO's and OPW's allows efficiently to use advantages both approaches.

We have to solve the Kohn-Sham equation for the electronic states in a periodic systems

$$[-\frac{1}{2}\nabla^2 + v_{\text{eff}}(\boldsymbol{r})]\,\varphi_n(\boldsymbol{k};\boldsymbol{r}) = \varepsilon_n(\boldsymbol{k})\,\varphi_n(\boldsymbol{k};\boldsymbol{r})\ . \qquad (3.66)$$

Here v_{eff} is the effective potential (3.55). The wave function $\varphi_n(\boldsymbol{k};\boldsymbol{r})$ describes the Kohn-Sham one-electron state with the wave vector $\boldsymbol{k}$ and the band index n. Without any restriction of generality the effective potential can be split into two parts: a lattice sum of single local potentials decreasing smoothly to zero at the muffin tin radii and a smooth Fourier transformed potential which is the difference between the total effective and the local potential. Thus, the effective potential can be represented as follows

$$v_{\text{eff}}(\boldsymbol{r}) = \sum_{\boldsymbol{R}\boldsymbol{S}} v_{\boldsymbol{S}}^{\text{loc}}(\boldsymbol{r} - \boldsymbol{R} - \boldsymbol{S}) + \sum_{\boldsymbol{G}} e^{i\boldsymbol{G}\cdot\boldsymbol{r}}\, v^{\text{ft}}(\boldsymbol{G})\,, \qquad (3.67)$$

where $\boldsymbol{R}$ and $\boldsymbol{G}$ are direct and reciprocal lattice vectors, and $\boldsymbol{S}$ is a site position in an unit cell. The local potential is decomposed into angular contributions and with our conditions has the form

$$v_{\boldsymbol{S}}^{\text{loc}}(\boldsymbol{r}) = \begin{cases} \sum_L v_{\boldsymbol{S}L}^{\text{loc}}(r)\,Y_L(\hat{\boldsymbol{r}}) & : \quad r \le r_{MT} \\ 0 & : \quad r > r_{MT}\ , \end{cases} \qquad (3.68)$$

where $\hat{\boldsymbol{r}}$ is a normal vector along the vector $\boldsymbol{r}$. Here $Y_L(\hat{\boldsymbol{r}})$ are spherical harmonic functions with the combined index $L = \{l, m\}$. This decomposition greatly simplifies the evaluation of the required matrix elements. One-electron wave functions are sought in the combined basis approach

$$\varphi_n(\boldsymbol{k};\boldsymbol{r}) = \sum_{\mu} A_{n\mu}(\boldsymbol{k})\,\phi_\nu(\boldsymbol{k};\boldsymbol{r}) + \sum_{\boldsymbol{G}} B_{n\boldsymbol{G}}(\boldsymbol{k})\,\phi_{\boldsymbol{G}}(\boldsymbol{k};\boldsymbol{r})\ . \qquad (3.69)$$

$\varphi_\mu(\boldsymbol{k};\boldsymbol{r})$ is the Bloch sum of localised site orbitals $\phi_\mu(\boldsymbol{r} - \boldsymbol{R} - \boldsymbol{S}_\mu)$. The μ−sum runs over both core and valence orbitals:$\mu = \{c, \nu\}$. The core orbital contributions in a valence state are needed to provide orthogonalization of the valence state to true core states. $\varphi_{\boldsymbol{G}}(\boldsymbol{k};\boldsymbol{r}) \sim e^{i(\boldsymbol{k}+\boldsymbol{G})\cdot\boldsymbol{r}}$ is a normalised plane wave. We use core orbital contributions and plane wave contributions seperately, and do not form orthogonalised plane waves (OPWs) at the outset.

In the basis set three types of functions are used: true core orbitals, squeezed local valence orbitals and plane waves. By our definition, true core orbitals are solutions of the Kohn-Sham equation (3.66) which have negligible nearest neighbour overlap among each other (typically less than 10^{-6}). The highest fully occupied shells of each angular momentum (e.g. $2s$ and $2p$ in aluminium, or $3s$ and $3p$ in a $3d$-metal) are treated like valence orbitals in most of our calculations, because their nearest neighbour overlap is not small enough to be neglected, if a larger number of plane waves is included. This is usually the reason for the over-completeness breakdown of OPW expansions. The local basis function (both core and valence) can be constructed from radial functions $\xi^\mu_{Snl}(r)$ which are solutions of the radial Schrödinger equation

$$\left[-\frac{1}{2r}\frac{\partial^2}{\partial r^2}r + \frac{l(l+1)}{2r^2} + v^\mu_S(r) \right] \xi^\mu_{Snl}(r) = \varepsilon^\mu_{Snl}\xi^\mu_{Snl}(r) \,. \tag{3.70}$$

Here, in the case of core electrons ($\mu = c$) the potential $v^c_S(r)$ is the crystal potential averaged around the center S. To obtain squeezed valence electrons ($\mu = \nu$) we use a specially prepared spherical potential by adding an artificial attractive potential to $v^c_S(r)$:

$$v^\nu_S(r) = v^c_S(r) + \left(\frac{r}{r_\nu} \right)^4 \tag{3.71}$$

with a parameter r_ν which serves to tune the radial expansion of the basis functions and can be found self-consistently on the total energy minimum condition. A useful local valence basis orbital should on the contrary rapidly die off outside the atomic volume of its centre, but smooth enough for the Bloch sums of those orbitals to provide a smooth and close approximant to the true valence Bloch wave function so that the remaining difference between the two may be represented by a few OPWs. A local basis function is denoted in the following manner

$$\eta_\mu(\boldsymbol{r}) = \xi^\mu_{Snl}(r)Y_L(\hat{\boldsymbol{r}}) \,, \tag{3.72}$$

where the lower index $\mu = \{c, \nu\}$ acts as multi-index $\{Snlm\}$ for core and valence basis functions respectively.

The third kind of basis functions are plane waves normalised to the crystal volume V:

$$\eta_{\boldsymbol{k}}(\boldsymbol{r}) = \frac{1}{\sqrt{V}}e^{i\boldsymbol{k}\cdot\boldsymbol{r}} \,, \quad V = NV_u \,, \tag{3.73}$$

where N is the number of unit cells and V_u is the unit cell volume. To summarise, the entries in the expansion (3.66) are the basis Bloch functions

$$\phi_c(\boldsymbol{k}; \boldsymbol{r}) \equiv (\boldsymbol{r}|\boldsymbol{k}c) = \sum_{\boldsymbol{R}} \eta_c(\boldsymbol{r} - \boldsymbol{R} - \boldsymbol{S}_c)\frac{1}{\sqrt{N}}e^{i\boldsymbol{k}\cdot(\boldsymbol{R}+\boldsymbol{S}_c)} \,, \tag{3.74}$$

$$\phi_\nu(\boldsymbol{k}; \boldsymbol{r}) \equiv (\boldsymbol{r}|\boldsymbol{k}\nu) = \sum_{\boldsymbol{R}} \eta_\nu(\boldsymbol{r} - \boldsymbol{R} - \boldsymbol{S}_\nu)\frac{1}{\sqrt{N}}e^{i\boldsymbol{k}\cdot(\boldsymbol{R}+\boldsymbol{S}_\nu)} \,, \tag{3.75}$$

$$\phi_{\boldsymbol{G}}(\boldsymbol{k}; \boldsymbol{r}) \equiv (\boldsymbol{r}|\boldsymbol{k}\boldsymbol{G}) = \eta_{\boldsymbol{k}+\boldsymbol{G}}(\boldsymbol{r}) \tag{3.76}$$

The flexibility of the basis (3.72) and (3.73) consists in a balance between the radial extension of the valence basis orbitals and the number of plane waves needed to converge the expansion (3.69). By reducing the parameters r_ν and/or the number of local basis orbitals, the number of multi-center integrals needed in the calculation is reduced at the price of slowing down the convergence speed with the number of plane waves included, and vice versa. Our approach provides a full interpolation between the LCAO and OPW approaches adopting pseudo-potential features.

After the expansion of the wave functions in (3.66) we have the following system of linear equations

$$\sum_\mu [(\boldsymbol{k}\mu'|H|\boldsymbol{k}\mu) - E\,(\boldsymbol{k}\mu'|\boldsymbol{k}\mu)]\,A^\mu(\boldsymbol{k}) = 0\,, \tag{3.77}$$

which gives us the sought band energies $E = E_n(\boldsymbol{k})\,, n = 1, ..., M$, where M gives the rank of the coefficient matrix and corresponds to the number of basis functions. For the expansion coefficients $A_n^\mu(\boldsymbol{k})$ there are M different solutions. The equation system poses the eigenvalue problem

$$(\mathcal{H} - E\mathcal{S})\mathcal{A} = 0 \tag{3.78}$$

with Hamilton matrix $\mathcal{H}$ and overlap matrix $\mathcal{S}$. To solve this eigenvalues problem the matrix elements of the Hamiltionan and overlap matrices must be calculated. This is a non-trivial problem because the basis consists of three types of functions. Moreover the local valence functions are extended in real space and overlap with each other and with the core orbitals. Because of this one needs to calculate multi-center integrals which are assumed to be independent of the wave vector. Due to the limited space we shall not discuss this problem in the paper and refer the reader to the literature [3.20, 3.17, 3.32] for more details. The secular equation (3.78) can be solved in the same manner as discussed in the previous paper by H. Eschrig. The orthogonality of the core orbitals allows us to restructure the matrix in (3.78) so that the eigenvalue problem will be reduced. According to this scheme the solution of the system of (3.78) can be carried out in the following manner. As the first step the matrices $\mathcal{S}_{c\lambda}$, $\mathcal{S}_{\lambda\lambda}$, $\mathcal{H}_{\lambda\lambda}$ ($\lambda = \{\nu, G\}$ is common index for valence basis functions) and the energies of core states ε_c are determined. Then the matrix

$$\mathcal{S}_{\lambda\lambda} - \mathcal{S}_{c\lambda}^\dagger \mathcal{S}_{c\lambda}$$

is decomposed into the product of left tridiagonal matrix $\mathcal{S}_{\lambda\lambda}^l$ and right tridiagonal matrix $\mathcal{S}_{\lambda\lambda}^{l\dagger}$ with the Cholesky method. The next step is the calculation of the inverse matrix $(\mathcal{S}_{\lambda\lambda}^l)^{-1}$. Using this matrix we can calculate the matrix

$$\tilde{\mathcal{H}}_{\lambda\lambda} = (\mathcal{S}_{\lambda\lambda}^l)^{-1\dagger} \left[\mathcal{H}_{\lambda\lambda} - \mathcal{S}_{c\lambda}^\dagger \mathcal{H}_{cc} \mathcal{S}_{c\lambda} \right] (\mathcal{S}_{\lambda\lambda}^l)^{-1}\,. \tag{3.79}$$

After diagonalisation of the matrix $\tilde{\mathcal{H}}_{\lambda\lambda}$ we get the matrix $U_{\lambda\lambda}$, which diagonalizes $\mathcal{H}$ and with this matrix the expansion coefficients $\mathcal{A}$ can be written as:

$$\mathcal{A}^T = \begin{pmatrix} 1 & -\mathcal{S}_{c\lambda} \left(\mathcal{S}_{\lambda\lambda}^r \right)^{-1} U_{\lambda\lambda} \\ 0 & \left(\mathcal{S}_{\lambda\lambda}^r \right)^{-1} U_{\lambda\lambda} \end{pmatrix}. \tag{3.80}$$

The first column of the matrix $\mathcal{A}$ corresponds to the expansion coefficients of core states. According to our assumption that the core states are completely occupied, they do not overlap and are independent of the crystal momentum $\boldsymbol{k}$. Therefore one of the coefficients $A^c(\boldsymbol{k})$ is equal to one, and all other $A^\mu(\boldsymbol{k})$ are zero. The upper block of this column is hence a unit matrix 1 with the dimension $M_c \times M_c$, where M_c is number of the core states, and the lower block is a zero matrix. The second column represents the expansion coefficients of valence functions, where the upper block includes the orthogonalization correction of the valence states to the core states, and the dimension is equal to $M_c \times M_\lambda$. Here M_λ is the number of the valence states. The preceding scheme corresponds to the orthogonalisation corrections of the valence basis due to the core states (e.g. see [3.21]). With this representation the Bloch wave function (3.69) has a convenient form, and the estimations of the matrix elements and the charge density are substantially simplified.

After the eigenvalue problem (3.78) is solved, we can estimate the Bloch function (3.69) and the electronic charge density, which in our method is treated in the same manner as the crystal potential (3.67). This approach provides a very accurate numerical representation of the charge density and the crystal potential, which can be calculated self-consistently within the local density approximation.

As it was already mentioned above the main advantage of the combined basis method is the flexibility of the basis which allows to optimize efficiently band structure calculations without losing the high numerical accuracy. This fact is illustrated in Fig. 3.4 where the self-consistent band energy for Cu is presented. In this picture we demonstrate the convergence of the band energy with the number of plane waves in the basis. The band energies have been calculated for different parameters r_ν in (3.71) which regulates the extension of the local basis functions. The overlap between the local orbitals on different centers is getting larger with increasing the parameter r_ν. Then one needs to calculate more multi-center integrals in matrix elements of the eigenvalue matrix. For smaller r_ν the overlap is reduced but one needs more plane waves to obtained an accurate band structure. Another advantage of the method is the completeness of the basis for large energy range. In many approaches, specially in linearized methods, the basis functions are appropriate only for valence bands. This is sufficient for a study of ground state properties, but makes the methods not applicable for the spectroscopy. In the combined basis method the plane waves fit adequately the high lying bands which enables accurate calculations of spectroscopic characteristics within the local density approximation. Figure 3.5 shows an example of the band structure of Si calculated by the combined basis method. The band energies are shown along symmetry lines in the Brillouin zone. The indirected gap is 0.64 eV

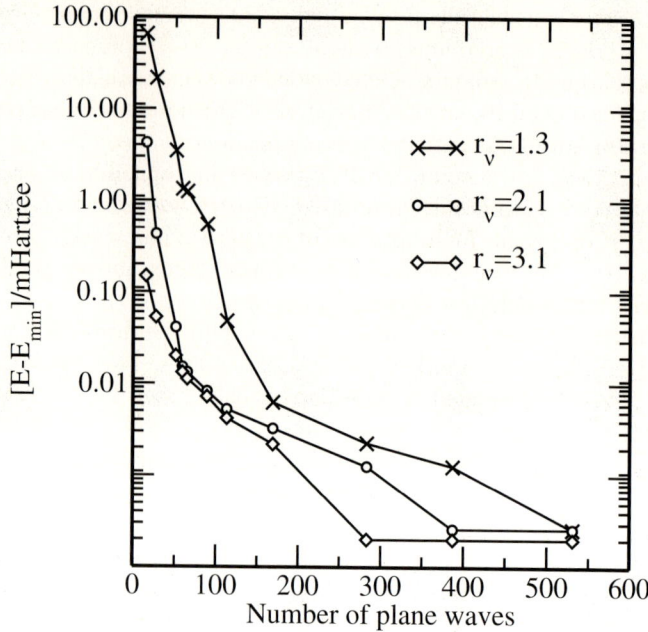

Fig. 3.4. Convergency of the self-consistent band energy of Cu for the different local basis extensions with the number of plane waves

Si

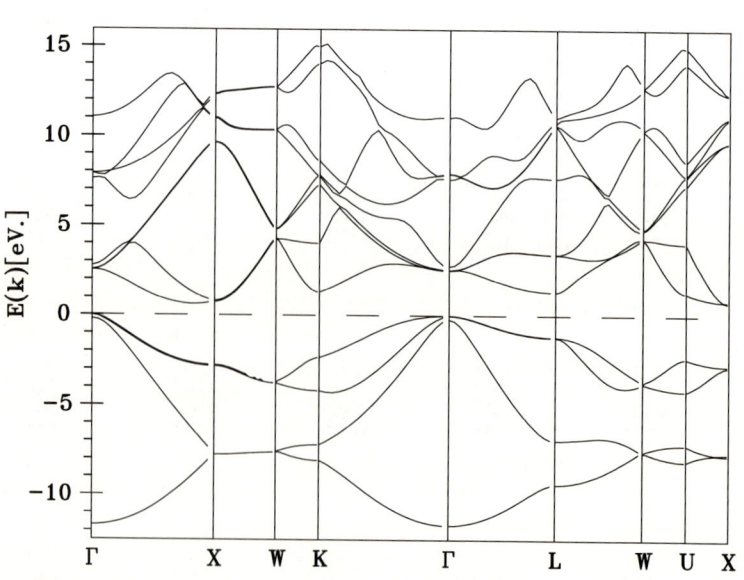

Fig. 3.5. Band structure for Si calculated with the combined basis method [3.17].

which corresponds to the expected LDA value. The band structure is in good agreement with very accurate APW and pseudo-potential calculations.

To summarize, the combined basis method is a high precision tool for the calculation of the electronic structure of solids. The basis set, consisting of localised valence orbitals and plane waves is appropriate for different types of electrons in solids. There are no shape restriction concerning both the electron density and the self-consistent potential. Typically 65 plane waves per atom yield an absolute accuracy of ≤ 1 mRyd for the total energy for a typical transition metal. Including orthogonalised plane waves into the basis set makes it possible to achieve the high precision for total energy calculations of both close packed and open systems.

3.3 GW Approximation

We shall now consider a way for approximating the self-energy using Hedin's set of (3.42)-(3.45). The most difficult part of these equations is the evaluation of the vertex function (3.45). The vertex function contains particle-hole correlation effects and is defined implicitly through the Bethe-Salpeter equation, which involves a two-particle Green's function. The functional derivative $\delta\Sigma/\delta G$ is not trivial to obtain, because the dependence of the self-energy with respect to the full Green's function is not explicitly known. If the electrons interact not too strongly, this functional derivative is small and the vertex function can be approximated by its zero order expression [3.27–3.29]:

$$\Gamma(1,2;3) \approx \delta(1-2)\delta(2-3). \tag{3.81}$$

This yields a simplified version of Hedin's set of equations:

$$\Sigma(1,2) = iW(1,2)G(1,2), \tag{3.82}$$

$$W(1,2) = v(1,2) + \int d(3,4)W(1,3)P(3,4)v(4,2), \tag{3.83}$$

$$P(1,2) = -iG(1,2)G(2,1). \tag{3.84}$$

In this approximation the self-energy is expressed as a product of the self-consistent single-particle propagator G and the self-consistent dynamically screened interaction W. This gives the name for the approximation: GW. The GW approximation (GWA) is consistent in the Baym-Kadanoff sense [3.9,3.8], i.e. it is a particle- and energy-conserving approximation. The GWA corresponds to the first iteration of the Hedin's equations and can be interpreted as the first order term of an expansion of the self-energy in terms of the screened interaction. The (3.41),(3.82)-(3.84) can be solved self-consistently, but in practice, such a calculation is computationally very expensive. Moreover, the experience with self-consistent GW implementations (see the review by F. Aryasetiawan and O. Gunnarson [3.5] and references therein) shows that in many cases the self-consistency even worsens the results in comparison with non-self-consistent calculations. The main reason for this is the neglect the vertex correction. Most existing GW calculations do not attempt self-consistency, but determine good approximations for the single-particle propagator G and the screened interaction separately, i.e., they adopt a "best G, best W" philosophy. The common choice for the single-particle propagator is usually the LDA or Hartree-Fock Green's function. Using this Green's function the linear response function is obtained via the (3.84), and afterward it is used for the calculation of the screened Coulomb interaction (3.83). The self-energy is then determined without further iteration. Nevertheless, with the first iteration of the *GW* approximation encouraging results have been achieved. In Fig. 3.6 we show a typical result for the energy bands of MgO within the GWA (dotted line) compared to a conventional LDA calculation. It is clear from this plot that the GWA gap is in much better agreement with the experimental value than the LDA results.

Below we describe briefly some existing implementations of the GW method. More details can be found in the original papers and in the reviews [3.5, 3.7, 3.44]. The GW integral equations (3.41),(3.82)-(3.84) can be represented in some basis set in the real or reciprocal space and solved by

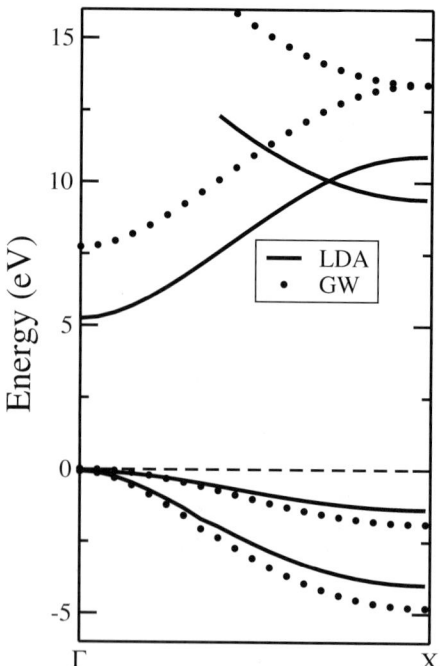

Fig. 3.6. Energy bands of MgO from KKR-LDA (solid) and GWA (dots) calculations. The LDA band gap is found to be 5.2 eV, the GWA band gap is 7.7 eV which is in good agreement with the experimental gap of 7.83 eV.

matrix inversion. The basis set should be appropriate to the symmetry of the particular problem and should be able to represent as accuratly as possible the quite distinctive behaviour of the functions, involved in the GW approach.

– *Plane wave methods*

Pseudo-potentials in conjunction with a plane wave basis set are widely used in computational condensed matter theory due to their ease of use and their systematic convergence properties. Because of simplicity of the GW equations in the plane wave basis, the implementation of the GWA is easy, and many pseudo-potential codes contain a GW part. Conventionally a pseudo-potential method can be applied for electronic structure study of systems with delocalised *sp*-electrons for which the plane wave basis converges rapidly. But due to recent development of the pseudo-potential technique new ultra-soft pseudo-potential methods [3.56] can also be applied to materials with localised *d*- and *f*-electrons. Most of existing pseudo-potential programs are well optimised and successfully used not only for bulk-systems but also for surfaces, interfaces, defects, and clusters. A disadvantage of the plane wave basis is bad convergence for systems with localised electrons. For transition metals or f-electron systems one needs several thousand plane waves. Also the number of basis functions needed for convergence is increasing with the volume of the system. Typically a plane wave GW calculation scales with N^4, that makes calculations of extended systems very expensive.

– *The Gaussian basis method*

Rohlfing, Krüger, and Pollman [3.41] have developed a GW method which combines a pseudo-potential basis and localised Gaussian orbitals. In this approach the LDA Green function is obtained by a conventional pseudo-potential method. Afterwards the Green's function and all GW equations are represented in localised Gaussian orbitals. This essentially reduces the size of the problem. Typically one needs 40-60 Gaussian functions. Another advantage of the method is that many of integrals can be calculated analytically. Because the pseudo-potential part is restricted for *sp*-electron compounds, the method can not be used for systems with localised electrons. A serious problem of the approach is the convergence of the Gaussian basis: while a Gaussian basis can systematically converge, the number of the basis functions needed for convergence can be quite various for different materials.

– *The linearised augmented plane wave (LAPW) method*

The LAPW method is one of the most popular methods for the electronic structure study. The basis consists of local functions, obtained from the Schrödinger equation for atomic-like potential in a muffin-tin sphere on some radial mesh, and plane waves, which describe the interstitial region. The local functions are matched on the sphere to plane waves. Such combination of two different kinds of basis functions makes the LAPW method extremely accurate for systems with localised or delocalised electrons. Also the plane waves are better suited for high energy states, which are usually badly represented by a conventional tight-binding method. All this makes the LAPW method attractive for the GW implementation. Hamada and coworkers developed a GW method with the LAPW [3.25] and applied it to Si. 45 basis functions per Si atom were needed which corresponds to a reduction by factor of five compared to plane wave calculations. But the computational afford is comparable with the pseudo-potential calculations because the evaluation of matrix elements is more expensive. Although a GW-LAPW realisation was successfully used also for Ni [3.4], the method did not become very popular because of the computational costs. With further development of computer technology this method may become very promising, as it was shown recently by Usuda and coworkers [3.55] in the GW-LAPW study in wurtzite ZnO.

– *The linearised muffin-tin orbitals (LMTO) method*
The LMTO is an all-electron method [3.1, 3.2] in which the wave functions are expanded as follows,

$$\psi_{n\boldsymbol{k}} = \sum_{\boldsymbol{R}L} \chi_{\boldsymbol{R}L}(\boldsymbol{r}; \boldsymbol{k}) b_{n\boldsymbol{k}}(\boldsymbol{R}L), \tag{3.85}$$

where χ is the LMTO basis, given in the atomic sphere approximation by

$$\chi_{\boldsymbol{R}L}(\boldsymbol{r}, \boldsymbol{k}) = \varphi_{\boldsymbol{R}L}(\boldsymbol{r}) + \sum_{\boldsymbol{R}'L'} \dot{\varphi}_{\boldsymbol{R}'L'}(\boldsymbol{r}) H_{\boldsymbol{R}L;\boldsymbol{R}'L'}(\boldsymbol{k}). \tag{3.86}$$

Here $\varphi_{\boldsymbol{R}L}(\boldsymbol{r}) = \phi_{\boldsymbol{R}L}(r) Y_L(\hat{\boldsymbol{r}})$ is a solution to the Schrödinger equation inside a sphere centred on an atom at site $\boldsymbol{R}$ for a certain energy ϵ_ν, $\dot{\varphi}_{\boldsymbol{R}'L'}(\boldsymbol{r})$ is the energy derivative of $\varphi_{\boldsymbol{R}L}(\boldsymbol{r})$ at the energy ϵ_ν, and $H_{\boldsymbol{R}L;\boldsymbol{R}'L'}(\boldsymbol{k})$ are the so-called LMTO structure constants. An advantage of the LMTO is that the basis functions do not depend on $\boldsymbol{k}$. The LMTO method is characterised by high computational speed, requirement of a minimal basis set (typically 9-16 orbitals per an atom), and good accuracy in the low energy range.

Aryasetiawan and Gunnarsson suggested to use a combination of the LMTO basis functions for solving the GW equations [3.6]. They showed that a set of products $\phi\phi$, $\phi\dot{\phi}$, and $\dot{\phi}\dot{\phi}$ forms a complete basis for the polarisation function (3.84) and the self-energy (3.82). This scheme allows accurate description of systems with any kinds of electrons typically with 60-100 product functions. A disadvantage of the approach is a bad representation of high energy states in the LMTO method, which are important for calculations of the polarisation function and the self-energy. Recently, Kotani and van Schilfgaarde developed a full-potential version of the LMTO product basis method [3.36], with an accuracy which is substantially better than that of the conventional GW-LMTO implementation.

– *The spacetime method*
Most of the existing implementations of the GWA are in the real fre-

quency / reciprocal space representation. In this approach the evaluation of the linear response function

$$P_{\boldsymbol{q}}(\omega) = -\frac{i}{(2\pi)^4} \int\limits_{-\infty}^{\infty} \mathrm{d}\varepsilon \int\limits_{\Omega_{BZ}} \mathrm{d}^3 k G_{\boldsymbol{k}}^{LDA}(\varepsilon) G_{\boldsymbol{k}-\boldsymbol{q}}^{LDA}(\varepsilon - \omega) \qquad (3.87)$$

and the self-energy

$$\Sigma_{\boldsymbol{q}}(\omega) = \frac{i}{(2\pi)^4} \int\limits_{-\infty}^{\infty} \mathrm{d}\varepsilon \int\limits_{\Omega_{BZ}} \mathrm{d}^3 k W_{\boldsymbol{k}}(\varepsilon) G_{\boldsymbol{k}-\boldsymbol{q}}^{LDA}(\varepsilon - \omega) \qquad (3.88)$$

involves very expensive convolutions. In the real-space/time representation both functions are simple products (3.84) and (3.82), which eliminates two convolutions in reciprocal and frequency space. The idea to chose different representations to minimise the computations is realized in the spacetime method [3.50]. In this scheme the LDA wave functions $\Phi_{n\boldsymbol{k}}(\boldsymbol{r})$ are calculated with a pseudo-potential method. Then the non-interacting Green's function is analytically continued from real to imaginary frequencies and Fourier transformed into the imaginary time:

$$G^{\mathrm{LDA}}(\boldsymbol{r}, \boldsymbol{r}'; i\tau) = \begin{cases} i \sum\limits_{n\boldsymbol{k}}^{\mathrm{occ.}} \Phi_{n\boldsymbol{k}}(\boldsymbol{r}) \Phi_{n\boldsymbol{k}}^*(\boldsymbol{r}') e^{\varepsilon_{n\boldsymbol{k}}\tau}, & \tau > 0 \\ -i \sum\limits_{n\boldsymbol{k}}^{\mathrm{unocc.}} \Phi_{n\boldsymbol{k}}(\boldsymbol{r}) \Phi_{n\boldsymbol{k}}^*(\boldsymbol{r}') e^{\varepsilon_{n\boldsymbol{k}}\tau}, & \tau < 0 \end{cases}. \qquad (3.89)$$

Here $\boldsymbol{r}$ denotes a point in the irreducible part of the real space unit cell while $\boldsymbol{r}'$ denotes a point in the "interaction cell" outside of which G^{LDA} is set to zero. The linear response function is calculated in the real-space and for imaginary time with the formula (3.84) and afterwards Fourier transformed from $i\tau$ to $i\omega$ and from real space to reciprocal one. The screened Coulomb interaction is evaluated as in a conventional plane wave method, and is then transformed into the real-space/imaginary time representation to obtain the self-energy according the (3.82). Further, the self-energy can be Fourier transformed into the imaginary frequency axis and reciprocal space, and analytically continued to real frequencies. The spacetime method decreases substantially the computational time and makes the calculation of large systems accessible. A main computational problem of the spacetime method is the storage of evaluated functions (G, P, W, and Σ) in both representations.

Acknowledgements

We thank V. Dugaev and Z. Szotek for many useful discussions. A. Ernst gratefully acknowledges support from the the DFG through the Forschergruppe 404 "Oxidic Interfaces" and the National Science Foundation under Grant No. PHY99-07949.

References

[3.1] O.K. Andersen. Linear methods in band theory. *Phys. Rev. B*, 12:3060, 1975.

[3.2] O.K. Andersen and O. Jepsen. Explicit, first-principles tight-binding theory. *Phys. Rev. Lett.*, 53:2571, 1984.

[3.3] V.N. Antonov, A.Y. Perlov, A.P. Shpak, and A.N. Yaresko. Calculation of the magneto-optical properties of ferromagnetic metals using the spin-polarised relativistic lmto method. *JMMM*, 146:205, 1995.

[3.4] F. Aryasetiawan. Self-energy of ferromagnetic nickel in the gw approxima-tion. *Phys. Rev. B*, 46:13051, 1992.

[3.5] F. Aryasetiawan and O. Gunnarson. The *GW* method. *Rep. Prog. Phys.*, 61:237, 1998.

[3.6] F. Aryasetiawan and O. Gunnarsson. Product-basis method for calculating dielectric matrices. *Phys. Rev. B*, 49:16214, 1994.

[3.7] W.G. Aulbur, L. Jönsson, and J.W. Wilkins. Quasiparticle calculations in solids. In H. Ehrenreich and F. Spaepen, editors, *Solid State Physics*, vol-ume 54. Academic, San Diego, 2000.

[3.8] G. Baym. Self-consistent approximations in many-body systems. *Phys. Rev.*, 127:1391, 1962.

[3.9] G. Baym and L. Kadanoff. Conservation laws and correlation functions. *Phys. Rev.*, 124:287, 1961.

[3.10] P. Blaha, K. Schwarz, P. Sorantin, and S. Trickey. Full-Potential, Linearized Augmented Plane Wave Programs for Crystalline Systems. *Comp. Phys. Commun.*, 59:399, 1990.

[3.11] F. Bloch. Über die Quantenmechanik der Elektronen in Kristallgittern. *Z. Phys.*, 52:555–600, 1928.

[3.12] P.E. Blöchel. Projector augmented-wave method. *Phys. Rev. B*, 50:17953, 1994.

[3.13] D.J. Ceperley and L. Alder. Ground state of the electron gas by a stochastic method. *Phys. Rev. Lett.*, 45:566, 1980.

[3.14] R.M. Dreizler and E.K.U. Gross. *Density Functional Theory*. Springer, Berlin, 1990.

[3.15] H. Dreyssé, editor. *Electronic Structure and Physical Properties of Solids. The Uses of the LMTO Method*. Springer, Berlin, 2000.

[3.16] H. Ebert and G. Schütz, editors. *Magnetic Dichroism and Spin Polarization in Angle-Resolved Photoemission*. Number 466 in Lecture Notes in Physics. Springer, Berlin, 1996.

[3.17] A. Ernst. full-potential-*Verfahren mit einer kombinierten Basis für die elek-tronische Struktur*. PhD thesis, Tu Dresden, 1997.

[3.18] A. Ernst, W.M. Temmerman, Z. Szotek, M. Woods, and P.J. Durham. Real-space angle-resolved photoemission. *Phil. Mag. B*, 78:503, 1998.

[3.19] Eschrig. *The Fundamentals of Density Functional Theory*. Teubner, Stuttgart, 1996.

[3.20] H. Eschrig. Mixed basis method. In E.K.U. Gross and R.M. Dreizler, editors, *Density Funtional Theory*, page 549. Plenum Press, New York, 19.

[3.21] H. Eschrig. *Optimized LCAO Method and the Electronic Structure of ex-tended systems*. Springer-Verlag Berlin, 1989.

[3.22] A. Fetter and J. Walecka. *Quantum Theory of Many-Particle Systems*. McGraw-Hill, New York, 1971.

[3.23] A. Gonis. *Green Functions for Ordered and Disordered Systems*, volume 4 of *Studies in Mathematical Physics*. North-Holland, Amsterdam, 1992.

[3.24] E. Gross and E. Runge. *Vielteilchentheorie*. Teubner, Stuttgart, 1986.

[3.25] N. Hamada, M. Hwang, and A.J. Freeman. Self-energy correction for the energy bands of silicon by the full-potential linearized augmented-plane-wave method: Effect of the valence-band polarization. *Phys. Rev. B*, 41:3620, 1990.

[3.26] D.R. Hamann, M. Sclüter, and C. Chiang. Norm-conserving pseudopotentials. *Phys. Rev. Lett*, 43:1494, 1979.

[3.27] L. Hedin. *Ark. Fys.*, 30:231, 1965.

[3.28] L. Hedin. New method for calculating the one-particle green's function with application to the electron-gas problem. *Phys. Rev.*, 139:A796, 1965.

[3.29] L. Hedin and S. Lundqvist. Effects of electron-electron and electron-phonon interactions on the one-electron states of solids. In F. Seitz and D. Turnbull, editors, *Solid State Physics*, volume 23. New York: Academic, 1969.

[3.30] P. Hohenberg and W. Kohn. Inhomogeneous electron gas. *Phys. Rev.*, 136:B864, 1964.

[3.31] R.O. Jones and O. Gunnarson. The density functional formalism, its applications and prospects. *Rev. Mod. Phys.*, 61:689, 1989.

[3.32] K. Koepernik and H. Eschrig. Full-potential nonorthogonal local-orbital minimum-basis band-structure scheme. *Phys. Rev. B*, 59:1743, 1999.

[3.33] W. Kohn and N. Rostoker. Solution of the Schrödinger equation in periodic lattices with an application to metallic lithium. *Phys. Rev.*, 94:1111, 1954.

[3.34] W. Kohn and L.J. Sham. Self-consistent equations including exchange and correlation effects. *Phys. Rev*, 140:A1133, 1965.

[3.35] J. Korringa. *Physica*, 13:392, 1947.

[3.36] T. Kotani and M. van Schilfgaarde. All-electron gw approximation with the mixed basis expansion based on the full-potential lmto method. *Solid State Communications*, 121:461, 2002.

[3.37] G.S. Krinchik and V.A. Artem'ev. Magneto-optical properties of ni, co, and fe in the ultraviolet, visible and infrared parts of the spectrum. *Soviet Physics JETP*, 26:1080–1968, 1968.

[3.38] M. Lüders, A. Ernst, W.M. Temmerman, Z. Szotek, and P.J. Durham. *Ab initio* angle-resolved photoemission in multiple-scattering formulation. *J. Phys.: Condens. Matt.*, 13:8587, 2001.

[3.39] J.M. Luttinger. Analytic properties of single-particle propagators for many-fermion systems. *Phys.Rev*, 121(4):942, 1961.

[3.40] J.M. Luttinger and J.C. Ward. Ground-state energy of a many-fermion system. ii. *Phys. Rev.*, 118:1417, 1960.

[3.41] P.K. M. Rohlfing and J. Pollmann. Quasiparticle band-structure calculations for c, si, ge, gaas, and sic using gaussian-orbital basis sets. *Phys. Rev. B*, 48:17791, 1993.

[3.42] R. Mattuck. *A Guide to Feynman Diagrams in the Many-Body Problem*. McGraw-Hill, New York, 1976.

[3.43] M. Milun, P. Pervan, B. Gumhalter, and D.P. Woodruff. Photoemission intensity oscillations from quantum-well states in the Ag/V(100) overlayer system. *Phys. Rev. B*, 59:5170, 1999.

[3.44] G. Onida, L. Reining, and A. Rubio. Electronic excitations: density-functional versus many-body green's-function approaches. *Reviews of Modern Physics*, 72(2):601, 2002.

[3.45] R.G. Parr and W. Yang. *Density-Functional Theory of Atoms and Molecules.* Oxford University Press, New York, 1989.

[3.46] J. Perdew. What do kohn-sham orbital energies mean? how do atoms dissociate? In R.M. Dreizler and da Providencia J., editors, *Density Functional Methods in Physics*, volume Physics 123 of *NATO ASI Series B*, page 265. Plenum, Press, New York and London, 1985.

[3.47] J.P. Perdew and Y. Wang. Accurate and simple analytic representation of the electron-gas correlation energy. *Phys. Rev. B*, 45:13 244, 1992.

[3.48] J.P. Perdew and A. Zunger. Self-interaction correction to density-functional approximations for many-electron systems. *Phys. Rev. B*, 23:5048, 1981.

[3.49] M. Petersilka, U.J. Gossmann, and E.K.U. Gross. Excitation energies from time-dependent density-functional theory. *Phys. Rev. Lett.*, 76:1212, 1996.

[3.50] H.N. Rojas, R.W. Godby, and R.J. Needs. Space-time method for ab initio calculations of self-energies and dielectric response functions of solids. *Phys. Rev. Lett.*, 74:1827, 1995.

[3.51] L.J. Sham and W. Kohn. One-particle properties of an inhomogeneous interacting electron gas. *Phys. Rev.*, 145:561, 1966.

[3.52] J.C. Slater. Damped Electron Waves in Crystals. *Phys. Rev.*, 51:840–846, 1937.

[3.53] A. Svane. Electronic structure of cerium in the self-interaction corrected local spin density approximatio. *Phys. Rev. Lett.*, 72:1248, 1996.

[3.54] Z. Szotek, W.M. Temmerman, and H. Winter. Self-interaction corrected, local spin density description of the gamma −> alpha transition in ce. *Phys. Rev. Lett.*, 72:1244, 1994.

[3.55] M. Usuda, N. Hamada, T. Kotani, and M. van Schilfgaarde. All-electron gw calculation based on the lapw method: Application to wurtzite zno. *Phys. Rev. B*, 66:125101, 2002.

[3.56] D. Vanderblit. Soft self-consistent pseudopotentials in a generalized eigenvalue formalism. *Phys. Rev. B*, 41:1990, 1990.

[3.57] U. von Barth and L. Hedin. A local exchange-correlation potential for the spin polarised case. *J. Phys. C: Sol. State Phys.*, 5:1629, 1972.

[3.58] S.H. Vosko, L. Wilk, and M. Nusair. Accurate spin-dependent electron liquid correlation energies for local spin density calculations: A critical analysis. *Can. J. Phys.*, 58:1200, 1980.

[3.59] P. Weinberger. *Electron Scattering Theory of Ordered and Disordered Matter.* Clarendon Press, Oxford, 1990.

[3.60] A.R. Williams, J. Kübler, and C.D. Gelatt. Cohesive properties of metallic compounds: Augmented-spherical-wave calculations. *Phys. Rev. B*, 19:6094, 1979.

4 A Solid-State Theoretical Approach to the Optical Properties of Photonic Crystals

K. Busch[1], F. Hagmann[1], D. Hermann[1], S.F. Mingaleev[1,2], and M. Schillinger[2]

[1] Institut für Theorie der Kondensierten Materie, Universität Karlsruhe, 76128 Karlsruhe, Germany
[2] Bogolyubov Institute for Theoretical Physics, 03143 Kiev, Ukraine

Abstract. In this chapter, we outline an efficient approach to the calculation of the optical properties of Photonic Crystals. It is based on solid state theoretical concepts and exploits the conceptual similarity between electron waves propagation in electronic crystals and electromagnetic waves propagation in Photonic Crystals. Based on photonic bandstructure calculations for infinitely extended and perfectly periodic systems, we show how defect structures can be described through an expansion of the electromagnetic field into optimally localized photonic Wannier functions which have encoded in themselves all the information of the underlying Photonic Crystals. This Wannier function approach is supplemented by a multipole expansion method which is well-suited for finite-sized and disordered structures. To illustrate the workings and efficiency of both approaches, we consider several defect structures for TM-polarized radiation in two-dimensional Photonic Crystals.

4.1 Introduction

The invention of the laser turned Optics into Photonics: This novel light source allows one to generate electromagnetic fields with previously unattainable energy densities and temporal as well as spatial coherences. As a result, researchers have embarked on a quest to exploit these properties through perfecting existing and creating novel optical materials with tailor made properties. A particular prominent example is the development of low-loss optical fibers which form the backbone of today's long-haul telecommunication systems [4.1]. With the recent advances in micro-fabrication technologies, another degree of freedom has been added to the flexibility in designing photonic systems: Microstructuring dielectric materials allows one to obtain control over the flow of light on lengths scales of the wavelength of light itself. For instance, the design of high-quality ridge waveguiding structures has facilitated the realization of functional elements for integrated optics such as beamsplitters and Mach-Zehnder interferometers [4.2].

The past two decades have witnessed a strongly increased interest in a novel class of micro-structured optical materials. Photonic Crystals (PCs) consist of a micro-fabricated array of dielectric materials in two or three spatial dimensions. A carefully engineered combination of microscopic scat-

K. Busch, F. Hagmann, D. Hermann, S.F. Mingaleev, and M. Schillinger, A Solid-State Theoretical Approach, Lect. Notes Phys. **642**, 55–74 (2004)

tering resonances from individual elements of the periodic array and Bragg scattering from the corresponding lattice leads to the formation of a photonic bandstructure. In particular, the flexibility in material composition, lattice periodicity, symmetry, and topology of PCs allows one to tailor the photonic dispersion relations to almost any need. The most dramatic modification of the photonic dispersion in these systems occurs when suitably engineered PCs exhibit frequency ranges over which the light propagation is forbidden irrespective of direction [4.3, 4.4]. The existence of these so-called complete Photonic Band Gaps (PBGs) allows one to eliminate the problem of light leakage from sharply bent optical fibers and ridge waveguides. Indeed, using a PC with a complete PBG as a background material and embedding into such a PC a circuit of properly engineered waveguiding channels permits to create an optical micro-circuit inside a perfect optical insulator, i.e. an optical analogue of the customary electronic micro-circuit. In addition, the absence of photon states for frequencies in a complete PBG allows one to suppress the emission of optically active materials embedded in PCs. Furthermore, the multi-branch nature of the photonic bandstructure and low group velocities associated with flat bands near a photonic band edge may be utilized to realize phase-matching for nonlinear optical processes and to enhance the interaction between electromagnetic waves and nonlinear and/or optically active material.

These prospects have triggered enormous experimental activities aimed at the fabrication of two-dimensional (2D) as well as three-dimensional (3D) PC structures for telecommunication applications with PBGs in the near infrared frequency range. Considering that the first Bragg resonance occurs when the lattice constant equals half the wavelength of light, fabrication of PCs with bandgaps in the near IR requires substantial technological resources. For 2D PCs, advanced planar microstructuring techniques borrowed from semiconductor technology can greatly simplify the fabrication process and high-quality PCs with embedded defects and waveguides have been fabricated in various material systems such as semiconductors [4.5–4.10], polymers [4.11, 4.12], and glasses [4.13, 4.14]. In these structures, light experiences PBG effects in the plane of propagation, while the confinement in the third direction is achieved through index guiding. This suggests that fabricational imperfections in bulk 2D PCs as well deliberately embedding defect structures such as cavities and waveguide bends into 2D PCs will inevitably lead to radiation losses into the third dimension. Therefore, it is still an open question as to whether devices with acceptable radiation losses can be designed and realized in 2D PCs. However, radiation losses can be avoided altogether if light is guided within the comlete PBG of 3D PCs and, therefore, the past years have seen substantial efforts towards the manufacturing of suitable 3D PCs. These structures include layer-by-layer structures [4.15, 4.16], inverse opals [4.17–4.19] as well as the fabrication of templates via laser holography [4.20, 4.21] and two-photon polymerization (sometimes also referred to as stereo-lithography) [4.22–4.24].

Given this tremendous flexibility in the fabrication of PCs (and the cost associated with most of the fabrication techniques), it is clear that modeling of the linear, nonlinear and quantum optical properties of PCs is a crucial element of PC research. In this manuscript, we would like to outline how the far-reaching analogies of electron wave propagation in crystalline solids and electromagnetic wave propagation in PCs can be utilized to obtain a theoretical framework for the quantitative description of light propagation in PCs. In Sect. 4.2, we describe an efficient method for obtaining the photonic bandstructure which is based on a Multi-Grid technique. The results of photonic bandstructure computations are the basis for the description of defect structures such as cavities and waveguides using a Wannier function approach (Sect. 4.3). As an example, we illustrate the design of a near optimal PC waveguide bend. Finally, in Sect. 4.4, we utilize this bend design and a multi-pole expansion technique to construct a PC beamsplitter within a finite-sized PC and discuss the role of fabricational tolerances on the performance of the device.

4.2 Photonic Bandstructure Computation

Photonic bandstructure computations determine the dispersion relation of infinitely extended defect-free PCs. In addition, they allow one to design PCs that exhibit PBGs and to accurately interpret measurements on PC samples. As a consequence, photonic bandstructure calculations represent an important predictive as well as interpretative basis for PC research and, therefore, lie at the heart of theoretical investigations of PCs. More specifically, the goal of photonic bandstructure computations is to find the eigenfrequencies and associated eigenmodes of the wave equation for the perfect PC, i.e., for an infinitely extended periodic array of dielectric material. For the simplicity of presentation we restrict ourselves in the remainder of this manuscript to the case of TM-polarized radiation propagating in the plane of periodicity (x, y)-plane of 2D PCs. In this case, the wave equation in the frequency domain (harmonic time dependence) for the z-component of the electric field reads

$$\frac{1}{\epsilon_p(\boldsymbol{r})} \left(\partial_x^2 + \partial_y^2\right) E(\boldsymbol{r}) + \frac{\omega^2}{c^2} E(\boldsymbol{r}) = 0. \tag{4.1}$$

Here c denotes the vacuum speed of light and $\boldsymbol{r} = (x, y)$ denotes a two-dimensional position vector. The dielectric constant $\epsilon_p(\boldsymbol{r}) \equiv \epsilon_p(\boldsymbol{r} + \boldsymbol{R})$ is periodic with respect to the set $\mathcal{R} = \{n_1 \boldsymbol{a}_1 + n_2 \boldsymbol{a}_2; (n_1, n_2) \in \mathcal{Z}^2\}$ of lattice vectors $\boldsymbol{R}$ generated by the primitive translations $\boldsymbol{a}_i$, $i = 1, 2$ that describe the structure of the PC. Equation (4.1) represents a differential equation with periodic coefficients and, therefore, its solutions obey the Bloch-Floquet theorem

$$E_{\boldsymbol{k}}(\boldsymbol{r} + \boldsymbol{a}_i) = e^{i \boldsymbol{k} \boldsymbol{a}_i} E_{\boldsymbol{k}}(\boldsymbol{r}), \tag{4.2}$$

where $i = 1, 2$. The wave vector $\boldsymbol{k} \in 1.\text{BZ}$ that labels the solution is a vector of the first Brillouin zone (BZ) known as the crystal momentum. As a result of this reduced zone scheme, the photonic bandstructure acquires a multi-branch nature that is associated with the backfolding of the dispersion relation into the 1. BZ. This introduces a discrete index n, the so-called band index, that enumarates the distinct eigenfrequencies and eigenfunctions at the same wave vector $\boldsymbol{k}$.

The photonic dispersion relation $\omega_n(\boldsymbol{k})$ gives rise to a photonic Density of States (DOS), which plays a fundamental role for the understanding of the quantum optical properties of active material embedded in PCs [4.25]. The photonic DOS, $N(\omega)$, is defined by "counting" all allowed states with a given frequency ω

$$N(\omega) = \sum_n \int_{1.\text{BZ}} d^2k \; \delta(\omega - \omega_n(\boldsymbol{k})). \qquad (4.3)$$

Other physical quantities such as group velocities $\boldsymbol{v}_n(\boldsymbol{k}) = \nabla_{\boldsymbol{k}} \omega_n(\boldsymbol{k})$ can be calculated through adaption of various techniques known from electron bandstructure theory. For details, we refer to [4.26] and [4.27].

A straightforward way of solving (4.1) is to expand all the periodic functions into a Fourier series over the reciprocal lattice $\mathcal{G}$, thereby transforming the differential equation into an infinite matrix eigenvalue problem, which may be suitably truncated and solved numerically. Details of this plane wave method (PWM) for isotropic systems can be found, for instance, in [4.26,4.28] and for anisotropic systems in [4.29]. While the PWM provides a straightforward approach to computing the bandstructure of PCs, it also exhibits a number of shortcomings such as slow convergence associated with the truncation of Fourier series in the presence of discontinuous changes in the dielectric constants. In particular, this slow convergence makes the accurate calculation of Bloch functions a formidable and resource-consuming task. Therefore, we have recently developed an efficient real space approach to computing photonic bandstructures [4.27]. Within this approach, the wave equation, (4.1), is discretized in a single unit cell in real space (defined through the set of space points $\boldsymbol{r} = r_1\boldsymbol{a}_1 + r_2\boldsymbol{a}_2$ with $r_1, r_2 \in [-1/2, 1/2]$), leading to a sparse matrix problem. The Bloch-Floquet theorem, (4.2), provides the boundary condition for the elliptic partial differential equation (4.1). In addition, the eigenvalue is treated as an additional unknown for which the normalization of the Bloch functions provides the additional equation needed for obtaining a well-defined problem. The solution of this algebraic problem is obtained by employing Multi-Grid (MG) methods which guarantee an efficient solution by taking full advantage of the smoothness of the photonic Bloch functions [4.27,4.30] (see also the chapter of G. Wittum in this volume). Even for the case of a naive finite difference discretization, the MG-approach easily outperforms the PWM and leads to a substantial reduction in CPU time. For instance, in the present case of 2D systems for which the Bloch functions are required we save one order of magnitude in CPU time as compared

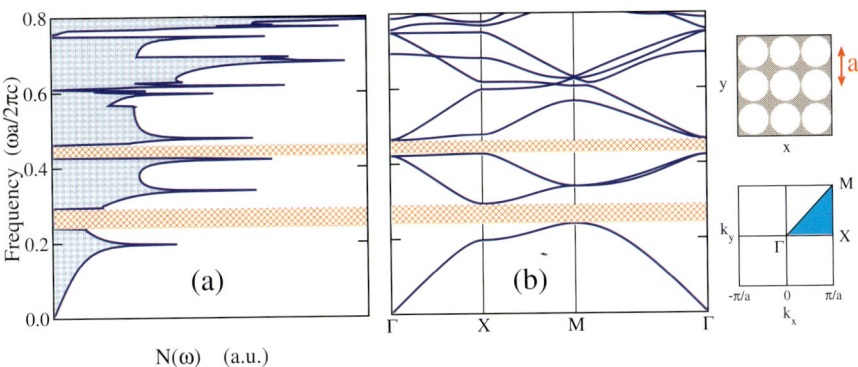

Fig. 4.1. Density of States (a) and photonic band structure (b) for TM-polarized radiation in a square lattice (lattice constant a) of cylindrical air pores of radius $R_{\text{pore}} = 0.475a$ in dielectric with $\varepsilon = 12$ (silicon). This PC exhibits a large fundamental gap extending from $\omega = 0.238 \times 2\pi c/a$ to $\omega = 0.291 \times 2\pi c/a$. A higher order band gap extends from $\omega = 0.425 \times 2\pi c/a$ to $\omega = 0.464 \times 2\pi c/a$.

to PWM. Additional refinements such as a finite element discretization will further increase the efficiency of the MG-approach.

In Fig. 4.1(b), we show the bandstructure for TM-polarized radiation in a 2D PC consisting of a square lattice (lattice constant a) of cylindrical air pores (radius $r_{\text{pore}} = 0.475a$) in a silicon matrix ($\varepsilon_{\text{p}} = 12$). This structure exhibits two 2D PBGs. The larger, fundamental bandgap (20% of the midgap frequency) extends between $\omega = 0.238 \times 2\pi c/a$ to $\omega = 0.291 \times 2\pi c/a$ and the smaller, higher order bandgap (8% of the midgap frequency) extends from $\omega = 0.425 \times 2\pi c/a$ to $\omega = 0.464 \times 2\pi c/a$. Furthermore, in Fig. 4.1(a) we depict the DOS for our model system, where the photonic band gaps are manifest as regions of vanishing DOS. Characteristic for 2D systems is the linear behavior for small frequencies as well as the logarithmic singularities, the so-called van Hove singularities, associated with vanishing group velocities for certain frequencies inside the bands (compare with Fig. 4.1(a)).

4.3 Defect Structures in Photonic Crystals

To date, the overwhelming majority of theoretical investigations of cavities and waveguiding in PCs has been carried out using Finite-Difference Time-Domain (FDTD) and/or Finite-Element (FE) techniques. However, applying general purpose methodologies such as FDTD or FE methods to defect structures in PCs largely disregards information about the underlying PC structure which is readily available from photonic bandstructure computation. As a result, only relatively small systems can be investigated and the physical insight remains limited.

4.3.1 Maximally Localized Photonic Wannier Functions

A more natural description of localized defect modes in PCs consists in an expansion of the electromagnetic field into a set of localized basis functions which have encoded into them all the information of the underlying PC. Therefore, the most natural basis functions for the description of defect structures in PCs are the so-called photonic Wannier functions, $W_{n\boldsymbol{R}}(\boldsymbol{r})$, which are formally defined through a lattice Fourier transform

$$W_{n\boldsymbol{R}}(\boldsymbol{r}) = \frac{V_{\mathrm{WSC}}}{(2\pi)^2} \int_{\mathrm{BZ}} d^2\boldsymbol{k}\, e^{-i\boldsymbol{k}\boldsymbol{R}}\, E_{n\boldsymbol{k}}(\boldsymbol{r}) \qquad (4.4)$$

of the extended Bloch functions, $E_{n\boldsymbol{k}}(\boldsymbol{r})$. The above definition associates the photonic Wannier function $W_{n\boldsymbol{R}}(\boldsymbol{r})$ with the frequency range covered by band n, and centers it around the corresponding lattice site $\boldsymbol{R}$. In addition, the completeness and orthogonality of the Bloch functions translate directly into corresponding properties of the photonic Wannier functions. Computing the Wannier functions directly from the output of photonic bandstructure programs via (4.4) leads to functions with poor localization properties and erratic behavior (see, for instance, Fig. 2 in [4.31]). These problems originate from an indeterminacy of the global phases of the Bloch functions. It is straightforward to show that for a group of N_{w} bands there exists, for every wave vector $\boldsymbol{k}$, a free unitary transformation between the bands which leaves the orthogonality relation of Wannier functions unchanged. A solution to this unfortunate situation is provided by recent advances in electronic bandstructure theory. Marzari and Vanderbilt [4.32] have outlined an efficient scheme for the computation of maximally localized Wannier functions by determining numerically a unitary transformation between the bands that minimizes an appropriate spread functional $\mathcal{F}$

$$\mathcal{F} = \sum_{n=1}^{N_{\mathrm{w}}} \left[\langle n\boldsymbol{0}|\, r^2\, |n\boldsymbol{0}\rangle - (\langle n\boldsymbol{0}|\, \boldsymbol{r}\, |n\boldsymbol{0}\rangle)^2 \right] = \mathrm{Min} . \qquad (4.5)$$

Here we have introduced a shorthand notation for matrix elements according to

$$\langle n\boldsymbol{R}|\, f(\boldsymbol{r})\, |n'\boldsymbol{R}'\rangle = \int_{\mathbb{R}^2} d^2 r\, W_{n\boldsymbol{R}}^*(\boldsymbol{r})\, f(\boldsymbol{r})\, \varepsilon_{\mathrm{p}}(\boldsymbol{r})\, W_{n'\boldsymbol{R}'}(\boldsymbol{r}) , \qquad (4.6)$$

for any function $f(\boldsymbol{r})$. For instance, the orthonormality of the Wannier functions in this notation read as

$$\langle n\boldsymbol{R}|\, |n'\boldsymbol{R}'\rangle = \int_{\mathbb{R}^2} d^2 r\, W_{n\boldsymbol{R}}^*(\boldsymbol{r})\, \varepsilon_{\mathrm{p}}(\boldsymbol{r})\, W_{n'\boldsymbol{R}'}(\boldsymbol{r}) = \delta_{nm}\delta_{\boldsymbol{R}\boldsymbol{R}'} , \qquad (4.7)$$

The field distributions of the optimized Wannier functions belonging to the six most relevant bands of our model system are depicted in Fig. 4.2 (see

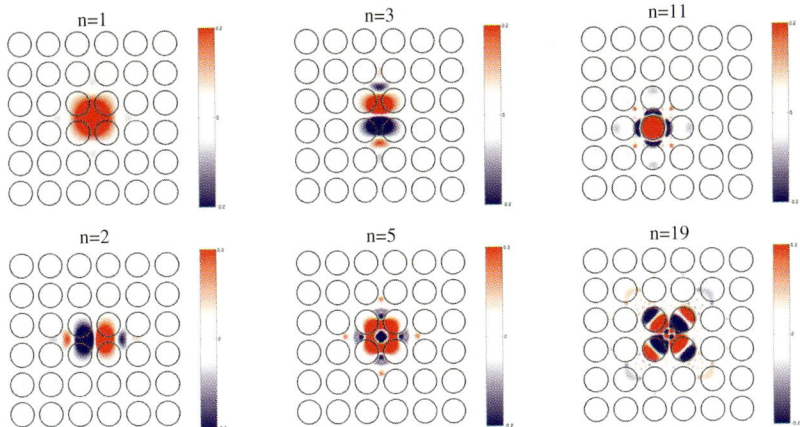

Fig. 4.2. Photonic Wannier functions, $W_{n\mathbf{0}}(\mathbf{r})$, for the six bands that are most relevant for the description of the localized defect mode shown in Fig. 4.3(a). These optimally localized Wannier functions have been obtained by minimizing the corresponding spread functional, (4.5). Note, that in contrast to the other bands, the Wannier center of the eleventh band is located at the center of the air pore. The parameters of the underlying PC are the same as those in Fig. 4.1.

also the discussion in Sect. 4.3.3). Their localization properties as well as the symmetries of the underlying PC structure are clearly visible. It should be noted that the Wannier centers of all calculated bands (except of the eleventh band) are located halfway between the air pores, i.e. inside the dielectric (see [4.32] for more details on the Wannier centers). In addition, we would like to point out that instead of working with the electric field [4.33, 4.31], (4.1), one may equally well construct photonic Wannier functions for the magnetic field, as recently demonstrated by Whittaker and Croucher [4.34].

4.3.2 Defect Structures via Wannier Functions

The description of defect structures embedded in PCs starts with the corresponding wave equation in the frequency domain

$$\nabla^2 E(\mathbf{r}) + \left(\frac{\omega}{c}\right)^2 \left(\varepsilon_{\mathrm{p}}(\mathbf{r}) + \delta\varepsilon(\mathbf{r})\right) E(\mathbf{r}) = 0 \ . \tag{4.8}$$

Here, we have decomposed the dielectric function into the periodic part, $\varepsilon_{\mathrm{p}}(\mathbf{r})$, and the contribution, $\delta\varepsilon(\mathbf{r})$, that describes the defect structures. Within the Wannier function approach, we expand the electromagnetic field according to

$$E(\mathbf{r}) = \sum_{n,\mathbf{R}} E_{n\mathbf{R}} \, W_{n\mathbf{R}}(\mathbf{r}) \ , \tag{4.9}$$

with unknown amplitudes $E_{n\mathbf{R}}$. Inserting this expansion into the wave equation (4.8) and employing the orthonomality relations, (4.7), leads to the basic equation for lattice models of defect structures embedded in PCs

$$\sum_{n',\mathbf{R}'} \left\{ \delta_{nn'}\delta_{\mathbf{R}\mathbf{R}'} + D_{\mathbf{R}\mathbf{R}'}^{nn'} \right\} E_{n'\mathbf{R}'} = \left(\frac{c}{\omega}\right)^2 \sum_{n',\mathbf{R}'} A_{\mathbf{R}\mathbf{R}'}^{nn'} E_{n'\mathbf{R}'} . \qquad (4.10)$$

The matrix $A_{\mathbf{R}\mathbf{R}'}^{nn'}$ depends only on the Wannier functions of the underlying PC and is defined through

$$A_{\mathbf{R}\mathbf{R}'}^{nn'} = - \int_{\mathbf{R}^2} d^2\mathbf{r} \ W_{n\mathbf{R}}^*(\mathbf{r}) \nabla^2 W_{n'\mathbf{R}'}(\mathbf{r}) . \qquad (4.11)$$

The localization of the Wannier functions in space leads to a very rapid decay of the magnitude of matrix elements with increasing separation $|\mathbf{R} - \mathbf{R}'|$ between lattice sites, effectively making the matrix $A_{\mathbf{R}\mathbf{R}'}^{nn'}$ sparse. Furthermore, it may be shown that the matrix $A_{\mathbf{R}\mathbf{R}'}^{nn'}$ is Hermitian and positive definite. Similarly, once the Wannier functions of the underlying PC are determined, the matrix $D_{\mathbf{R}\mathbf{R}'}^{nn'}$ depends solely on the overlap of these functions, mediated by the defect structure:

$$D_{\mathbf{R}\mathbf{R}'}^{nn'} = \int_{\mathbf{R}^2} d^2\mathbf{r} \ W_{n\mathbf{R}}^*(\mathbf{r}) \, \delta\varepsilon(\mathbf{r}) W_{n'\mathbf{R}'}(\mathbf{r}) . \qquad (4.12)$$

As a consequence of the localization properties of both the Wannier functions and the defect dielectric function, the Hermitian matrix $D_{\mathbf{R}\mathbf{R}'}^{nn'}$, too, is sparse. In the case of PCs with inversion symmetry, $\varepsilon_{\mathrm{p}}(\mathbf{r}) \equiv \varepsilon_{\mathrm{p}}(-\mathbf{r})$, the Wannier functions can be chosen to be real. Accordingly, both matrices, $A_{\mathbf{R}\mathbf{R}'}^{nn'}$ and $D_{\mathbf{R}\mathbf{R}'}^{nn'}$ become real symmetric ones.

Depending on the nature of the defect structure, we are interested in (i) frequencies of localized cavity modes, (ii) dispersion relations for straight waveguides, or (iii) transmission and reflection through waveguide bends and other, more complex defect structures. In the following, we consider each of these cases separately.

4.3.3 Localized Cavity Modes

As a first illustration of the Wannier function approach, we consider the case of a simple cavity created by infiltrating a single pore at the defect site $\mathbf{R}_{\mathrm{def}}$ with a material with dielectric constant $\varepsilon_{\mathrm{def}}$, as shown in the inset of Fig. 4.3(a). In this case, we directly solve (4.10) as a generalized eigenvalue problem for the cavity frequencies that lie within the PBG, and reconstruct the cavity modes from the corresponding eigenvectors. In Fig. 4.3(a) we compare the frequencies of these cavity modes calculated from (4.10) with corresponding calculations using PWM-based super-cell calculations.

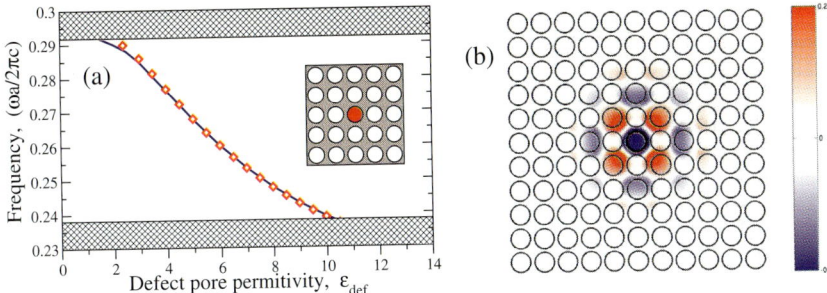

Fig. 4.3. (a) Frequencies of localized cavity modes created by infiltrating a single defect pore with a material with dielectric constant ε_{def} (see inset). The results of the Wannier function approach (diamonds) using $N_{\text{W}} = 10$ Wannier functions per unit cell are in complete agreement with numerically exact results of the super-cell calculations (full line). The parameters of the underlying PC are the same as those in Fig 4.1. (b) Electric field distribution for the cavity mode with frequency $\omega = 0.290 \times 2\pi c/a$, created by infiltrating the pore with a polymer with $\varepsilon_{\text{def}} = 2.4$.

Upon increasing ε_{def}, a non-degenerate cavity mode with monopole symmetry emerges from the upper edge of the bandgap. The results of the Wannier function approach using the $N_{\text{W}} = 10$ most relevant Wannier functions *per unit cell* in (4.10) are in complete agreement with numerically exact results of the super-cell calculations. In Fig. 4.3(b), we depict the corresponding mode

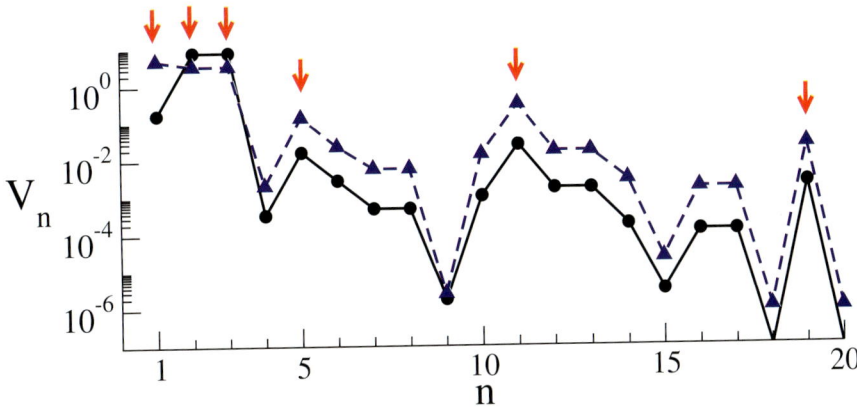

Fig. 4.4. The strength V_n of the individual contributions from the Wannier functions of the lowest 20 bands (index n) to the formation of the cavity modes depicted in Fig. 4.3. The Wannier functions with $V_n \leq 10^{-3}$ may be safely leaved out of account. Arrows indicate the six must relevant Wannier functions depicted in Fig. 4.2. The parameters of the underlying PC are the same as those in Fig. 4.1.

structure for a monopole cavity mode created by infiltration of a polymer with $\varepsilon_{\text{def}} = 2.4$ into the pore. The convergence properties of the Wannier function approach should depend on the nature and symmetry properties of the cavity modes under consideration. To discuss this issue in greater detail, it is helpful to define a measure V_n of the strength of the contributions to a cavity mode from the individual Wannier function associated with band n via $V_n = \sum_{\boldsymbol{R}} |E_{n\boldsymbol{R}}|^2$. In Fig. 4.4 we display the dependence of the parameter V_n on the band index n for the cavity modes shown in Fig. 4.3, for two values of the defect dielectric constant, $\varepsilon_{\text{def}} = 2.4$ (solid line) and $\varepsilon_{\text{def}} = 8$ (dashed line), respectively. In both cases, the most relevant contributions to the cavity modes originate from the Wannier functions belonging to bands $n = 1, 2, 3, 5, 11$, and 19, and all contributions from bands $n > 20$ are negligible. These most relevant Wannier functions for our model system are shown in Fig. 4.2. In fact, fully converged results are obtained when we work with the 10 most relevant Wannier functions per unit cell (for a comparison with numerically exact super-cell calculations see Fig. 4.3(a)).

4.3.4 Dispersion Relations of Waveguides

The efficiency of the Wannier function approach is particularly evident when considering defect clusters consisting of several defect pores. In this case the defect dielectric function, $\delta\varepsilon(\boldsymbol{r})$, can be written as a sum over positions, $\boldsymbol{R}_m$, of the individual defect pores, so that (4.12) reduces to a sum

$$D_{\boldsymbol{R}\boldsymbol{R}'}^{nn'} = \sum_m D(m)_{\boldsymbol{R}-\boldsymbol{R}_m,\boldsymbol{R}'-\boldsymbol{R}_m}^{nn'} , \qquad (4.13)$$

over the matrix elements $D(m)_{\boldsymbol{R},\boldsymbol{R}'}^{nn'}$ of the individual defects (see discussion in [4.31] for more details). Therefore, for a given underlying PC structure, it becomes possible to build up a database of matrix elements, $D(m)_{\boldsymbol{R},\boldsymbol{R}'}^{nn'}$, for different geometries (radii, shapes) of defect pores, which allow us highly efficient defect computations through simple matrix assembly procedures. This is in strong contrast to *any* other computational technique known to us.

Arguably the most important types of defect clusters in PCs are one or several adjacent straight rows of defects. Properly designed, such defect rows form a PC waveguide which allows the efficient guiding of light for frequencies within a PBG [4.35, 4.36]. Due to the one-dimensional periodicity of such a waveguide, its guided modes, $E^{(p)}(\boldsymbol{r}\,|\,\omega) = \sum_{n,\boldsymbol{R}} E_{n\boldsymbol{R}}^{(p)}(\omega)\, W_{n\boldsymbol{R}}(\boldsymbol{r})$, obey the 1D Bloch-Floquet theorem

$$E_{n\boldsymbol{R}+\boldsymbol{s}_{\text{w}}}^{(p)}(\omega) = e^{i\boldsymbol{k}_p(\omega)\boldsymbol{s}_{\text{w}}}\, E_{n\boldsymbol{R}}^{(p)}(\omega) , \qquad (4.14)$$

and thus they can be labeled by a wave vector, $\boldsymbol{k}_p(\omega)$, parallel to the waveguide director, $\boldsymbol{s}_{\text{w}} = w_1\boldsymbol{a}_1 + w_2\boldsymbol{a}_2$, where $\boldsymbol{a}_1 = (a, 0)$ and $\boldsymbol{a}_2 = (0, a)$ are the primitive lattice vectors of the PC, and integers w_1 and w_2 define the direction

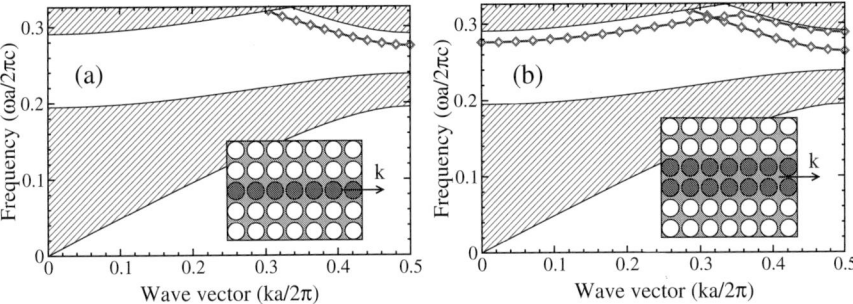

Fig. 4.5. Dispersion relations of the propagating guided mode for PC waveguides consisting of **(a)** one row and **(b)** two rows of defect pores infiltrated with a polymer with $\varepsilon_{\mathrm{def}} = 2.4$. The calculations within the Wannier function approach (diamonds), based on (4.10) and (4.14) in which we accounted for the interaction of 5 nearest pores along the waveguide and used 10 most relevant Wannier functions per unit cell, are in complete agreement with the results of supercell calculations (solid lines). The gray areas represent the projected band structure of the underlying model PC (see Fig. 4.1). The red circles in the insets indicate the positions of the infiltrated pores.

of the waveguide (for instance, an x-axis directed W1-waveguide is described through $w_1 = 1$ and $w_2 = 0$). Commonly, investigations of PC waveguides consist of calculations of the dispersion relations, $\mathbf{k}_p(\omega)$, of all the guided modes, which can be obtained by substituting (4.14) into (4.10) as is described in details in [4.31].

To date, investigations of straight PC waveguides have concentrated on the calculation of dispersion relations for *propagating guided modes* with real wave vectors, $k_{\mathrm{p}}(\omega)$, only. Such calculations can be accurately carried out also by employing the supercell technique. In Fig. 4.5 we display the dispersion relations for the propagating guided modes of the W1 and W2 waveguides created by infiltrating a polymer into one row and two rows of pores, calculated within the Wannier function approach. The results of these calculations are fully converged and in complete agreement with the results of plane-wave based supercell computations. Similar to the calculations of complex cavity structures, the calculations of waveguide dispersion relations within the Wannier function approach require fairly minimal computational resources in comparison with the supercell technique.

We would like to emphasize that, in contrast to the supercell technique, the Wannier function approach enables us to also obtain the dispersion relations for *evanescent guided modes* with complex wave vectors $k_{\mathrm{p}}(\omega)$. Since such modes grow or decay along the waveguide direction, they are largely irrelevant in perfectly periodic straight waveguides. However, they start to play an important role as soon as the perfect periodicity of the waveguide is broken either through imperfections due to fabricational tolerances, or through the deliberate creation of deviations from periodicity such as bends or coupled cavity-waveguide systems for Wavelength Division Multiplexing (WDM) applications. In such cases, these *evanescent* guided modes give rise to light localization effects and determine the non-trivial transmission and reflection properties of PC circuits [4.31, 4.37] as we will discuss below.

4.3.5 Photonic Crystal Circuits

In this section we demonstrate that the Wannier function approach provides an efficient simulation tool for the description of light through PC circuits which allows one to overcome most of the limitations related to FDTD or FE methods. As an illustration, we consider light propagation through two-port PC circuits such as waveguide bends or coupled cavity-waveguide systems. The common feature of these devices is that two semi-infinite straight PC waveguides act as leads that are connected through a finite-sized region of defects. In this case, light propagation through the device at frequency ω is governed by (4.10), which should be truncated (to obtain equal number of equations and unknowns) by prescribing certain values to the expansion coefficients, $E_{n\boldsymbol{R}}$, at some sites inside the waveguiding leads. Since these values determine the amplitudes of the incoming light, it is physically more transparent to express the expansion coefficients $E_{n\boldsymbol{R}}$ within the leads through a superposition of the guided modes $\boldsymbol{\Phi}^{(p)}(\omega)$ with wave vectors $k_{\mathrm{p}}(\omega)$ of the corresponding infinite straight waveguide. In a numerical implementation this is facilitated by replacing the expansion coefficients $E_{n\boldsymbol{R}}$ for all lattice sites $\boldsymbol{R}$ inside each waveguiding lead, W_i, $i = 1, 2$, according to

$$E_{n\boldsymbol{R}}^{\mathrm{w}_i} = \sum_{p=1}^{N} u_{\mathrm{w}_i}^{(p)}(\omega) E_{n\boldsymbol{R}}^{(p)}(\omega) + \sum_{p=N+1}^{2N} d_{\mathrm{w}_i}^{(p)}(\omega) E_{n\boldsymbol{R}}^{(p)}(\omega) , \qquad (4.15)$$

where $u_{\mathrm{w}_i}^{(p)}$ and $d_{\mathrm{w}_i}^{(p)}$ are amplitudes of the guided modes, and we assume that all $2N$ guided modes are ordered in the following way: $p = 1$ to N are occupied by the propagating guided modes with $\mathrm{Re}[k_{\mathrm{p}}] > 0$ and evanescent guided modes with $\mathrm{Im}[k_{\mathrm{p}}] > 0$, whereas $p = N+1$ to $2N$ are occupied by the propagating guided modes with $\mathrm{Re}[k_{\mathrm{p}}] < 0$ and evanescent guided modes with $\mathrm{Im}[k_{\mathrm{p}}] < 0$. Assuming that the amplitudes, $u_{\mathrm{w}_1}^{(p)}$ and $d_{\mathrm{w}_2}^{(p)}$, of all the propagating (evanescent) guided modes which propagate (decay) in the direction of the device are known (they depend on the purpose of our calculation or on the experimental setup), we can now substitute (4.15) into (4.10) and, solving the resulting system of coupled equations, find the unknown expansion coefficients $E_{n\boldsymbol{R}}$ for the sites $\boldsymbol{R}$ inside the domain of the device (which can be used, e.g., for visualization of the field propagation through the device), and the amplitudes, $u_{\mathrm{w}_2}^{(p)}$ and $d_{\mathrm{w}_1}^{(p)}$ of all *outgoing propagating* and *growing evanescent* guided modes. In [4.31] some of us have demonstrated, by comparison with the FDTD calculations [4.35], that the results of such transmission calculations based on the Wannier function approach are indeed very accurate and

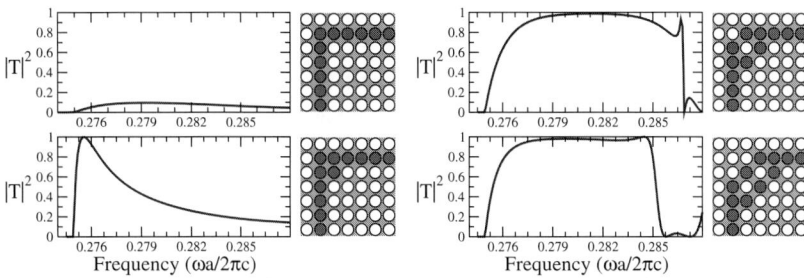

Fig. 4.6. Transmission spectra, $|T(\omega)|^2$, for four different bend geometries embedded in our 2D model PC. The results of the Wannier function approach are obtained with $N_R = 5$, $L = 5$, and the $N_W = 10$ most relevant Wannier functions. The parameters of the underlying PC are the same as those in Fig. 4.1.

agree extremely well with FDTD calculations. Now, in Fig. 4.6, we present the results of Wannier function calculations of the transmission spectra for four different bend geometries with attached single-mode waveguide leads (see Fig. 4.5) that are embedded in our model PC. The improvement of the optimal design (lower right in Fig. 4.6) over the naive bend (upper left in Fig. 4.6) is apparent: For the optimal design we find a wide frequency range of nearly perfect transmission as opposed to maximal 10% transmission in the naive design. In Sect. 4.4, we will utilize this optimized bend design for the construction of a beamsplitter and a discussion of the influence of fabricational tolerances on the performance of the device.

The efficiency of the Wannier function approach for transmission calculations becomes apparent when considering that – once the Wannier functions for the underlying PC have been obtained – the calculation of a single data point in the reflection spectra of Fig. 4.6 reduces to the solution of a single sparse system of some 800 equations, which even on a laptop computer takes only a few seconds. Therefore, the Wannier function approach outlined above will (i) enable a reverse engineering of defect structures with prescribed functionality and (ii) allow detailed studies regarding the robustness of successful designs with respect to fabricational tolerances. Moreover, the Wannier function approach can be straightforwardly applied, with comparable efficiency, to investigations of the transmission spectra through PC circuits made from highly dispersive and/or nonlinear materials. Of paramount importance is the fact that, in contrast to the FDTD or FE methods, the Wannier function approach permits one to accurately and efficiently calculate the complete scattering matrices of PC devices [4.31], allowing us to construct a PC circuit theory in which *individual devices are replaced by simple equivalent scattering matrices* which are assembled by simple scattering matrix multiplication rules to form the scattering matrix of *large-scale circuits* [4.38]. We would like to emphasize that in some sense these scattering matrices can be regarded as the optical analogue of the impedance matrices associated with multi-port devices in microwave technology [4.39].

4.4 Finite Photonic Crystals

The Wannier function approach described above is particularly useful for large systems, and has no restriction in the shape of the elements conforming the system. Extensions of this approach have to include the description of coupling in and out of PCs, in order words to treat finite systems and sources [4.40]. If we restrict ourselves to cylindrical symmetry, a complementary and also quite efficient tool can be found in the recently developed multipole expansion technique [4.41, 4.42]. Apart from the field pattern generated by a source, this approach allows one to calculate the local density of states (LDOS): For applications to quantum optical experiments in PCs it is necessary to investigate not only the (overall) availability of modes with frequency ω but also the local coupling strength of an emitter at a certain position $\boldsymbol{r}$ in the PC to the electromagnetic environment provided by the PC. Consequently, it is the overlap matrix element of the emitters dipole moment to the eigenmodes (Bloch functions) that is determining quantum optical properties such as decay rates etc. [4.25]. This may be combined into the local DOS (LDOS), $N(\boldsymbol{r}, \omega)$, which, for an infinite system, is defined as

$$N(\boldsymbol{r}, \omega) = \sum_n \int_{BZ} d^2k \, |E_{n\boldsymbol{k}}(\boldsymbol{r})|^2 \, \delta(\omega - \omega_n(\boldsymbol{k})). \tag{4.16}$$

Similar to the DOS, the LDOS of an infinite system vanishes for frequencies lying in the band gap, revealing the suppression of light emission at those frequencies. However, actual devices are not of infinite extent, and, therefore, in these finite-sized structures the LDOS will be very small but non vanishing. For finite systems, the LDOS can be obtained by extracting the imaginary part of Green's tensor $\mathcal{G}(\boldsymbol{r}, \boldsymbol{r}; \omega)$

$$N(\boldsymbol{r}; \omega) = -\frac{2\omega n_{Si}^2}{\pi c^2} \, \mathrm{Im}\mathrm{Tr}[\mathcal{G}(\boldsymbol{r}, \boldsymbol{r}; \omega)] \,, \tag{4.17}$$

where n_{Si} represents the index of refraction of the background where the air pores are embedded. The Green's tensor $\mathcal{G}(\boldsymbol{r}, \boldsymbol{r}_s; \omega)$ represents the field distribution at an observation point $\boldsymbol{r}$ generated by δ-source at $\boldsymbol{r}_s$. For infinite systems, it is straightforward to show that (4.17) agrees with (4.16). In the present case of a TM-polarized radiation in a 2D PC only the G_{zz} component of the Green's tensor is needed and satisfies the wave equation

$$\left(\partial_x^2 + \partial_y^2\right) G_{zz} + \frac{\omega^2}{c^2}\epsilon(\boldsymbol{r})G_{zz} = \delta(\boldsymbol{r} - \boldsymbol{r}_s). \tag{4.18}$$

The multipole method consists in expanding Green's function G_{zz} in cylindrical harmonics both outside and inside the pores that comprise the PC. Subsequently the corresponding expansion coefficients are obtained by imposing appropriate continuity conditions across the pore surfaces and the Sommerfeld radiation condition as a boundary condition at infinity [4.41, 4.42]. In particular, we consider a total of N_c pores embedded in a silicon background. Inside the lth cylinder and in a coordinate system centered around the pore center r_l, the G_{zz} is given by

$$G_{zz,l}^{\text{int}}(r, r_s) = \frac{1}{4i} \chi_l^{\text{int}}(r_s) H_0^{(1)}(kn_l|r - r_s|)$$

$$+ \sum_{m=-\infty}^{\infty} C_m^l J_m(kn_l|r - r_l|) e^{im \arg(r - r_l)} . \tag{4.19}$$

Here, $k = n_{\text{Si}}\omega/c$ is the wavenumber in the background material (silicon), r_l is the cylinder position, and the value of χ_l^{int} indicates whether r_s lies inside or outside of the lth cylinder ($\chi_l^{\text{int}} = 1$ or $\chi_l^{\text{int}} = 0$, respectively). Finally, $\arg(r - r_l)$ denotes the polar angle of the vector $r - r_l$, and $H_0^{(1)}$ and J_m denote Hankel and Bessel functions, respectively. A similar expression for G_{zz} at an observation point r in proximity to but outside of pore q centered at r_q can be written as

$$G_{zz}^{\text{ext}}(r, r_s) = \frac{1}{4i} \chi^{\text{ext}}(r_s) H_0^{(1)}(k|r - r_s|)$$

$$+ \sum_{q=1}^{N_c} \sum_{m=-\infty}^{\infty} B_m^q H_m^{(1)}(k|r - r_s|) e^{im \arg(r - r_s)} . \tag{4.20}$$

Again χ^{ext} accounts for the position of the source.

The coefficients C_m^l of (4.19) and B_m^l in (4.20) for the same pore l are linked through continuity conditions across the pore surface. These continuity conditions together with the requirement of consistency of the various expansions centered around different pores give rise to the full multiple scattering problem and determine a system of linear equations. Once this system is solved, the Green's function $G_{zz}^{\text{ext}}(r, r_s)$ can be reconstructed and field distributions and LDOS may be obtained (for details of the calculations as well as the implementation, we refer to [4.42]).

To demonstrate the feasibility of this approach as a modeling tool, we discuss a beamsplitter based on the optimized bend designed within the Wannier function approach (see Sect. 4.3.5). In Fig. 4.7 we display the field distribution of a plane wave (frequency $\omega = 0.282 \times 2\pi c/a$) impinging on a trial beamsplitter based on the naive beamsplitter (upper left in Fig. 4.6). Although the wave couples into the device through the input waveguide, this beamsplitter is unable to guide any radiation to any of the output ports and all the incoming radiation is reflected back. The failure of this naive example of a beamsplitter manifests the need of a more thorough investigation of the parameters to construct working devices. A more complicated beamsplitter based on the optimized bend design (lower right in Fig. 4.6) is depicted in Fig. 4.8. The good operation characteristics of the device for the same freqquncy ($\omega = 0.282 \times 2\pi c/a$) are apparent. More precisely, a analysis of the

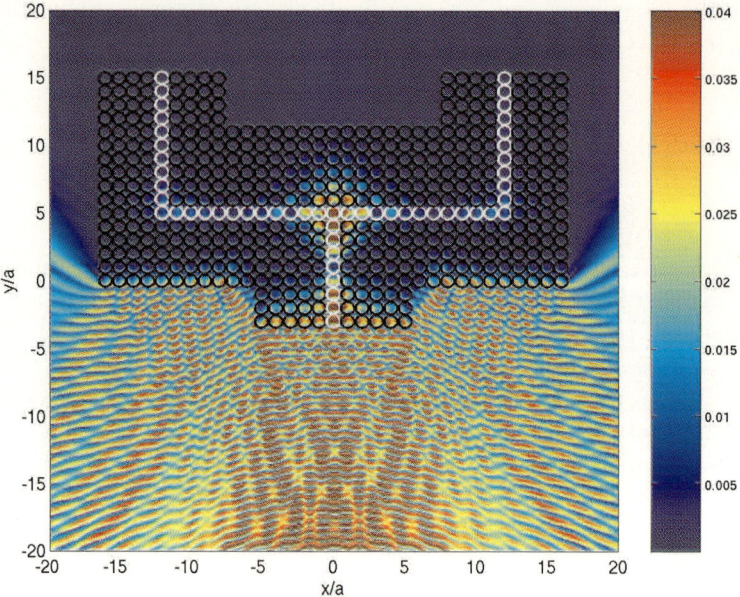

Fig. 4.7. Field distribution of a beamsplitter design based on the naive beamsplitter (upper left in Fig. 4.6). A point source far away from the PC structure emits at $\omega = 0.282 \times 2\pi c/a$, so that effectively a plane wave impinges on the PC structure. Clearly visible is the coupling into the central waveguide structure (Defect pores infilled with polymer are indicated through a white circle). Little intensity is transfered to the arms of the beamsplitter and practically nothing is transmitted around the bend. The parameters of the underlying PC are the same as those in Fig. 4.1.

Poynting vector in the input waveguide and the two output waveguides reveals that about 92% of the intensity are transmitted, 46% in each arm of the beamsplitter [4.43].

So far, we have been considering devices built within perfect lattices. This, unfortunately, is far from the experimental situation. Defects or imperfections are always present and they greatly influence the response of any actual device. It is thus important to characterize the effects of disorder on the device performance. As an illustration, we consider the optimized beamsplitter of Fig. 4.8 as the "perfect" device and suppose that during the fabrication process fabricational tolerances lead to a random variation of the pore diameter ranging from $r/a = 0.46$ to $r/a = 0.48$ (corresponding roughly to 3% radial disorder). The resulting performance for the operating frequency $\omega = 0.282 \times 2\pi c/a$ of the "perfect" device is depicted in Fig. 4.9. Clearly, the performance is compromised for even this moderate degree of disorder of about 3%. Although this speaks for itself, we would like to emphasize the im-

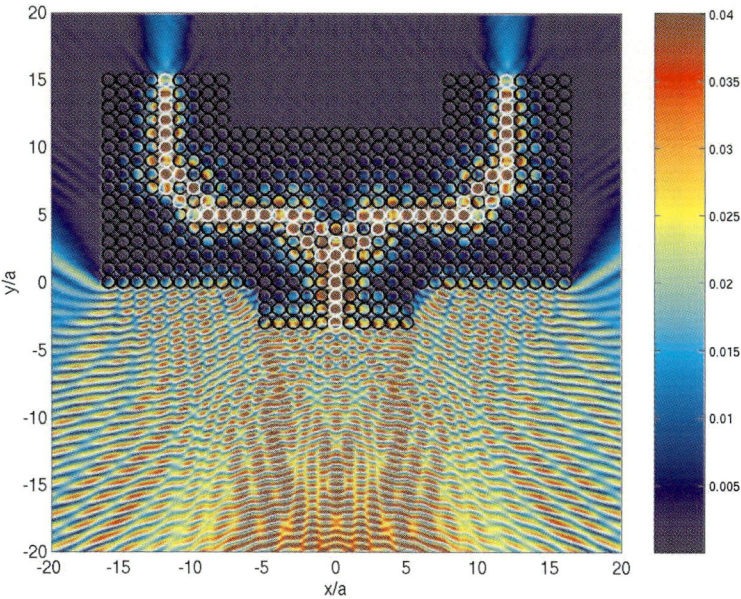

Fig. 4.8. Field distribution of a beamsplitter design based on the optimized beam-splitter (lower right in Fig. 4.6). A point source far away from the PC structure emits at $\omega = 0.282 \times 2\pi c/a$, so that effectively a plane wave impinges on the PC structure. Clearly visible is the coupling into the central waveguide structure (Defect pores infilled with polymer are indicated through a white circle). Substantial intensity is transfered to the arms of the beamsplitter and is fully transmitted around the bend, resulting in an effective beamsplitter. The parameters of the underlying PC are the same as those in Fig. 4.1.

portance of systematic investigations of the effects of fabricational tolerances on the performance of PC-based devices. This area of research has received very little attention until now.

4.5 Conclusions and Outlook

In summary, we have outlined a framework based on solid-state theoretical methods that allows one to qualitatively and quantitatively treat electro-magnetic wave propagation in PCs. Photonic bandstructure computations for infinitely extended PCs provides photonic bandstructures and other physical quantities such as DOS and group velocities [4.26,4.27]. Furthermore, the input of bandstructure calculations facilitate the construction of maximally localized photonic Wannier functions which allow one to efficiently obtain the properties of defect structures embedded in PCs. In particular, the efficiency

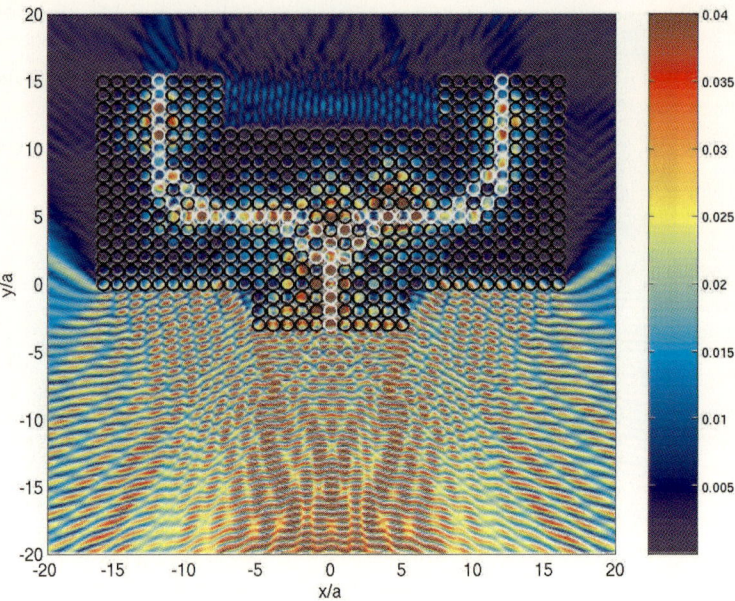

Fig. 4.9. Field distribution of a beamsplitter including fabricational tolerances. The beamsplitter is based on the optimized design of Fig. 4.8 and a random variation of the pore diameter ranging from $r/a = 0.46$ to $r/a = 0.48$ (roughly 3% radial disorder) has been added in order to model fabricational imperfections. A point source far away from the PC structure emits at $\omega = 0.282 \times 2\pi c/a$, so that effectively a plane wave impinges on the PC structure (Defect pores infilled with polymer are indicated through a white circle). The rather poor performance of the device is evident when comparing with the simulation for the perfect structure in Fig. 4.8. The parameters of the underlying PC are the same as those in Fig. 4.1.

of the Wannier function approach allows one to investigate large-scale PC circuits which, to date, are beyond the reach of standard simulation techniques such as FDTD or FE methods. Perhaps even more important is the fact that using the Wannier function approach facilitates the efficient exploration of huge parameter spaces for the design of defect structures embedded in a given PC basis structure.

The Wannier function approach is complemented by a multipole expansion technique which are well-suited for the investigation of finite-sized PCs. The usefulness of this multipole expansion manifests itself when we are considering efficient designs for actual finite-sized devices. A judicious approach that combines the results of optimizations via Wannier function studies with the multipole expansion technique has allowed us to desgin a realistic beamsplitter. Unfortunately, the experimental situation is far from ideal and the lattice where the design is realized is not defect-free. We introduced realistic fabricational tolerances in into the optimized beamsplitter and analyzed its response. Under these conditions the device clearly lost its functionality, showing the importance of finding designs that are as robust as possible under the influence of fabricational imperfections.

Acknowledgments

This work was supported by the Center for Functional Nanostructures (CFN) of the Deutsche Forschungsgemeinschaft (DFG) within project A 1.2. The research of K.B. is further supported by the DFG under grant Bu 1107/2-2 (Emmy-Noether program). The work of M.S. is funded in the framework of the DFG Research Training Group 786 *Mixed Fields and Nonlinear Interactions* at the University of Karlsruhe.

References

[4.1]   G.P. Agrawal: *Nonlinear Fiber Optics*, 3rd edn (Academic Press, San Diego Dan Francisco New York Boston London Sydney Tokyo 2001)
[4.2]   R. März: *Integrated Optics: Design and Modeling*, (Artech House, 1995)
[4.3]   E. Yablonovitch, Phys. Rev. Lett. **58**, 2059 (1987)
[4.4]   Phys. Rev. Lett. **58**, 2486 (1987)
[4.5]   A. Birner, R.B. Wehrspohn, U.M. Gösele, and K. Busch, Adv. Mater. **13**, 377 (2001)
[4.6]   T.F. Krauss and R.M. de la Rue, Prog. Quantum Electron. **23**, 51 (1999)
[4.7]   A. Forchel et al., Microelectron. Eng. **53**, 21 (2000)
[4.8]   M. Loncar, T. Doll, J. Vuckovic, and A. Scherer, J. Lightwave Technol. **18**, 1402 (2000)
[4.9]   H. Benisty et al., IEEE J. Quantum Electron. **38**, 770 (2002)
[4.10] S. Noda, M. Imada, A. Chutinan, and N. Yamamoto, Opt. Quantum Electron. **34**, 723 (2002)
[4.11] C. Liguda et al., Appl. Phys. Lett. **78**, 2434 (2001)
[4.12] A.C. Edrington et al., Adv. Mater. **13**, 421 (2001)
[4.13] A. Rosenberg, R.J. Tonucci, H.B. Lin, and E.L. Shirley, Phys. Rev. B **54**, R5195 (1996)
[4.14] O.J.A. Schueller et al., Appl. Opt. **38**, 5799 (1999)
[4.15] S.-Y. Lin et al., Nature **394**, 251 (1998)
[4.16] S. Noda, K. Tomoda, N. Yamamoto, and A. Chutinan, Science **289**, 604 (2000)
[4.17] J.E.G.J. Wijnhoven and W.L. Vos, Science **281**, 802 (1998)
[4.18] A. Blanco et al., Nature **405**, 437 (2000)
[4.19] Y.A. Vlasov, X.Z. Bo, J.C. Sturm, and D.J. Norris, Nature **414**, 289 (2001)
[4.20] M. Campbell et al., Nature **404**, 53 (2000)
[4.21] Y.V. Miklyaev et al., Appl. Phys. Lett. **82**, 1284 (2003)
[4.22] H.B. Sun, S. Matsuo, and H. Misawa, Appl. Phys. Lett. **74**, 786 (1999)
[4.23] H.B. Sun et al., Appl. Phys. Lett. **79**, 1 (2001)

[4.24] M. Straub and M. Gu, Opt. Lett. **27**, 1824 (2002)

[4.25] N. Vats, S. john, and K. Busch, Phys. Rev. A **65**, 043808 (2002)

[4.26] K. Busch and S. John, Phys. Rev. E **58**, 3896 (1998)

[4.27] D. Hermann, M. Frank, K. Busch, and P. Wölfle, Optics Express **8**, 167 (2001)

[4.28] K.-M. Ho, C.T. Chan, and C.M. Soukoulis Phys. Rev. Lett. **65**, 3152 (1990)

[4.29] K. Busch and S. John, Phys. Rev. Lett. **83**, 967 (1999)

[4.30] A. Brandt, S. McCormick, and J. Ruge, SIAM J. Sci. Stat. Comput. **4**, 244 (1983)

[4.31] K. Busch, S.F. Mingaleev, A. Garcia-Martin, M. Schillinger, D. Hermann, J. Phys.: Condens. Matter **15**, R1233 (2003)

[4.32] N. Marzari and D. Vanderbilt, Phys. Rev. B **56**, 12847 (1997)

[4.33] A. Garcia-Martin, D. Hermann, K. Busch, and P. Wölfle, Mater. Res. Soc. Symp. Proc. **722**, L 1.1 (2002)

[4.34] D.M. Whittaker and M.P. Croucher, Phys. Rev. B **67**, 085204 (2003)

[4.35] A. Mekis et al., Phys. Rev. Lett. **77**, 3787 (1996)

[4.36] J.D. Joannopoulos, P.R. Villeneuve, and S.H. Fan, Nature **386**, 143 (1997)

[4.37] S.F. Mingaleev and Y.S. Kivshar, Opt. Lett. **27**, 231 (2002)

[4.38] S.F. Mingaleev and K. Busch, Opt. Lett. **28**, 619 (2003)

[4.39] H. Brand: *Schaltungslehre linearer Mikrowellennetze*, (Hirzel, Stuttgart 1995)

[4.40] D. Hermann, A. Garcia-Martin, J.J. Saenz, S.F. Mingaleev, M. Schillinger, and K. Busch, in preparation

[4.41] A.A. Asatryan et al., Phys. Rev. E **63**, 046612 (2001)

[4.42] A.A. Asatryan et al., Waves in Random Media **13**, 9 (2003)

[4.43] F. Hagmann and K. Busch, unpublished

5 Simulation of Active and Nonlinear Photonic Nano-Materials in the Finite-Difference Time-Domain (FDTD) Framework

A. Klaedtke, J. Hamm, and O. Hess

Advanced Technology Institute, School of Electronics and Physical Sciences, University of Surrey, Guildford, Surrey, GU2 7XH, UK

Abstract. A numerical method is presented that unites three-dimensional finite-differnce time-domain (FDTD) computer simulations of active, nonlinear photonic nano-materials with optical Bloch equations describing their microscopic spatio-temporal dynamics. The constituent equations are derived and the algorithm is discussed. Computationally simulated Rabi oscillations closely correspond to analytic results. Fully three-dimensional simulations reveal the nonlinear spatio-temporal dynamics of high-finesse whispering gallery modes in microdisk lasers.

5.1 Introduction

With Moore's famous law still holding at the time of writing of this chapter, processors are getting faster and the main memory capacity of computers is increasing steadily. This opens up the possibility to refine physical models to a detailed level unthinkable several decades ago. Closely linked to this remarkable progress in the development of nano-structuring technologies in micro- and nano-electronics, novel photonic nano-materials and structures can be fabricated today that even carry functionalities on sub-wavelength scales.

Theoretical models for and computer simulation of photonic nano-materials often rely on well-known simplifications that profit from the assumption that all optical effects may be described on spatial scales that correspond to several times the optical wavelength. However, in the novel nano-photonic materials the typical structural variation are significantly smaller than the wavelength. In these cases (and even close to the wavelength) these simplifications can longer be made any more. Indeed, Maxwell's equations describing the spatio-temporal evolution of the electromagnetic field have to be considered in full detail. Compared with simpler models this enormously increases the computational complexity. Ironically, the steadily shrinking scales of novel electronic and photonic nano-structures make on the one hand computer simulations of their optical and materials properties more tedious but on the other hand provide the basis for the conception and realization of increasingly more powerful computing platforms on which the simulations run.

A. Klaedtke, J. Hamm, and O. Hess, Simulation of Active and Nonlinear Photonic Nano-Materials in the Finite-Difference Time-Domain Framework, Lect. Notes Phys. **642**, 75–101 (2004)
http://www.springerlink.com/

In this chapter we will first describe a well established combination of algorithms to solve the Maxwell curl equations with boundary conditions and discuss their implementation. The code snippets shown are in a pseudo programming language similar to C or C++. Subroutines are called without the global variables as parameters. In the second part of this report, a method is demonstrated which couples the optical Bloch equations in a special form to the electromagnetic field. The optical Bloch equations describe the quantum behaviour of a system interacting with the electric field on the basis of a dipole interaction Hamiltonian. With the resulting time-domain full vectorial Maxwell Bloch equations, the interplay of light with optically active nano-materials can be computationally investigated.

5.2 Finite-Difference in Time-Domain

For the calculations involving the electromagnetic fields we will formulate the Maxwell curl equations in the so called Heaviside-Lorentz form. This unit system is conveniant to equalize the magnitude of the numerical values of the electric and magnetic field allowing higher precision when additions are performed on digital microprocessors using floating point numbers. Note that if we were using the MKSA system of units in contrast, we would introduce an imbalance with the dielectric constant of vacuum ϵ_0 and the permeability of vacuum μ_0 being of significantly different magnitude. Furthermore, we conveniently set the speed of light in vacuum c to 1.

Frequently, the Maxwell curl equations can be simplified taking the specific properties of the nano-materials into account. Considering non-magnetic materials results in the equivalence of the magnetic induction and the magnetic field $\boldsymbol{H}$ (in the Heaviside-Lorentz system of units). Furthermore we assume that the backround dielectric constant ϵ_r should not vary in time (only be spatially dependent) and of scalar nature. These assumptions lead to the following equations describing the coupled spatio-temporal dynamics of the three fields $\boldsymbol{E}$ (the electric field), $\boldsymbol{D}$ (the electric displacement field) and $\boldsymbol{H}$ (the magnetic field):

$$\nabla \times \boldsymbol{H}(\boldsymbol{r},t) = c^{-1} \frac{\partial \boldsymbol{D}(\boldsymbol{r},t)}{\partial t}, \qquad \nabla \times \boldsymbol{E}(\boldsymbol{r},t) = -c^{-1} \frac{\partial \boldsymbol{H}(\boldsymbol{r},t)}{\partial t} \qquad (5.1)$$

$$\boldsymbol{D}(\boldsymbol{r},t) = \epsilon_r(\boldsymbol{r}) \boldsymbol{E}(\boldsymbol{r},t) + \boldsymbol{p}(\boldsymbol{r},t) \qquad (5.2)$$

The polarisation density $\boldsymbol{p}$ is the link to the (spatio-temporal) material properties. The implementation of those will be discussed in the second part.

In 1966, Yee [5.2] presented a formulation for solving the Maxwell curl equations 5.1 in a discretised way. The algorithm named after him is highly efficient and still rather accurate. To translate the differentials in the Maxwell equations, Yee made use of a straightforward approach for the difference approximation of second order accuracy in δs, exemplified in 5.3. The differential at the position p is approximated by the difference quotient of the two neighbouring values at half-step distances

$$\left.\frac{\partial f(s)}{\partial s}\right|_{s=p} \to \lim_{\delta s \to 0} \frac{f(p + \delta s/2) - f(p - \delta s/2)}{\delta s}. \tag{5.3}$$

In these (from the point of view of the Maxwell equations "auxiliary") equations for the description of material properties, the use of an averaging rule is necessary, as a field value is not calculated for a certain time step. The following equation will be used in this case:

$$f(t) \to \lim_{\delta t \to 0} \frac{f(t + \delta t/2) + f(t - \delta t/2)}{2} \tag{5.4}$$

When carrying out the differencing it is essential to avoid as many averaging procedures as possible to achieve a high numerical stability.

The field values in the equations in discrete form will be labeled in a special way to make the appearance short and concise. Therefore we introduce a notation for field values F at certain positions in Cartesian space x, y, z and time t. The discrete field values should be pinned on an evenly spaced grid with a spatial stepping of δs and a temporal stepping of δt. Offsets can be changed as necessary.

$$F|_{i,j,k}^{m} := F(i\,\delta s, j\,\delta s, k\,\delta s, m\,\delta t) = F(x, y, z, t) \tag{5.5}$$

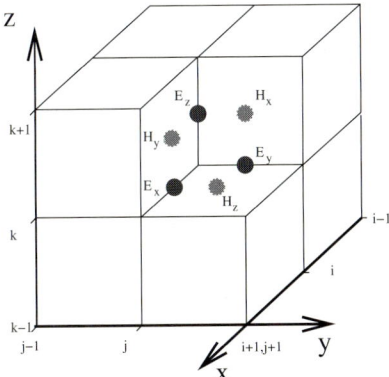

Fig. 5.1. The "Yee cube" showing the distribution of the field vector components in one grid cell at space point i, j, k. Shown are the Cartesian components of the electric (E) and the magnetic (H) fields.

```
void calcH () {
  for (int i = isc; i < iec; i++) {
    for (int j = jsc; j < jec; j++) {
      for (int k = ksc; k < kec; k++) {
        H_x(i,j,k) -= cdtds * ( E_z(i,j+1,k) - E_z(i,j,k) -
                                E_y(i,j,k+1) + E_y(i,j,k) );
        H_y(i,j,k) -= cdtds * ( E_x(i,j,k+1) - E_x(i,j,k) -
                                E_z(i+1,j,k) + E_z(i,j,k) );
        H_z(i,j,k) -= cdtds * ( E_y(i+1,j,k) - E_y(i,j,k) -
                                E_x(i,j+1,k) + E_x(i,j,k) );
      }
    }
  }
}
```

Fig. 5.2. Pseudo code realisation of the algorithm calculating the evolution of the magnetic field. The global constants `isc`, `jsc` and `ksc` define the lower bound of the core region of the simulation. `iec`, `jec` and `kec` define the upper bounds of the core region. The global constant `cdtds` is the factor $c \cdot \delta t / \delta s$. The global grid variables `H` and `E` represent the electric and magnetic fields.

The result of the translation to difference form of the Maxwell curl equations 5.1 is given in 5.6. For brevity, we only show the equation for the first Cartesian component x of the updating equation for the electric displacement field $\boldsymbol{D}$. The other five scalar equations that complete the curl equation set are of similar form and can be easily deduced.

$$
D_x\big|_{i+\frac{1}{2},j,k}^{m+\frac{1}{2}} = D_x\big|_{i+\frac{1}{2},j,k}^{m-\frac{1}{2}} + \frac{c\,\delta t}{\delta s}\left[\left(H_z\big|_{i+\frac{1}{2},j+\frac{1}{2},k}^{m} - H_z\big|_{i+\frac{1}{2},j-\frac{1}{2},k}^{m} \right) \right.
$$
$$
\left. - \left(H_y\big|_{i+\frac{1}{2},j,k+\frac{1}{2}}^{m} - H_y\big|_{i+\frac{1}{2},j,k-\frac{1}{2}}^{m} \right) \right]
\tag{5.6}
$$

Translating this into a pseudo programming language is straight forward. The algorithm is composed of two parts. One part is the calculation of the new electric displacement field (Fig. 5.3) and the other results in the new magnetic field (Fig. 5.2).

A very important aspect to consider in practice is the order of storage of the three dimensional arrays in memory. The innermost loop (k in the examples) should belong to the fastest index in the array. This means consecutive iterations for the index of this dimension should access neighbouring positions in memory. The cache architecture of the computer can then preload several array elements from the slower main memory needed in successive calculations all at once. The following iterations can then be executed with fast cache access.

An interesting detail of the Yee algorithm is the spatial distribution of the grid points as well as their temporal distribution. These spatial positions are

```
void calcD () {
  for (int i = isc; i < iec; i++) {
    for (int j = jsc; j < jec; j++) {
      for (int k = ksc; k < kec; k++) {
        D_x(i,j,k) += cdtds * ( H_z(i,j,k) - H_z(i,j-1,k) -
                                H_y(i,j,k) + H_y(i,j,k-1) );
        D_y(i,j,k) += cdtds * ( H_x(i,j,k) - H_x(i,j,k-1) -
                                H_z(i,j,k) + H_z(i-1,j,k) );
        D_z(i,j,k) += cdtds * ( H_y(i,j,k) - H_y(i-1,j,k) -
                                H_x(i,j,k) + H_x(i,j-1,k) );
      }
    }
  }
}
```

Fig. 5.3. Pseudo code realisation of the algorithm calculating the evolution of the electric displacement field. In addition to the global parameters introduced in calcH (Fig. 5.2), the global grid D represents the electric displacement field.

usually visualized on the so called Yee cube (Fig. 5.1), a three dimensional rectangular grid displaying the positions of the field values.

The electric displacement D and the magnetic induction B are at the positions of their appertaining fields E and H, which are shown in the Yee cube. Similarly, the polarisation density p is situated at the position of the electric field and the electric displacement field.

The discretisation of the material equation 5.2, relating the electric field E to the electric displacement field D and the polarisation density p represents no further difficulty, as all field values are given at the same positions in space and time. A pseudo code implementation is shown in Fig. 5.4.

```
void calcE () {
  for (int i = isc; i < iec; i++) {
    for (int j = jsc; j < jec; j++) {
      for (int k = ksc; k < kec; k++) {
        E_x(i,j,k) = InvEps_x(i,j,k) * D_x(i,j,k);
        E_y(i,j,k) = InvEps_y(i,j,k) * D_y(i,j,k);
        E_z(i,j,k) = InvEps_z(i,j,k) * D_z(i,j,k);
      }
    }
  }
}
```

Fig. 5.4. Pseudo code realisation of the algorithm calculating the evolution of the electric field. The global grid InvEps represents the inverse of the ϵ_r material index. The different Cartesian indices x, y and z account for the spatial offset according to the Yee-cube.

We should point out that the use of three different grids for the inverse ϵ_r field is not mandatory. Only one grid is of real significance. Values of ϵ_r at the half-step points in space can be calculated by interpolation of neighbouring points as needed.

The Yee algorithm is highly efficient as only few and computationally fast operations (i.e. multiplications and sums) have to be processed. And it is still a rather accurate method to calculate the evolution of the electromagnetic fields in spite of using only approximations of second order accuracy in the differences. This fact can be attributed to the type of algorithm, called a half-time stepping or leap-frog type. This means that the algorithm leaps from the calculation of the new electric field, using the old magnetic field, to the calculation of the new magnetic field using the thus calculated magnetic field, and so on.

It can be shown that in d spatial dimensions the algorithm requires to satisfy the Courant condition

$$c \cdot \delta t \leq d^{-\frac{1}{2}} \cdot \delta s \tag{5.7}$$

in order to be stable. Another restriction has to be made for the spatial stepping δs. The wavelengths in the material with the highest diffraction index, which are to be investigated, should be represented by at least 12 grid points. This is a more strict application of the Nyquist criterion which has proven to be a good rule of thumb.

It should be mentioned here that there is a drawback of the method. Energy should be conserved by the basic Maxwell equations in free space. However, it turns out that the Yee algorithm is only energy conserving with respect to averaged values over longer times [5.5]. If strict energy conservation is required at all times higher orders of approximation for the differentials have been applied in simulations (see Taflove [5.6]).

5.3 Uniaxial Perfectly Matching Layers (UPML) Boundary Conditions

As the computational grid for the fields can not be infinite, they have to be terminated somewhere. The equations describing the termination are called boundary conditions. There are boundary conditions representing all kinds of real world systems. Metallic or closed boundary conditions are simple to implement, as the field value is just assumed to be vanishing at the position of the border. Electromagnetic waves are then totally reflected. A pseudo code implementation is given in Fig. 5.5. Periodic boundaries represent a certain kind of Bloch condition.

In our simulations we use open boundaries to model free space. Energy transmitted away from the object of interest by electromagnetic waves has to be lost. Moreover, there should be no feedback of energy from the borders

```
void calcPCW () {
  for (int k = 0; k < kmax; k++) {
    for (int i = 0; i < imax; i++) {
      E_z(i,0,k) = 0.;
      E_x(i,0,k) = 0.;
    }
    for (int j = 0; j < jmax; j++) {
      E_z(0,j,k) = 0.;
      E_y(0,j,k) = 0.;
    }
  }
  for (int j = 0; j < jmax; j++) {
    for (int i = 0; i < imax; i++) {
      E_y(i,j,0) = 0.;
      E_x(i,j,0) = 0.;
    }
  }
}
```

Fig. 5.5. Pseudo code realisation of the metallic boundary conditions. The tangential field values of the electric fields are set to 0. It is only necessary to do this on the lower bounds of the grid dimensions, as only these values are accessed. The global parameters imax, jmax and kmax are the upper bounds of the grid dimensions.

of the simulation region. It turned out that the so called Uniaxial Perfectly Matching Layers (U-PML) boundary condition is best suited for us.

The basic idea behind PML boundaries is to set up a region of conducting material which absorbs electromagnetic waves without reflection. As this region with the special material properties has to be terminated on the edge of

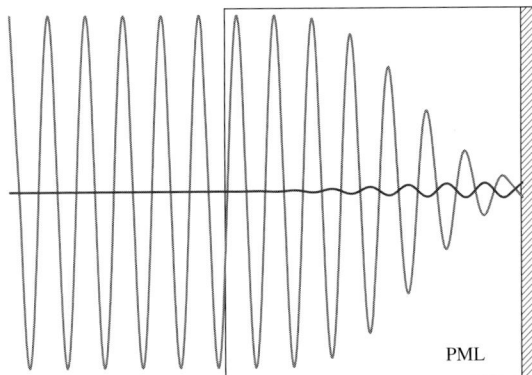

Fig. 5.6. A wave coming from the left enters the PML region and is dampened. It is reflected on the right by a perfectly conducting wall and passes the PML region again being still further absorbed.

the computational cube it is terminated with a totally reflecting boundary. The workings of such an enclosing PM layer is shown in Fig. 5.6. A wave coming from the left enters the PML. On the right side, the wave is totally reflected and crosses the PML again being further reduced in magnitude.

To describe such a behaviour in terms of the Maxwell equations in frequency space, the dielectric constant and the permeability have to be tensorial quantities. It turns out that both can be separated into a scalar and a tensorial part, with the tensorial part s being the same for both.

$$
\begin{aligned}
\nabla \times \boldsymbol{H} &= -\imath\omega\,\epsilon_r \boldsymbol{s} \cdot \boldsymbol{E} \\
\nabla \times \boldsymbol{E} &= \imath\omega\,\mu_r \boldsymbol{s} \cdot \boldsymbol{H}
\end{aligned}
\tag{5.8}
$$

The ansatz requiring a tensor s that absorbs an incoming wave without reflecting it leads to 5.9.

$$
\boldsymbol{s} = \begin{bmatrix} \frac{s_y s_z}{s_x} & & 0 \\ & \frac{s_x s_z}{s_y} & \\ 0 & & \frac{s_x s_y}{s_z} \end{bmatrix} \quad \cdot\,, \; s_i = \kappa_i - \frac{\sigma_i}{\imath\omega}
\tag{5.9}
$$

The parameters s_i — with i being one of x, y or z — can be separated into a real (κ_i) and an imaginary part (σ_i/ω). The real part degrades evanescent waves. The imaginary part absorbs energy.

The conversion of the differential equation 5.8 from frequency space to time-domain requires the introduction of two new fields $\boldsymbol{D}$ and $\boldsymbol{B}$ as shown in 5.10 to avoid the calculation of a convolution integral.

$$
\boldsymbol{D} = \epsilon_r \begin{bmatrix} \frac{s_z}{s_x} & & 0 \\ & \frac{s_x}{s_y} & \\ 0 & & \frac{s_y}{s_z} \end{bmatrix} \cdot \boldsymbol{E}; \qquad \boldsymbol{B} = \mu_r \begin{bmatrix} \frac{s_z}{s_x} & & 0 \\ & \frac{s_x}{s_y} & \\ 0 & & \frac{s_y}{s_z} \end{bmatrix} \cdot \boldsymbol{H}
\tag{5.10}
$$

With this substitution, it is possible to separate the real and imaginary parts occurring on the right side of 5.8. The transformation to time-domain is then simply the substitution of $-\imath\omega$ by the time derivative.

$$
\begin{aligned}
\nabla \times \boldsymbol{H} &= \begin{bmatrix} \kappa_y & & 0 \\ & \kappa_z & \\ 0 & & \kappa_x \end{bmatrix} \cdot \frac{\partial \boldsymbol{D}}{\partial t} + \begin{bmatrix} \sigma_y & & 0 \\ & \sigma_z & \\ 0 & & \sigma_x \end{bmatrix} \cdot \boldsymbol{D} \\
\nabla \times \boldsymbol{E} &= -\begin{bmatrix} \kappa_y & & 0 \\ & \kappa_z & \\ 0 & & \kappa_x \end{bmatrix} \cdot \frac{\partial \boldsymbol{B}}{\partial t} - \begin{bmatrix} \sigma_y & & 0 \\ & \sigma_z & \\ 0 & & \sigma_x \end{bmatrix} \cdot \boldsymbol{B}
\end{aligned}
\tag{5.11}
$$

$$\begin{bmatrix} \kappa_x & & 0 \\ & \kappa_y & \\ 0 & & \kappa_z \end{bmatrix} \cdot \frac{\partial\,D}{\partial t} + \begin{bmatrix} \sigma_x & & 0 \\ & \sigma_y & \\ 0 & & \sigma_z \end{bmatrix} \cdot D =$$

$$\epsilon_r \left\{ \begin{bmatrix} \kappa_z & & 0 \\ & \kappa_x & \\ 0 & & \kappa_y \end{bmatrix} \cdot \frac{\partial\,E}{\partial t} + \begin{bmatrix} \sigma_z & & 0 \\ & \sigma_x & \\ 0 & & \sigma_y \end{bmatrix} \cdot E \right\}$$

$$\begin{bmatrix} \kappa_x & & 0 \\ & \kappa_y & \\ 0 & & \kappa_z \end{bmatrix} \cdot \frac{\partial\,B}{\partial t} + \begin{bmatrix} \sigma_x & & 0 \\ & \sigma_y & \\ 0 & & \sigma_z \end{bmatrix} \cdot B =$$

$$\mu_r \left\{ \begin{bmatrix} \kappa_z & & 0 \\ & \kappa_x & \\ 0 & & \kappa_y \end{bmatrix} \cdot \frac{\partial\,H}{\partial t} + \begin{bmatrix} \sigma_z & & 0 \\ & \sigma_x & \\ 0 & & \sigma_y \end{bmatrix} \cdot H \right\}$$

(5.12)

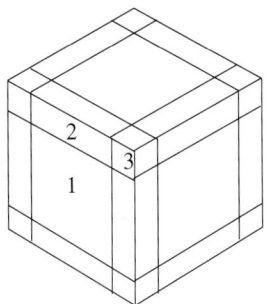

Fig. 5.7. The faces (1), edges (2) and corners (3) a PML has to be separated into.

An exemplary assembly of regions for the simulation is shown in Fig. 5.7. The inner core is surrounded by three types of PML layers which are finally terminated on the surface of the cube. As only that part of a wave travelling in the direction away from the center of the simulation should be absorbed, the fields components which are perpendicular to the faces of the cube should be effected. The PML has got a finite thickness, therefore edges and corners have to have more than one absorption coefficient to be present. On the faces of the cube, the waves should only be absorbed in the perpendicular direction, along the edges in both transverse directions. And in the corners the waves are to be absorbed in all directions. The conductivities and absorption coefficients σ_i and κ_i have to be set accordingly.

Discretisation of 5.11 and 5.12 is straightforward with respect to the discretisation rule (5.3). The results for the x component of the electric field and the displacement field update are shown in 5.13 and 5.14. Other components and field updates are of the same form and can easily be derived.

$$D_x\big|_{i+\frac{1}{2},j,k}^{m+\frac{1}{2}} = \frac{\kappa_y - \frac{1}{2}\sigma_y\,\delta t}{\kappa_y + \frac{1}{2}\sigma_y\,\delta t} D_x\big|_{i+\frac{1}{2},j,k}^{m-\frac{1}{2}} + \frac{\delta t}{\kappa_y + \frac{1}{2}\sigma_y\,\delta t} \cdot$$

$$\left\{ \frac{H_z\big|_{i+\frac{1}{2},j+\frac{1}{2},k}^{m} - H_z\big|_{i+\frac{1}{2},j-\frac{1}{2},k}^{m}}{\delta s} - \right.$$

$$\left. \frac{H_y\big|_{i+\frac{1}{2},j,k+\frac{1}{2}}^{m} - H_y\big|_{i+\frac{1}{2},j,k-\frac{1}{2}}^{m}}{\delta s} \right\}$$

(5.13)

$$E_x\big|_{i+\frac{1}{2},j,k}^{m+\frac{1}{2}} = \frac{\kappa_z - \frac{1}{2}\sigma_z\,\delta t}{\kappa_z + \frac{1}{2}\sigma_z\,\delta t} E_x\big|_{i+\frac{1}{2},j,k}^{m-\frac{1}{2}} + \frac{1}{\epsilon_r}\left(\kappa_z + \frac{1}{2}\sigma_z\,\delta t\right)^{-1}$$

$$\left\{ \left(\kappa_x + \frac{1}{2}\sigma_x\,\delta t\right) D_x\big|_{i+\frac{1}{2},j,k}^{m+\frac{1}{2}} - \right.$$

(5.14)

$$\left. \left(\kappa_x - \frac{1}{2}\sigma_x\,\delta t\right) D_x\big|_{i+\frac{1}{2},j,k}^{m-\frac{1}{2}} \right\}$$

The pseudo code realisation of the above described and discretized algorithm is shown in Fig. 5.8. We restrict the presentation to the electric field part and leave the programming of the magnetic field calculation to the reader. What is missing in this code is the setup of the coefficient arrays ce_p (5.15 with $p \in \{x, y, z\}$) which is presented in Fig. 5.9. A similar setup has to be worked out for the magnetic field coefficient arrays.

$$ce_p(0, i) = \kappa_p(i) + \frac{1}{2}\sigma_p(i)\,\delta t$$

$$ce_p(1, i) = \left(\kappa_p(i) + \frac{1}{2}\sigma_p(i)\,\delta t\right)^{-1}$$

$$ce_p(2, i) = \kappa_p(i) - \frac{1}{2}\sigma_p(i)\,\delta t$$

(5.15)

$$ce_p(3, i) = \frac{\kappa_p(i) - \frac{1}{2}\sigma_p(i)\,\delta t}{\kappa_p(i) + \frac{1}{2}\sigma_p(i)\,\delta t}$$

The implementation of just one highly absorbing sheath with a high absorption coefficient σ results in a large numerical error at the interface to the inner region. This error expresses itself in an artificial reflection. It turns out that the use of several layers with a smoothly increasing absorption coefficient resolves the problem. The same applies for the coefficient κ. A potential rising has proven to be a good approach and is shown in 5.16.

$$\sigma(x) = \sigma_{\max}\left(\frac{x}{d}\right)^p \qquad \kappa(x) = 1 + (\kappa_{\max} - 1)\left(\frac{x}{d}\right)^p$$

(5.16)

The parameter d is the thickness in layers of the boundary. The exponent p and the values for $\sigma_{\max}$ and $\kappa_{\max}$ can be freely chosen. Taflove shows in his excellent book on FDTD methods [5.6] that a choice for p between 3 and 4 gives reasonable results.

```
void calcPML_E () {
 calcPartPML_E(0, isc, 0, jmax, 0, kmax);
 calcPartPML_E(iec, imax, 0, jmax, 0, kmax);
 calcPartPML_E(isc, iec, 0, jsc, 0, kmax);
 calcPartPML_E(isc, iec, jec, jmax, 0, kmax);
 calcPartPML_E(isc, iec, jsc, jec, 0, ksc);
 calcPartPML_E(isc, iec, jsc, jec, kec, kmax);
}

void calcPartPML_E (int is, int ie,
                    int js, int je,
                    int ks, int ke) {
 float Dold_x, Dold_y, Dold_z;

 for (int i = is; i < ie; i++) {
  for (int j = js; j < je; j++) {
   for (int k = ks; k < ke; k++) {
    Dold_x = D_x(i,j,k);
    Dold_y = D_y(i,j,k);
    Dold_z = D_z(i,j,k);

    D_x(i,j,k) = ce_y(3,j) * Dold_x + dtds * ce_y(1,j) *
                 ( H_z(i,j,k) - H_z(i,j-1,k) -
                 H_y(i,j,k) + H_y(i,j,k-1) );
    D_y(i,j,k) = ce_z(3,k) * Dold_y + dtds * ce_z(1,k) *
                 ( H_x(i,j,k) - H_x(i,j,k-1) -
                 H_z(i,j,k) + H_z(i-1,j,k) );
    D_z(i,j,k) = ce_x(3,i) * Dold_z + dtds * ce_x(1,i) *
                 ( H_y(i,j,k) - H_y(i-1,j,k) -
                 H_x(i,j,k) + H_x(i,j-1,k) );

    E_x(i,j,k) = ce_z(3,k) * E_x(i,j,k) + ce_z(1,k)*InvEps_x(i,j,k) *
                 ( ce_x(0,i) * D_x(i,j,k) - ce_x(2,i) * Dold_x );
    E_y(i,j,k) = ce_x(3,i) * E_y(i,j,k) + ce_x(1,i)*InvEps_y(i,j,k) *
                 ( ce_y(0,j) * D_y(i,j,k) - ce_y(2,j) * Dold_y );
    E_z(i,j,k) = ce_y(3,j) * E_z(i,j,k) + ce_y(1,j)*InvEps_z(i,j,k) *
                 ( ce_z(0,k) * D_z(i,j,k) - ce_z(2,k) * Dold_z );
   }
  }
 }
}
```

Fig. 5.8. One part of a pseudo code realisation of the U-PML boundary condition. The global parameter dtds is $\delta t/\delta s$. The global parameter arrays ce_x, ce_y and ce_z have to be set up once initially.

He also shows that the reflection coefficient R (which is dependent on the angle φ of the incoming wave) can be approximated by 5.17.

$$R(\varphi) = \exp\left[-2\eta_{\mathrm{fs}}\,\epsilon_{\mathrm{r}}\,\cos\varphi \int_0^d \sigma(x)dx\right] \qquad (5.17)$$

```
void setupPMLcoeff_E (vector ce, vector cm,
                      int ls, int le, int lmax) {
 vector val1(-1, PMLMAX-1, 1.);
 vector val2(-1, PMLMAX-1, 1.);
 vector val1p(-1, PMLMAX-1, 1.);
 vector val2p(-1, PMLMAX-1, 1.);

 for (int l = 0; l < pmlmax; l++) {
  x = float(l) / float(pmlmax);
  sigma = SigmaMax * pow(x,POTPML);
  kappa = 1. + (KappaMax - 1.) * pow(x,POTPML);
  val1(l) = kappa + 0.5 * sigma * dt;
  val2(l) = kappa - 0.5 * sigma * dt;

  x = (float(l) + 0.5) / float(PMLMAX-1)
  sigma = SigmaMax * pow(x, POTPML);
  kappa = 1. + (KappaMax - 1.) * pow(x,POTPML);
  val1p(l) = kappa + 0.5 * sigma * dt;
  val2p(l) = kappa - 0.5 * sigma * dt;
 }

 for (int l = 0; l < ls; l++) {
  ce(0,l) = val1p(ls - l - 2); ce(2,l) = val2p(ls - l - 2);
  ce(1,l) = 1. / val1(ls - l - 1);
  ce(3,l) = val2(ls - l - 1) / val1(ls - l - 1);

  cm(0,l) = val1(ls - l - 1); cm(2,l) = val2(ls - l - 1);
  cm(1,l) = 1. / val1p(ls - l - 2);
  cm(3,l) = val2p(ls - l - 2) / val1p(ls - l - 2);
 }

 for (int l = le; l < lmax; l++) {
  ce(0,l) = val1p(l - le); ce(2,l) = val2p(l - le);
  ce(1,l) = 1. / val1(l - le);
  ce(3,l) = val2(l - le) / val1(l - le);

  cm(0,l) = val1(l - le); cm(2,l) = val2(l - le);
  cm(1,l) = 1. / val1p(l - le);
  cm(3,l) = val2p(l - le) / val1p(l - le);
 }
}
```

Fig. 5.9. The setup routine of the ce_i and cm_i coefficient arrays. It has to be called for each dimension (x, y and z) with the matrices of coefficients. Both coefficient matrices ce and cm for each dimension have to have a range from 0 to 3 for the first index and the second index should be defined from 0 to the number of grid points in that dimension. The parameters ls, le and lmax are the starting, the end grid points of the core region, and the number of grid points respectively.

The impedance in free space is $\eta_{fs} \approx 733\Omega$. This reflectivity has to be minimised leading to an optimal value for σ_{max}. The aim is to adjust the parameters of the PMLs in such a way that the wave is damped as much as possible while keeping the numerical reflection error low.

5.4 Time-Domain Full Vectorial Maxwell-Bloch Equations

The coupling of material properties to the equations describing the behaviour of the electromagnetic fields in the framework discussed here is established by the functional dependence of the electric displacement D on the electric field E. Here we will restrict ourselves to a simple two level atomic system that is described by the dipolar interaction Hamiltonian with the electromagnetic field. The quantum mechanical behaviour of such an atomic system is described by complex pseudo-spin equations. Maxwell's (real valued) equations 5.1 have to be coupled to these. This was done in an approach first published by Ziolkowski *et al.* [5.3]. Here, we will present an alternative approach that follows the ideas that have been suggested by Nagra and York in 1998 [5.1]. As we will show in Appendix A, both models share the same functionality.

In the case of an atom with a dipolar interaction Hamiltonian, the functional dependence of D on E reads:

$$D(\omega) = (1 + \chi_{\text{Lorentz}}(\omega))\, \epsilon_{\text{r}}\, E(\omega) \tag{5.18}$$

The factor $\chi_{\text{Lorentz}}(\omega)$ represents the Lorentz line shape.

The susceptibility χ_{Lorentz} can be separated into two parts, being either entirely real χ' or imaginary χ''. They are given in 5.19. Absorption is described by the imaginary part (as can be seen in Fig. 5.10); the associated phase shift specified by the real part of χ is shown in Fig. 5.12.

$$
\begin{aligned}
\chi'(\omega) &\sim \frac{\omega_0 - \omega}{\frac{1}{\omega}\left(\omega_0^2 - \omega^2\right)^2 + 4\omega\gamma^2} \\
\chi''(\omega) &\sim \frac{2\gamma}{\left(\frac{\omega_0^2}{\omega} - \omega\right)^2 + 4\gamma^2}
\end{aligned}
\tag{5.19}
$$

The differential equation that describes this functional dependence is of the form of a damped harmonic oscillator equation (5.20).

$$\ddot{p} + 2\gamma\,\dot{p} + \omega_0^2\,p = \underbrace{2\frac{\omega_0^2}{\Omega}\frac{1}{\hbar}\cdot|d_{\text{cv}}|^2}_{\alpha}\cdot E\cdot(n + n_{\text{a}}N_0) \tag{5.20}$$

It is enhanced by a highly nonlinear coupling to the inversion density n. n_{a} is the density of atoms in the material. They are characterised by the damping factor γ which is related to the line width and the resonance frequency Ω. Both are related to the frequency ω_0 by

$$\Omega^2 = \omega_0^2 - \gamma^2. \tag{5.21}$$

The driving term on the right hand side of 5.20 that also characterises the atomic light field interaction, contains a coupling constant α, the electric

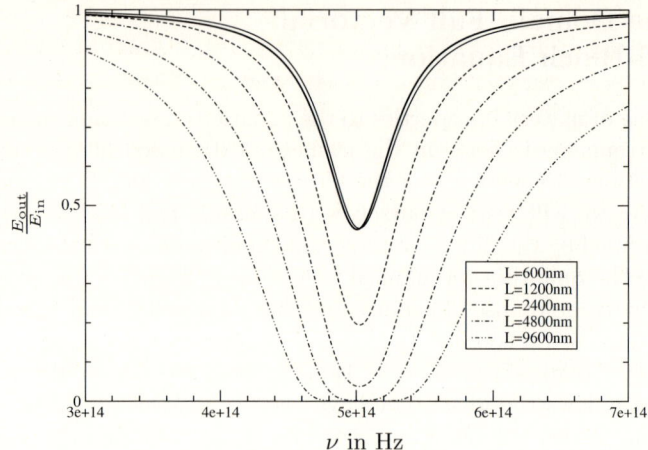

ν in Hz

Fig. 5.10. The absorption spectra of a pulse travelling different distances in an initially absorbing active Lorentzian medium. See text.

field strength E at the position of the atom and the atomic inversion density n. The coupling constant $\boldsymbol{\alpha}$ includes the dipole matrix element $|\boldsymbol{d}_{cv}|^2$ that determines the strength and spatial orientation of the polarisation to the light-field coupling. The atomic inversion density n

$$n = n_a \left(\rho_{aa} - \rho_{bb}\right) \tag{5.22}$$

is related to the diagonal density matrix elements ρ_{aa} and ρ_{bb} of the quantum mechanical system with a lower level a and an upper level b. It represents the occupation of these.

N_0 is related to the equilibrium occupation ρ_0 of the upper, more energetic level, in the following way:

$$N_0 = 1 - 2\rho_0 \tag{5.23}$$

An equation of a form similar to the Einstein rate equations can be rigidly obtained from quantum mechanically derived optical Bloch equations for a two level system. Equation 5.24 describes the temporal evolution of the damped inversion density for a pumped two level system interacting with the electric field.

$$\dot{n} = -2n_a\Lambda - \gamma_{nr}n - 2\frac{\Omega}{\hbar\omega_0^2}\,\dot{\boldsymbol{p}} \cdot \boldsymbol{E} \tag{5.24}$$

A pumping term with pump strength Λ is included in the rate equation for the inversion density n. The term with the relaxation constant γ_{nr} represents a non-radiative relaxation of excited systems.

The term proportional to $\dot{\boldsymbol{p}} \cdot \boldsymbol{E}$ stands for the work of the electromagnetic field done on the active material in quanta of approximately $\hbar\omega_0$.

Equations 5.20 and 5.24 together with 5.1 and 5.2 form the already mentioned "auxiliary differential equation" (ADE) FDTD algorithm to simulate active material interacting with the electromagnetic field. The difference form of the material equations are given in 5.25 and 5.26.

As a full tensorial expression for $\boldsymbol{\alpha}$ would make the presentation very long and nontransparent we will assume in the following a diagonal form of $\boldsymbol{\alpha}$. Note that the spatial centering of the grids given in the equation below for the inversion density is not fully correct. In principle, the field values for $\boldsymbol{p}$ and $\boldsymbol{E}$ would have to be averaged on a central point in space. And this value should then be averaged back to the space point that is required in the polarisation equation. As the error due to this incorrect centering is of minor importance in most simulations, we simplify the given equations.

$$
\begin{aligned}
p_x|^{m+\frac{1}{2}} = \ & \frac{2 - \omega^2\,\delta t^2}{1 + \gamma\,\delta t} \cdot p_x|^{m-\frac{1}{2}} \\
& - \frac{1 - \gamma\,\delta t}{1 + \gamma\,\delta t} \cdot p_x|^{m-\frac{3}{2}} \\
& + \alpha_x \frac{\delta t^2}{1 + \gamma\,\delta t} \cdot E_x|^{m-\frac{1}{2}} \cdot \left(n|^{m-\frac{1}{2}} + n_a N_0\right)
\end{aligned}
\tag{5.25}
$$

$$
\begin{aligned}
n|^{m+\frac{1}{2}} = \ & \frac{2 - \gamma_{\mathrm{nr}}\delta t}{2 + \gamma_{\mathrm{nr}}\delta t} \cdot n|^{m-\frac{1}{2}} \\
& - 4\frac{\delta t}{2 + \gamma_{\mathrm{nr}}\delta t} \cdot n_a\Lambda \\
& - 2\frac{\Omega}{(2 + \gamma_{\mathrm{nr}}\delta t)\,\hbar\omega^2}\left\{ \left(p_x|^{m+\frac{1}{2}} - p_x|^{m-\frac{1}{2}}\right) \cdot \right. \\
& \qquad\qquad \left(E_x|^{m+\frac{1}{2}} + E_x|^{m-\frac{1}{2}}\right) + \\
& \qquad\qquad \left(p_y|^{m+\frac{1}{2}} - p_y|^{m-\frac{1}{2}}\right) \cdot \\
& \qquad\qquad \left(E_y|^{m+\frac{1}{2}} + E_y|^{m-\frac{1}{2}}\right) + \\
& \qquad\qquad \left(p_z|^{m+\frac{1}{2}} - p_z|^{m-\frac{1}{2}}\right) \cdot \\
& \qquad\qquad \left. \left(E_z|^{m+\frac{1}{2}} + E_z|^{m-\frac{1}{2}}\right) \right\}
\end{aligned}
\tag{5.26}
$$

To complete the model the electric field values E from the Maxwell equations in Heaviside-Lorentz units have to be divided by ϵ_0 to link them with the above equations for the material system in the MKSA unit system.

In the pseudo code implementation shown in Fig. 5.11 the constant values PFA, PFB, PFC, NFA, NFB and NFC represent the values given in 5.27.

$$PFA = \frac{2 - \omega^2 \delta t^2}{1 + \gamma \delta t}$$

$$PFB = \frac{1 - \gamma \delta t}{1 + \gamma \delta t}$$

$$PFC_i = \alpha_i \frac{\delta t^2}{1 + \gamma \delta t} \frac{1}{\epsilon_0}$$

$$NFA = \frac{2 - \gamma_{\mathrm{nr}} \delta t}{2 + \gamma_{\mathrm{nr}} \delta t}$$ \hfill (5.27)

$$NFB = -4 \frac{\delta t}{2 + \gamma_{\mathrm{nr}} \delta t} \cdot n_a \Lambda$$

$$NFC = -2 \frac{\Omega}{(2 + \gamma_{\mathrm{nr}} \delta t) \epsilon_0 \hbar \omega^2}$$

The auxiliary equations do not impose any further restrictions on the stability rules for the simulation.

5.5 Computational Costs

The computing resources that one can use for the simulation of physical systems place a few restrictions on the system geometries and time scales to be investigated. Memory (especially the fast memory) is only available in restricted quantities. And processors always seem to be too slow to satisfy the demands of huge grid sizes.

To calculate a rough estimate of the amount of memory that is required for a simulation run we may resort to the following equation that gives an estimate of the required space S in bytes.

$$S \approx w \cdot \left[\underbrace{N^3 \cdot f_c}_{\text{core}} + \underbrace{12 \cdot p \cdot N^2}_{\text{PML}} \right] \tag{5.28}$$

The parameter w should be the number of bytes for one word, meaning the size of a floating point number. N stands for the number of grid points of one side of the simulation cube, with f_c required values per grid point in the core region. The perfectly matching boundary should be made of p layers with f_p values per grid point.

In the boundary region there have to be 12 arrays for the 4 physical quantities and their 3 Cartesian components. In the core region, the 6 components of the electric and magnetic fields come along with 3 components for the electric displacement field. If active material is present, 1 value per grid point for the inversion density and 6 values for the present and former polarisation components (as it is a differential equation of second order) have to be stored. In the pseudo code implementation 3 additional values for the old electric field components have to be stored. If the background dielectric

```
void calcPN () {
 float Ps_x, Ps_y, Ps_z;
 for (int i = isa; i < iea; i++) {
  for (int j = jsa; j < jea; j++) {
   for (int k = ksa; k < kea; k++) {
    Ps_x = P_x(i,j,k); Ps_y = P_y(i,j,k); Ps_z = P_z(i,j,k);

    P_x(i,j,k) = PFA * P_x(i,j,k) -
                 PFB * Po_x(i,j,k) +
                 PFC_x * Eo_x(i,j,k) * ( N(i,j,k) - N_0 );
    P_y(i,j,k) = PFA * P_y(i,j,k) -
                 PFB * Po_y(i,j,k) +
                 PFC_y * Eo_y(i,j,k) * ( N(i,j,k) - N_0 );
    P_z(i,j,k) = PFA * P_z(i,j,k) -
                 PFB * Po_z(i,j,k) +
                 PFC_z * Eo_z(i,j,k) * ( N(i,j,k) - N_0 );

    Po_x(i,j,k) = Ps_x; Po_y(i,j,k) = Ps_y; Po_z(i,j,k) = Ps_z;

    E_x(i,j,k) -= InvEps_x(i,j,k) * P_x(i,j,k);
    E_y(i,j,k) -= InvEps_y(i,j,k) * P_y(i,j,k);
    E_z(i,j,k) -= InvEps_z(i,j,k) * P_z(i,j,k);

    N(i,j,k) = NFA * N(i,j,k) +
               NFB +
               NFC * ( ( P_x(i,j,k) - Po_x(i,j,k) ) *
                       ( E_x(i,j,k) + Eo_x(i,j,k) ) +
                       ( P_y(i,j,k) - Po_y(i,j,k) ) *
                       ( E_y(i,j,k) + Eo_y(i,j,k) ) +
                       ( P_z(i,j,k) - Po_z(i,j,k) ) *
                       ( E_z(i,j,k) + Eo_z(i,j,k) ) );
   }
  }
 }
}
```

Fig. 5.11. The routine calculating the material behaviour. The active region is bounded by `isa` to `iea-1`, `jsa` to `jea-1` and `ksa` to `kea-1`. The global grids P and Po are the current and former time step field representations for the polarisation density. Eo is the old electric field that is defined globally. N represents the inversion density. The other values are also globally defined constant parameters defined in the text.

constant ϵ varies in space then either 1 or 3 values per grid point — depending on the accuracy needed — have to be added. In total, f_c will thus range from 6 to 22 values per spatial grid point. Consequently, the amount of memory needed for such a simulation is enormous. The computer has to access each value every time step at least once. Unfortunately, modern computers have the problem that it is much more time consuming to access main memory than it takes to perform an operation. Therefore it is obvious that a computer with fast memory access is much more important than the time it takes to perform a multiplication or addition.

5.6 Test Runs

Various test runs were performed using on a one dimensional system with 3000 spatial grid points and a spatial cell length of $\delta s = 6$nm. The time stepping δt was chosen to obey equation 5.7.

In our computer simulations, the active medium has a resonance frequency of $\Omega = 5 \times 10^{14}$Hz and a line width of $\Gamma = 2\gamma = \pi \times 10^{14}$s^{-1}. The dipole matrix element in this cases is $|d_{cv}|^2 \approx 3.83 \times 10^{-58}$A s m and the density of the atoms $n_a = 1 \times 10^{26}$m^{-3}. There is no pumping ($\Lambda = 0$) and the non-radiative recombination rate is set to $\gamma_{nr} = 5 \times 10^7$s^{-1}. Temporally narrow pulses are travelling through an active medium layer of different thicknesses L.

Initially, at $t = 0$, the medium is in its ground state ($n = n_a$) with the thermal equilibrium occupation of the upper state ρ_0 being 0. A pulse with an amplitude of 2.5×10^8V/m small enough that the population of the atomic states is not noticeably changed is sent through an absorbing medium of different lengths. The system then responds in a linear way. The center frequency of the pulse with a Gaussian envelope was exactly the resonance frequency of the Lorentzian medium. The distance the pulses had to travel through the Lorentzian medium were $L \in \{600, 1200, 2400, 4800, 9600\text{nm}\}$ respectively.

The absorption spectra of the pulse for the 5 different path lengths are shown in Fig. 5.10. The theoretical absorption coefficient is given by:

$$\frac{E_{\text{out}}}{E_{\text{in}}} = e^{-\frac{1}{2}\frac{\omega}{c}\chi''(\omega)L} \tag{5.29}$$

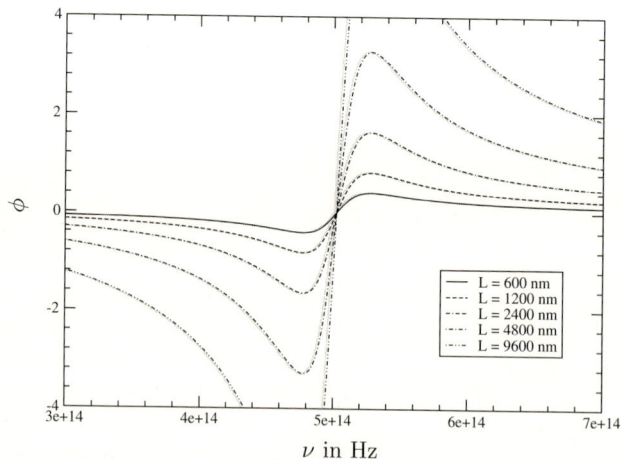

ν in Hz

Fig. 5.12. The phase shift spectra of a pulse travelling different distances in an initially unexcited active Lorentzian medium.

The spectrum of the smallest absorption, belonging to the 600 nm path length is compared to the theoretical prediction of the absorption behaviour. A slight shift to higher frequencies of about 3 parts in a thousand occurs in the case of the numerically calculated result. The phase shift $\phi(\omega)$ of the pulse is shown in 5.12 for the same set of lengths L. The theoretical prediction to which it is compared reads:

$$\phi = -\frac{1}{2}\frac{\omega}{c}\chi'(\omega)L \tag{5.30}$$

With the length of the absorbing medium increasing, the dip of the absorption curve gets deeper and the phase shift larger. Eventually the dip reaches 0, which means total absorption. The phase shift spectra are compared to the theoretical prediction of the absorption behaviour. As Fig. 5.12 shows, the analytic expression corresponds very well with our numerical results.

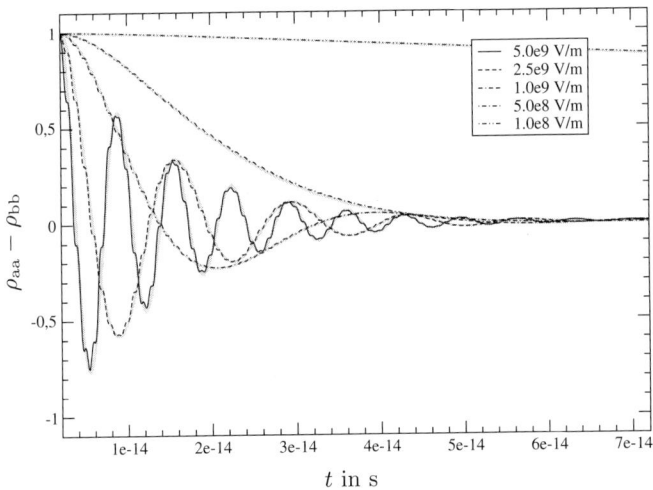

Fig. 5.13. Electromagnetic waves of different amplitudes drive a decaying Rabi oscillation to a steady state where neither light is emitted nor absorbed by the medium.

To further test the auxiliary difference equations we analyze Rabi oscillations on the basis of the rate equation 5.24 describing the atomic inversion density. Rabi oscillations (see for example R. Loudon, "The quantum theory of Light" [5.4]) refer to an oscillating behaviour of the occupation probability of the atomic states interacting with coherent EM waves of one single frequency (Fig. 5.13). In our test-setup of an atomic system we assume that initially the atoms were unperturbed and entirely in their lower states. A wave of a certain amplitude $\hat{E}$, but with the resonance frequency Ω of the quantum mechanical two-level system, is then switched on. This simulation is repeated for 5 different amplitudes of the probing wave. The result is presented in Fig. 5.13.

A theoretical investigation of the occupation probability results in the prediction of the frequency and the decay rate for the Rabi oscillation given in 5.31.

$$
\begin{aligned}
N(t) =& \rho_{\mathrm{aa}}(t) - \rho_{\mathrm{bb}}(t) \\
=& \begin{cases} e^{-\Gamma_{\mathrm{R}} t} \cdot \left[\cos \Omega_{\mathrm{R}} t \; + \frac{\Gamma_{\mathrm{R}}}{\Omega_{\mathrm{R}}} \sin \Omega_{\mathrm{R}} t \right] & \forall \, 4\omega_{\mathrm{R}}^2 > (\gamma - \gamma_{\mathrm{nr}})^2 \\ e^{-\Gamma_{\mathrm{R}} t} \cdot \left[\cosh \Omega_{\mathrm{R}} t + \frac{\Gamma_{\mathrm{R}}}{\Omega_{\mathrm{R}}} \sinh \Omega_{\mathrm{R}} t \right] & \forall \, 4\omega_{\mathrm{R}}^2 \le (\gamma - \gamma_{\mathrm{nr}})^2 \end{cases} \\
\Omega_{\mathrm{R}}^2 =& \frac{1}{4} \left(\gamma - \gamma_{\mathrm{nr}} \right)^2 - \omega_{\mathrm{R}}^2 \\
\Gamma_{\mathrm{R}} =& \frac{1}{2} \left(\gamma + \gamma_{\mathrm{nr}} \right) \\
\omega_{\mathrm{R}} =& \frac{\alpha \Omega}{2\hbar \omega_0^2} \hat{E}
\end{aligned}
\tag{5.31}
$$

So there are two cases. An oscillating one that is dampened and an over damped one that is not oscillating, but slowly relaxing to the steady state. In Fig. 5.13 this case is presented by the $1 \times 10^8 \mathrm{V/m}$ and the $5 \times 10^8 \mathrm{V/m}$ amplitudes, as opposed to the other 3 amplitudes.

A comparison of this theory with the numerical results shows good agreement. The fast periodic oscillations we obtain can be attributed to the simulation of the full dynamics in the algorithm versus the slowly varying amplitude approximation on which the analytic derivation is based.

As a further test we check the behaviour of a detuning in the frequency of the probing wave. In the simulations, we artificially reduce the line width

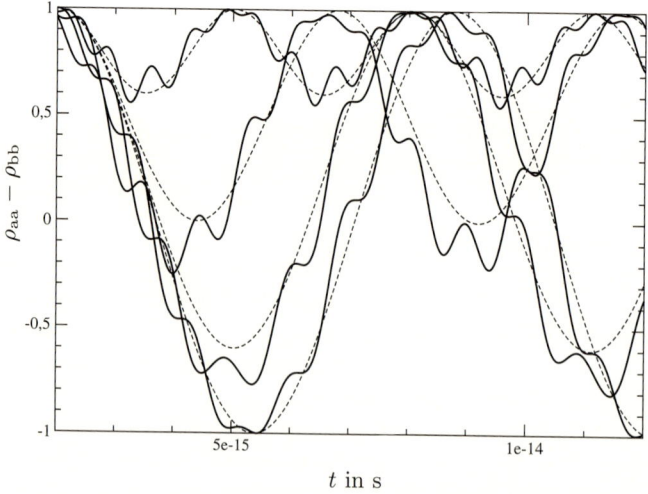

Fig. 5.14. Rabi flopping behaviour with 5 levels of frequency detuning.

of the dipole transition and the non radiative recombination rate γ_{nr} to zero. Thus the decay rate of the Rabi oscillation Γ_R vanishes. The amplitude of the probing wave is set to $\hat{E} = 5 \times 10^9 \text{V/m}$. In the computer simulations, five different ratios of detuning from the resonance frequency $|\Omega - \omega|/\omega_R$, $(0, 0.5, 1, 2)$ were compared with the theoretical/analytical prediction. The simulation results (dashed) are shown in Fig. 5.14. Similar to the previous example, the results match very well the analytically predicted behaviour.

Figure 5.15 shows as a nice result of all of the above that the application of pulses with the length of half the period of a Rabi oscillation (a π pulse) inverts the occupation. Likewise, in the case of applying a pulse of the length of an entire Rabi oscillation period (2π pulse) the population of the lower level goes through a cycle of depopulation an repopulation.

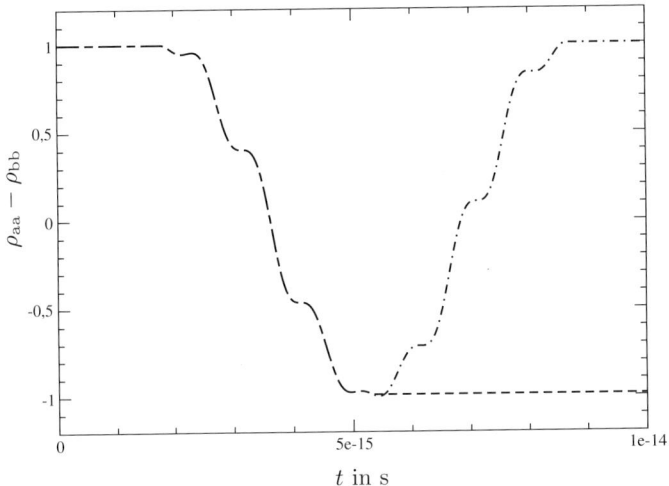

Fig. 5.15. Demonstration of the effect of so called π and 2π pulses (pulses of certain lengths) on the inversion density of the material system.

5.7 Microdisk Laser Dynamics

In contrast to the examples that we have considered so far (all of the above results were achieved using a one-dimensional code as presented by Nagra and York [5.1]) the microdisk resonator/laser (Fig. 5.16) that we will discuss in the following is (with respect to the computer simulation) a truly three-dimensional system. Figure 5.16 shows the essential features of a whispering gallery mode excited in a microdisk. These modes are the most intriguing attribute of electromagnetic fields in these cylindrical structures. They have their maximum field values on the circumference of the disk and can be

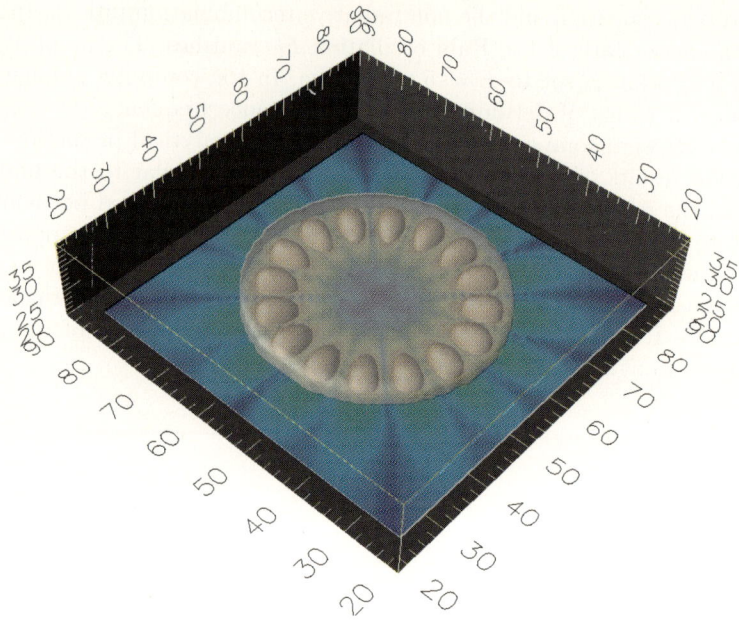

Fig. 5.16. A microdisk (hollow cylinder) showing the essential features of a HEM$_{8,1,0}$ whispering gallery mode. The features represent the isosurface of the z component of the magnetic field. The color plot shows component on a plane intersecting the disk in the middle.

characterised by the number of nodes in the azimuthal, the radial and the axial dimension. The mode shown in Fig. 5.16 is therefore a HEM$_{8,1,0}$ mode having 8 nodes around the disk, 1 node in the middle and 0 nodes in the axial direction.

Associated with every one of these modes in the disk is an eigen-frequency. If an active material inside the disk will support this frequency, strong emission of light could be achieved when pumping the respective quantum mechanical transition involved. The fact that these modes have high quality factors implies that they emit radiation very slowly. This makes them highly interesting for low threshold lasing devices.

Such a lasing device was simulated using the described ADE-FDTD algorithm in 3 spatial dimensions. The resonance frequency of the active Lorentzian medium was chosen to match the eigen-frequency of the HEM$_{4,1,0}$ mode of a microdisk with a radius of 105nm and a height of 30nm. The background material was chosen to have an ϵ_r of 11.1556. Pumping was homogeneous over the whole disk.

The result for the inversion density over time is plotted in Fig. 5.17. The upper curve gives the spatially averaged inversion density and the lower curve the inversion density at one point in the disk. The averaged inversion is higher than the inversion at the position of the whispering gallery mode as inversion is accumulating in the center due to the lack of a lasing center mode. After a few short relaxation oscillations, lasing sets in and the calculation reveals a rotation of the mode around the center of the disk.

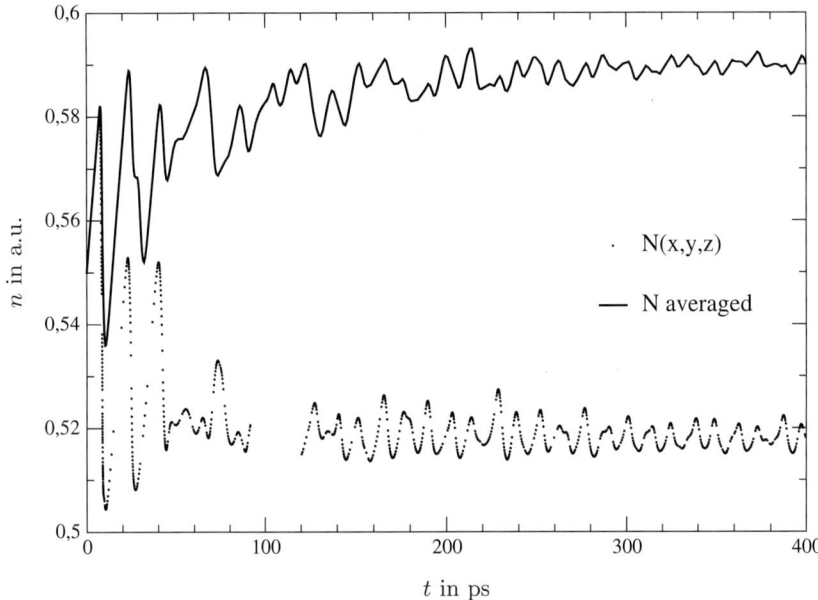

Fig. 5.17. Temporal inversion density plot from the simulation of lasing behaviour in a microdisk resonator filled with an excited two level atomic system active material. The inversion density is given in artificial units.

Figure 5.18 shows the inversion density n at an arbitrary time step in the simulation. Lasing at the eigen-frequency takes place and the associated $HEM_{4,1,0}$ whispering gallery mode spatially consumes the inversion density at the rim of the disk. It is thus embossing the mode structure into the density profile. In the middle, the pumping piled up an inversion hill which is not touched by the whispering gallery modes. Occasionally, the inversion density in the middle is high enough to support a highly lossy $HEM_{0,0,0}$ mode which then occurs for a very short period in time until the inversion density is again reduced below threshold.

As the "modes" in a nonlinear system are not fixed in space they can change their shape and position. With the inversion being, due to spatial hole burning, higher at the nodes of the modes the $HEM_{4,1,0}$ mode starts to rotate in order to maximize its depletion of the available inversion.

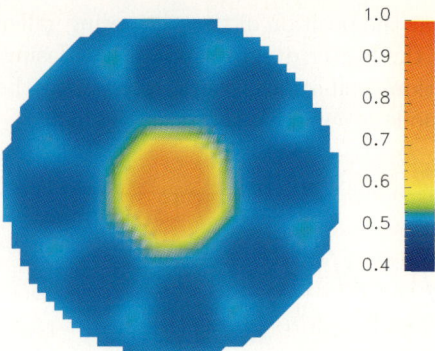

Fig. 5.18. Inversion density plot from a simulation of the lasing behaviour of a microdisk resonator filled with a pumped two level atomic system active material. The inversion density is given in artificial units.

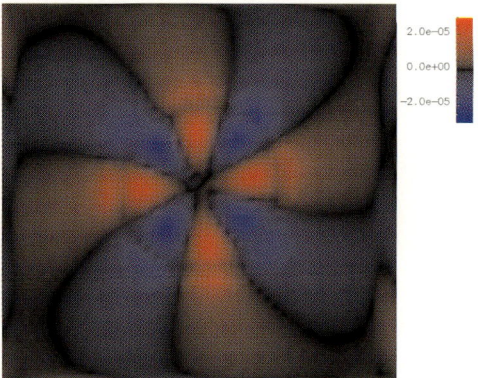

Fig. 5.19. Electric field color plot from a simulation of the lasing behaviour of a microdisk resonator filled with a pumped two level atomic system active material. The electric field strength is given in artificial units.

5.8 Conclusion

We have discussed a numerical method that unites finite-differnce time-domain (FDTD) computer simulations of active, nonlinear photonic nano-materials with optical Bloch equations describing their microscopic spatio-temporal dynamics. Computationally simulated Rabi oscillations closely correspond to analytic results. Fully three-dimensional simulations reveal the nonlinear spatio-temporal dynamics of high-finesse whispering gallery modes in microdisk lasers. These are, in particular, dynamic, fan-like oscillation of the electromagnetic fields and inversion density of the WGM driven by spatial holeburning effects.

Appendix A: Relations

In this section, the relation between the Lorentz oscillator equation to the polarisation pseudo-spin equation of a electric dipole transitions will be shown. For simplicity we restrict the discussion to the scalar form of the equations, as the important aspects of the amplitude dynamics are not related to the vectorial character.

The harmonic oscillator differential equation is of second order with real values, whereas the pseudo-spin differential equations are of first order with complex values. Equation 5.32 is the optical Bloch equation (see for example in Loudon [5.4]) for the off-diagonal part of the density matrix ρ of a two level system, which is related to the polarisation. The upper level is labeled b, the lower one a.

$$\partial_t \rho_{ba} = -\left(\gamma + \imath\,\Omega\right)\rho_{ba} + \imath\frac{1}{\hbar} d \cdot E \cdot (\rho_{aa} - \rho_{bb}) \qquad (5.32)$$

To demonstrate the relation of these two sets of equations, the Lorentz (or harmonic oscillator) 5.20 is first transfered into a system of differential equations of first order by introducing the new real quantity Q.

$$\begin{aligned} \partial_t P &= & Q \\ \partial_t Q &= -\omega_0^2 P - 2\gamma\,Q & + \alpha E N \end{aligned} \qquad (5.33)$$

Subsequently, we then diagonalise this coupled system of first order differential equations with real variables P and Q and constants γ and ω_0. This leads to two uncoupled differential equations of first order with complex variables Π and Π^*, shown in 5.34. The two decoupled equations are complex conjugates of each other. We introduce a new frequency $\Omega = \sqrt{\omega_0^2 - \gamma^2}$ being a real quantity for $\omega_0 > \gamma$.

$$\begin{aligned} \partial_t \Pi &= -\left(\gamma + \imath\,\Omega\right)\Pi - \frac{\imath}{2\eta\Omega}\left(\gamma + \imath\,\Omega\right)\alpha E N \\ \partial_t \Pi^* &= -\left(\gamma - \imath\,\Omega\right)\Pi^* + \frac{\imath}{2\eta\Omega}\left(\gamma - \imath\,\Omega\right)\alpha E N \end{aligned} \qquad (5.34)$$

The complex quantity Π is related to P and Q through the transformation matrix used in the diagonalisation. By normalising the eigen-vector of the equation system, an arbitrary constant η is introduced. To relate the value of Π to the polarisation P, η is chosen so that the real part of Π equals P and thus still represents the polarisation.

$$\Pi = -\frac{\imath\omega_0^2}{2\eta\Omega}\left(P + \frac{\gamma + \imath\,\Omega}{\omega_0^2}Q\right) \qquad (5.35)$$

In the limit of $\gamma/\omega_0 \to 0$ and the particular choice of $\eta = -\imath\omega_0^2/(2\Omega)$, the diagonalised Lorentz differential system of 5.34 reduces to a form which closely matches the optical Bloch equation for the polarisation (5.32).

$$\partial_t \Pi = - (\gamma + i \Omega) \, \Pi - i \frac{\Omega}{\omega_0^2} \alpha E N \qquad (5.36)$$

So the factor α is given by comparing the quantum mechanically derived equation to 5.36.

$$\alpha = \frac{\omega_0^2}{\hbar \Omega} d \qquad (5.37)$$

The second optical Bloch equation (5.38) describing the inversion $N :=$ $\rho_{aa} - \rho_{bb}$ of the two level system can then be directly translated to the real value space. Surely the system must be in any of the two levels, the sum of the probabilities of the system being in either one is normalised to $\text{Tr} \, [\rho] = \rho_{aa} + \rho_{bb} := 1$.

$$\partial_t N = - \frac{4}{\hbar} d \cdot E \cdot \Im [\rho_{ba}] \qquad (5.38)$$

With the identification of Π with ρ_{ba}, 5.35 and the norm $\eta = - i \omega_0^2 / (2 \Omega)$, the imaginary part of the off diagonal density matrix part is:

$$\Im [\rho_{ba}] = \frac{\Omega}{\omega_0^2} \partial_t P \qquad (5.39)$$

So the temporal inversion evolution 5.38 transforms to the following form.

$$\partial_t N = - 4 \frac{\Omega}{\hbar \omega_0^2} d \cdot E \cdot \partial_t P \qquad (5.40)$$

We will now finally introduce a few phenomenological terms. The inversion N should relaxate with a non radiative relaxation constant γ_{nr}, so that the steady state occupation probability ρ_0 of the upper level is reached. And there should be a possibility to non radiativly pump the upper level with a pumping rate Λ.

The non-radiative relaxation of the upper niveau for an unperturbed two-level system, with the relaxation rate γ_{nr} is given by the following rate equation.

$$\partial_t \, \rho_{bb} \Big|_{nr} = - \gamma_{nr} \cdot (\rho_{bb} - \rho_0) \qquad (5.41)$$

It can easily be deduced, that the evolution of the inversion N for this relaxation behaviour is as follows:

$$\partial_t \, N \Big|_{nr} = - \gamma_{nr} \cdot (N - (1 - 2\rho_0)) = - \gamma_{nr} \cdot (N - N_0) \qquad (5.42)$$

The pumping of the upper niveau with a rate of Λ can be expressed in the below equation for the inversion N:

$$\partial_t \, N \Big|_{pump} = - 2\Lambda \qquad (5.43)$$

Combining those two additional terms and inserting them in 5.40 results in the following evolution equation for the inversion N.

$$\partial_t N = -4\frac{\Omega}{\hbar\omega_0^2}d \cdot E \cdot \partial_t P - 2\Lambda - \gamma_{\mathrm{nr}} \cdot (N - N_0) \tag{5.44}$$

References

[5.1] Amit S. Nagra and Robert A. York. FDTD Analysis of Wave Propagation in Nonlinear Absorbing and Gain Media. *IEEE Transactions on Antennas and Propagation*, 46(3):334–340, March 1998. 1d adefdtd, 4 lvl atomic system, rate equations, pulse propagation.

[5.2] K. S. Yee. Numerical solution of initial boundary value problems involving Maxwell's equations in isotropic media. *IEEE Trans. Antennas Propag.*, 14:302–307, May 1966.

[5.3] Richard W. Ziolkowski, John M. Arnold, and Daniel M. Gogny. Ultrafast pulse interactions with two-level atoms. *Physical Review A*, 52(4):3082–3094, October 1995.

[5.4] Rodney Loudon. *The Quantum Theory of Light*. Oxford Science Publications, 1 edition, 1973.

[5.5] A. Taflove. *Computational Electrodynamics: The Finite-Difference Time-Domain Method*. Norwood, MA: Artech House, 1995.

[5.6] A. Taflove. *Advances in Computational Electrodynamics: The Finite-Difference Time-Domain Method*. Norwood, MA: Artech House, 1998.

6 Symmetry Properties
of Electronic and Photonic Band Structures

W. Hergert, M. Däne, and D. Ködderitzsch

Martin-Luther-University Halle-Wittenberg, Department of Physics,
Von-Seckendorff-Platz 1, 06120 Halle, Germany

Abstract. Group theoretical investigations have a huge potential to simplify calculations in solid state theory. We will discuss the application of group theory to electronic and photonic band structures. The symmetry properties of the Schrödinger equation and Maxwell's equations as well will be investigated. We have developed methods to simplify group theoretical investigations based on the computer algebra system *Mathematica*.

6.1 Introduction

The majority of physical systems exhibit intrinsic symmetries which can be used to simplify the solution of the equations governing these systems. Group theory as a mathematical tool plays an important role to classify the solutions within the context of the underlying symmetries. Extensive use of group theory has been made to simplify the study of electronic structure or vibrational modes of solids or molecules. There exists a number of excellent books illustrating the use of group theory. [6.1–6.4]

Similarities between the solution of the Schrödinger equation, or the Kohn-Sham equations in the framework of density functional theory for a crystal, and the solution of Maxwell's equations have been already pointed out by Joannopoulos *et al.*. [6.5] Therefore, it is clear that the same group theoretical concepts like in the theory of electronic band structures including two-dimensional and three-dimensional structures, surfaces as well as defects should be applicable to photonic band structure calculations, if we take into account the vectorial nature of the electromagnetic field.

There are several publications about group theoretical investigations of photonic crystals in the literature. Sakoda has extensively studied the symmetry properties of two-dimensional and three-dimensional photonic crystals (cf. [6.6–6.9]), starting from a plane wave representation of the electromagnetic fields. Group theoretical investigations are done also by Ohtaka and Tanabe. [6.10] They investigate the symmetry properties of photonic crystals represented by an array of dielectric spheres. In this case a series expansion in terms of vector spherical harmonics is used.

The problem at the end for electronic and the photonic band structure calculations is: How to apply group theory in the actual research work, if one goes away from all the textbook examples. The aim of the paper is to show,

W. Hergert, M. Däne, and D. Ködderitzsch, Symmetry Properties of Electronic and Photonic
Band Structures, Lect. Notes Phys. **642**, 103–125 (2004)
http://www.springerlink.com/

that computer algebra tools are appropriate to simplify group theoretical discussions connected with the calculation of electronic and photonic band structures.

After a short discussion of the usefulness of computer algebra systems, we will introduce basic concepts of group theory. Representation theory will be discussed next, followed by the analysis of the symmetry properties of the Schrödinger equation and of Maxwell's equations. We want to solve the electronic and photonic problem for a lattice periodic situation. In the one case we have a periodic potential $V(\mathbf{r}) = V(\mathbf{r}+\mathbf{R})$ in the other case a periodic dielectric constant $\epsilon(\mathbf{r}) = \epsilon(\mathbf{r} + \mathbf{R})$. The consequences of that periodicity will be considered. A simplification of the solution of both kinds of problems is possible, if symmetry-adapted basis functions are used. This is discussed in more detail for the electronic problem. At the end we apply group theory to the calculation of photonic band structures.

6.2 Group Theory Packages for Computer Algebra Systems

Computer algebra (CA) systems like *Mathematica* or *Maple* have been developed to allow formal mathematical manipulations. Nowadays those systems are complete in such a sense, that formal manipulations, numerical calculations, as well as graphical representations are possible in an easy and intuitive way with the same software. Apart from such general purpose CA systems, there exist also systems developed for special applications in mathematics. Group theoretical considerations, although conceptionally easy, lead very often to time-consuming algebraic calculations, which are error-prone. Therefore group theory is an excellent field for the application of CA systems. Some systems for abstract group theory are available [6.11], but they are not very helpful for considerations in solid state theory. K. Shirai developed a *Mathematica* package for group theory in solid state physics [6.12]. We followed similar ideas but tried to make the package more easy to use.

We have constructed a package for the CA system *Mathematica* which allows to do group theoretical manipulations which occur in solid state theory. [6.13] The software allows basic considerations with point groups, contains tight-binding theory for the electronic structure of solids, but also special applications to photonic crystals. The package is accompanied by an on line help, which is integrated in *Mathematica*'s help system. All considerations discussed in this paper can be found in a *Mathematica* notebook which is part of our package. We will give references to the commands implemented in the package throughout the paper.

6.3 Basic Concepts in Group Theory

In this section we will introduce the basic concepts of group theory. We will focus on definitions which will be necessary for the following discussions. To illustrate the concepts, let us discuss the symmetry group of a square in two dimensions, i.e. we are interested in all transformations which transform the square into itself. (cf. Fig. 6.1) This example will be of later use for the discussion of photonic band structures.

A set $\mathcal{G}$ of elements A,B,C ... is called a *group* if the following four axioms are fulfilled: i) There exists an operation, often called multiplication, which associates every pair of elements of $\mathcal{G}$ with another element of $\mathcal{G}$: $A \in \mathcal{G}, B \in \mathcal{G} \rightarrow A \cdot B = C, C \in \mathcal{G}$, ii) The associative law is valid: $A, B, C \in \mathcal{G} \rightarrow (A \cdot B) \cdot C = A \cdot (B \cdot C) = A \cdot B \cdot C$, iii) in the set exists an identity element: $A, E \in \mathcal{G} \rightarrow A \cdot E = E \cdot A = A$, iv) for all $A \in \mathcal{G}$ exists an inverse element $A^{-1} \in \mathcal{G}$ with $A \cdot A^{-1} = A^{-1} \cdot A = E$.

The symmetry group of the square consists of rotations of $\pi/2$ around the z-axis and mirror operations. The normal vectors of the mirror planes are the x and y-axis, $\overrightarrow{Oa}$ and $\overrightarrow{Ob}$. A mirror symmetry may be expressed as a twofold rotation, followed by an inversion. All the symmetry operations of the group of the square, named C_{4v} are:

$$C_{4v} = \left\{ E, C_{2z}, C_{4z}, C_{4z}^{-1}, IC_{2x}, IC_{2y}, IC_{2a}, IC_{2b} \right\} \tag{6.1}$$

If the group theory package is included in a *Mathematica* notebook by means of the command `Needs["GroupTheory'Master'"]`[1], the newly defined command `c4v=InstallGroup["C4v"]` will install the group in terms of the rotation matrices of the elements. The matrices are stored in the list `c4v`. The command `c4vs=GetSymbol[c4v]` will transform the elements of the group into symbolic form (cf.(6.1)). The symbols will be stored in the list `c4vs`. Operations on the group can be done in both representations of the group elements. The group multiplication is implemented in our *Mathematica* package by a redefinition of the infix-operator $\oplus$ (see [6.25]).

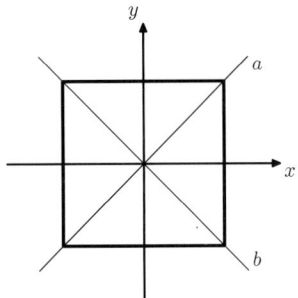

Fig. 6.1. Symmetry of a square.

---

[1] *Mathematica* commands are marked using this particular font.

For the graphical user interface of *Mathematica* we designed two additional palettes. The palette `SymmetryElements` contains all the symmetry elements of the 32 point groups. The palette `PointGroup` contains all the operations defined in the package.

The inspection of the multiplication table (`MultiplicationTable[c4v]`) of the group shows, that the table is not symmetric, indicating that the group multiplication is not commutative in this group. Therefore the group is not *abelian* (`AbelianQ[c4vs]` returns `False`.). All elements of the group can be also generated by successive multiplication of only two elements of the group, the so called *generators*. For our example we find the generators C_{4z}, IC_{2x}. (`Generators[group_]`) [2]

Any set of elements of a group which obeys all the group postulates is called a *subgroup* of the group. Examples for subgroups are: $\{E, C_{2z}, C_{4z}, C_{4z}^{-1}\}$ and $\{E, IC_{2y}\}$ (`SubGroupQ[group_,subgroup_]` performs a test).

Another structure of the group is introduced by the definition of *conjugate elements*. An element B of the group $\mathcal{G}$ is said to be conjugate to A if their exists a group element X such that $B = XAX^{-1}$. A *class* (`Classes[c4vs]`) is a collection of mutually conjugate elements of a group. The classes of the group C_{4v} are: (E), (C_{2z}), (C_{4z}, C_{4z}^{-1}), (IC_{2x}, IC_{2y}), (IC_{2a}, IC_{2b}). It is easy to prove that: i) E always forms a class on its own, ii) every group element in $\mathcal{G}$ is a member of some class of $\mathcal{G}$, iii) no element can be the member of two classes of $\mathcal{G}$, iv) if $\mathcal{G}$ is an abelian group, every element forms a class on its own.

We have already seen, that the group in the example can be represented by the abstract symbols or the rotation matrices. The relation between groups is formulated more rigorously in terms of *homomorphism* and *isomorphism*. Two groups $\mathcal{G} = \{A, B, C, \dots\}$ and $\mathcal{G}' = \{A', B', C', \dots\}$ are called homomorphic, if i) to each element of $\mathcal{G}$ corresponds one and only one element of $\mathcal{G}'$ ii) for all elements holds $A \to A', B \to B' \hookrightarrow AB \to A'B'$. The set of elements which are mapped to $E' \in \mathcal{G}'$ in a homomorphism is called *kernel of the homomorphism* $\mathcal{N}_K$. In contrast to the homomorphism the mapping in the *isomorphism* is one-to-one. Whereas the mapping of the symbols on the rotation matrices constitutes an isomorphism, the following mapping $E, C_{2z}, C_{4z}, C_{4z}^{-1} \Rightarrow +1$ and $IC_{2x}, IC_{2y}, IC_{2a}, IC_{2b} \Rightarrow -1$ constitutes a homomorphism.

6.4 Representation Theory

6.4.1 Matrix Representations of Groups

Matrix representations of symmetry groups are the essential tools to investigate symmetry properties of solutions of field equations as Schrödinger's equation or Maxwell's equations.

---

[2] The formal arguments of the commands indicate, which type of information has to be plugged in. For more information use the help system or the example notebook.

A group of square matrices, with the matrix multiplication as relation between the elements, which is homomorphic to a group $\mathcal{G}$ is called a matrix representation of $\mathcal{G}$. Each element $A \in \mathcal{G}$ corresponds to a matrix $\Gamma(A)$:

$$\Gamma(A)\Gamma(B) = \Gamma(C) \quad \forall \quad A, B, C \in \mathcal{G}$$
$$\Gamma(E) = E \quad \text{(identity matrix)} \tag{6.2}$$
$$\Gamma(A^{-1}) = \Gamma(A)^{-1}$$

It is obvious that an l-dimensional representation consisting of a set of $(l \times l)$-matrices can be transformed in another representation by means of a similarity transformation $\Gamma'(A) = S^{-1}\Gamma(A)S$ using a non-singular $(l \times l)$-matrix S.

The rotation matrices of a finite symmetry group form a three-dimensional matrix representation in that sense. The mapping of the group elements of C_{4v} onto the numbers $+1, -1$ represents also, in this case a one-dimensional, matrix representation of the group.

From two matrix representations Γ^1 and Γ^2 of $\mathcal{G}$ with the dimensions l_1 and l_2, a $(l_1 + l_2)$-dimensional representation Γ can be built, forming the block matrices

$$\Gamma(A) = \left(\begin{array}{c|c} \Gamma^1(A) & 0 \\ \hline 0 & \Gamma^2(A) \end{array} \right). \tag{6.3}$$

The representation Γ is called the *direct sum* of the representations Γ^1 and Γ^2: $\Gamma = \Gamma^1 \oplus \Gamma^2$. Representations which can be transformed into direct sums by similarity transformations are called *reducible*. If such a transformation is not possible, the representation is called *irreducible*. The rotation matrices of the group C_{4v} are given by:

$$\Gamma(E) = \begin{pmatrix} 1 & 0 & 0 \\ 0 & 1 & 0 \\ 0 & 0 & 1 \end{pmatrix} \qquad \Gamma(C_{2z}) = \begin{pmatrix} -1 & 0 & 0 \\ 0 & -1 & 0 \\ 0 & 0 & 1 \end{pmatrix} \qquad \Gamma(C_{4z}) = \begin{pmatrix} 0 & 1 & 0 \\ -1 & 0 & 0 \\ 0 & 0 & 1 \end{pmatrix}$$

$$\Gamma(IC_{4z}^{-1}) = \begin{pmatrix} 0 & -1 & 0 \\ 1 & 0 & 0 \\ 0 & 0 & 1 \end{pmatrix} \qquad \Gamma(IC_{2x}) = \begin{pmatrix} -1 & 0 & 0 \\ 0 & 1 & 0 \\ 0 & 0 & 1 \end{pmatrix} \qquad \Gamma(IC_{2y}) = \begin{pmatrix} 1 & 0 & 0 \\ 0 & -1 & 0 \\ 0 & 0 & 1 \end{pmatrix}$$

$$\Gamma(IC_{2a}) = \begin{pmatrix} 0 & -1 & 0 \\ -1 & 0 & 0 \\ 0 & 0 & 1 \end{pmatrix} \qquad \Gamma(IC_{2b}) = \begin{pmatrix} 0 & 1 & 0 \\ 1 & 0 & 0 \\ 0 & 0 & 1 \end{pmatrix}$$

The rotation matrices are block matrices and therefore reducible into a two-dimensional and a one-dimensional matrix representation of C_{4v}.

An important quantity of a matrix representation which will not change under a similarity transformation is the trace of the rotation matrix. This trace is called the *character* χ of the representation matrix.

$$\chi(A) = \sum_{j=1}^{l} \Gamma(A)_{jj} \tag{6.4}$$

The character system of a group is called the character table. The character table can be generated automatically using `CharacterTable[c4vs]`. We will get:

	C_1	C_2	$2 C_3$	$2C_4$	$2C_5$
Γ^1	1	1	1	1	1
Γ^3	1	1	1	-1	-1
Γ^4	1	1	-1	1	-1
Γ^2	1	1	-1	-1	1
Γ^5	2	-2	1	0	0

$C_1 = (E)$
$C_2 = (C_{2z})$
$C_3 = (C_{4z}, C_{4z}^{-1})$
$C_4 = (IC_{2x}, IC_{2x})$
$C_5 = (IC_{2a}, IC_{2b})$

The character table represents a series of character theorems, which are important for later use: i) the number of inequivalent irreducible representations of a group $\mathcal{G}$ is equal to the number of classes of $\mathcal{G}$, ii) the characters of the group elements in the same class are equal, iii) the sum of the squares of the dimensions of all irreducible representations is equal to the order of the group, iv) two representations are equivalent if their character systems are equivalent. Especially important are the following theorems:

– A representation Γ is irreducible, if:

$$\sum_{T \in \mathcal{G}} |\chi(T)|^2 = g . \tag{6.5}$$

(The sum of the squares of the characters of the rotation matrices of C_{4v} exceeds the group order $g = 8$, indicating that this representation cannot be irreducible.)
– The number n, how often an irreducible representation Γ^i, or a representation equivalent to Γ^i, is contained in the reduction of the reducible representation Γ, is given by:

$$n = \frac{1}{g} \sum_{T \in \mathcal{G}} \chi(T)\chi^i(T)^* \tag{6.6}$$

(g- order of the group, $\chi(T), \chi^i(T)$ - character of the group element in the representation Γ and Γ^i)
– Orthogonality theorems for characters

$$\sum_k \chi^i(C_k)^* \chi^j(C_k) N_k = g\,\delta_{ij} \tag{6.7}$$

$$\sum_i \chi^i(C_k)^* \chi^i(C_l) N_k = g\,\delta_{kl} \tag{6.8}$$

$(\chi^i(C_k), \chi^j(C_k)$ characters of elements in class C_k in irreducible representations Γ^i and Γ^j, N_k - number of elements in C_k. In the first relation the summation runs over all classes of the group. In the second relation it runs over all inequivalent irreducible representations of the group.)

The character theorems can be easily checked for all the point groups using *Mathematica*.

6.4.2 Basis Functions of Irreducible Representations

Our final goal is the investigation of the symmetry properties of the solutions of Maxwell's equation. Therefore we have to associate the symmetry expressed by the symmetry group, i.e. the point group C_{4v} in our example, with the symmetry properties of scalar and vector fields.

We define a transformation operator to a symmetry element T as $\hat{P}(T)$. For a scalar function it holds:

$$\hat{P}(T)f(\boldsymbol{r}) = f(T^{-1}\boldsymbol{r}) \tag{6.9}$$

The operators $\hat{P}(T)$ form a group of linear unitary operators. If we discuss Maxwell's equations, we have to deal with vector fields in general. Instead of (6.9) the following transformation has to be applied to the vector field $\boldsymbol{F}$:

$$\hat{P}(T)\boldsymbol{F}(\boldsymbol{r}) = T\boldsymbol{F}(T^{-1}\boldsymbol{r}) \tag{6.10}$$

Now we define the *basis functions* of an irreducible representation (IR) and concentrate on scalar fields. If a set of l-dimensional matrices $\Gamma(T)$ forms a representation of the group $\mathcal{G}$ and $\phi_1(\boldsymbol{r}), \ldots, \phi_l(\boldsymbol{r})$ is a set of linear independent functions such that

$$\hat{P}(T)\phi_n(\boldsymbol{r}) = \sum_{m=1}^{l} \Gamma(T)_{mn}\phi_m(\boldsymbol{r}) \qquad n = 1, 2, \ldots, l \tag{6.11}$$

then functions $\phi_n(\boldsymbol{r})$ are called partners in a set of basis functions of the representation Γ. ϕ_n is said to transforms like the nth row of the representation. We can write any function $\phi(\boldsymbol{r})$, which can be normalized, as a sum of basis functions of the irreducible representations Γ^p of the group.

$$\phi(\boldsymbol{r}) = \sum_{p} \sum_{n=1}^{l_p} \phi_n^p(\boldsymbol{r}) \tag{6.12}$$

Note, that this is not an expansion like the Fourier series. We don't expand a function with respect to an orthonormal complete set of functions here. The set of functions $\phi_n^p(\boldsymbol{r})$ depends on $\phi(\boldsymbol{r})$ itself. The functions $\phi_n^p(\boldsymbol{r})$ in (6.12) can be found by means of the so called *projection operators*.

$$\mathcal{P}^p_{mn} = \left(\frac{l_p}{g}\right) \sum_{T \in \mathcal{G}} \Gamma^p(T)^*_{mn} \hat{P}(T), \quad \mathcal{P}^p_{mn} \phi^q_i(\boldsymbol{r}) = \delta_{pq} \delta_{ni} \, \phi^p_m(\boldsymbol{r}) \quad (6.13)$$

$$\mathcal{P}^p = \left(\frac{l_p}{g}\right) \sum_{T \in \mathcal{G}} \chi^p(T)^* \hat{P}(T) \tag{6.14}$$

The projection operators are implemented in the package.
(cf. `CharacterProjectionOperator[classes_,chars_,func_]`.) As an example, we investigate the symmetry properties of a function $\phi(\boldsymbol{r}) = (a\,x + b\,y)g(r)$ (a and b are constants). We apply the projection operators connected to the group C_{4v}, to the function and get the following result

$$\phi^5_1(\boldsymbol{r}) = \mathcal{P}^5_{11}\,\phi(\boldsymbol{r}) = a\,x\,g(r) \quad \mathcal{P}^5_{12}\,\phi(\boldsymbol{r}) = b\,x\,g(r)$$

$$\mathcal{P}^5_{21}\,\phi(\boldsymbol{r}) = a\,y\,g(r) \qquad \phi^5_2(\boldsymbol{r}) = \mathcal{P}^5_{22}\,\phi(\boldsymbol{r}) = b\,y\,g(r) \tag{6.15}$$

It is impossible to project out parts of $\phi(\boldsymbol{r})$ transforming like other IRs of C_{4v}. Therefore our trial function $\phi(\boldsymbol{r}) = (a\,x + b\,y)g(r)$ is a sum of functions, transforming like the representation Γ^5 (E).

If we consider three-dimensional photonic crystals we cannot resort to scalar fields anymore. We have to take into account the full vectorial nature of the fields also in symmetry considerations. A more detailed discussion of the subject can be found in [6.14–6.16].

6.5 Symmetry Properties of Schrödinger's Equation and Maxwell's Equations

Here we want to investigate the symmetry properties of scalar or vector fields, which we get as solutions of Schrödinger's equation or Maxwell's equations. The operators $\hat{P}(T)$ form a group of linear unitary operators. This group is isomorphic to the group of symmetry elements T. The Hamilton-Operator $\hat{H}(\boldsymbol{r})$ of the time-independent Schrödinger equation $\hat{H}\psi(\boldsymbol{r}) = E\,\psi(\boldsymbol{r})$ is given by

$$\hat{H}(\boldsymbol{r}) = -\frac{\hbar^2}{2m}\frac{\partial^2}{\partial r^2} + V(\boldsymbol{r}) \ . \tag{6.16}$$

For an arbitrary transformation T the transformation behavior of the Hamiltonian is given by

$$\hat{H}(\boldsymbol{r}) = \hat{P}(T)\hat{H}(T\boldsymbol{r})\hat{P}(T)^{-1} \ . \tag{6.17}$$

For transformations which leave $\hat{H}$ invariant, i.e. $\hat{H}(T\boldsymbol{r}) = \hat{H}(\boldsymbol{r})$ we get:

$$[\hat{H}, \hat{P}(T)] = 0 \ . \tag{6.18}$$

All transformations, which let $\hat{H}$ invariant, form a group. The corresponding operators $\hat{P}(T)$ form an isomorphic group, the group of the Schrödinger equation. All elements of the group of the Schrödinger equation commute with $\hat{H}$. Because the operator of the kinetic energy is invariant under all rotations forming the group $O(3)$, the symmetry of the Hamiltonian is determined exclusively by the potential $V(\boldsymbol{r})$.

We want to consider photonic crystals. Therefore we have to extend the analysis to Maxwell's equations.

$$\nabla \boldsymbol{D}(\boldsymbol{r},t) = 0 \qquad\qquad \nabla \times \boldsymbol{E}(\boldsymbol{r},t) = -\tfrac{\partial}{\partial t}\boldsymbol{B}(\boldsymbol{r},t)$$
$$\nabla \boldsymbol{B}(\boldsymbol{r},t) = 0 \qquad\qquad \nabla \times \boldsymbol{H}(\boldsymbol{r},t) = \tfrac{\partial}{\partial t}\boldsymbol{D}(\boldsymbol{r},t) \tag{6.19}$$

We assume that we have no free charges and currents. Using the materials equations $\boldsymbol{D} = \epsilon\epsilon_0\boldsymbol{E}, \boldsymbol{B} = \mu_0\boldsymbol{H}$ and a harmonic time dependence for the fields, i.e. $\boldsymbol{E}(\boldsymbol{r},t) = \boldsymbol{E}(\boldsymbol{r})\exp(i\omega t)$ the basic equations are given by

$$\hat{\Xi}_E \boldsymbol{E}(\boldsymbol{r}) = \frac{1}{\epsilon(\boldsymbol{r})}\,\nabla \times (\nabla \times \boldsymbol{E}(\boldsymbol{r})) = \left(\frac{\omega}{c}\right)^2 \boldsymbol{E}(\boldsymbol{r}) \tag{6.20}$$

$$\hat{\Xi}_H \boldsymbol{H}(\boldsymbol{r}) = \nabla \times \frac{1}{\epsilon(\boldsymbol{r})}\,\nabla \times \boldsymbol{H}(\boldsymbol{r}) = \left(\frac{\omega}{c}\right)^2 \boldsymbol{H}(\boldsymbol{r}) \tag{6.21}$$

If we assume a two-dimensional photonic crystal, i.e. the dielectric constant varies in the x-y-plane $(\boldsymbol{r}_{||})$ we can resort to the solution of scalar equations for the field components in z-direction.

$$\hat{\Xi}_E^{2D} E_z(\boldsymbol{r}_{||}) = -\frac{1}{\epsilon(\boldsymbol{r}_{||})}\left(\frac{\partial^2}{\partial x^2} + \frac{\partial^2}{\partial y^2}\right)E_z(\boldsymbol{r}_{||}) = \left(\frac{\omega}{c}\right)^2 E_z(\boldsymbol{r}_{||}) \tag{6.22}$$

$$\hat{\Xi}_H^{2D} H_z(\boldsymbol{r}_{||}) = -\left(\frac{\partial}{\partial x}\frac{1}{\epsilon(\boldsymbol{r}_{||})}\frac{\partial}{\partial x} + \frac{\partial}{\partial y}\frac{1}{\epsilon(\boldsymbol{r}_{||})}\frac{\partial}{\partial y}\right)H_z(\boldsymbol{r}_{||})$$
$$= \left(\frac{\omega}{c}\right)^2 H_z(\boldsymbol{r}_{||}) \tag{6.23}$$

Corresponding to (6.18) we have to find all the transformations T which leave the operators $\hat{\Xi}_E, \hat{\Xi}_H$ for the three-dimensional case or $\hat{\Xi}_E^{2D}, \hat{\Xi}_H^{2D}$ for the two-dimensional case invariant. It can be shown that the group of Maxwell's equations is formed by the space group of the photonic crystal, i.e. consists of all translations and rotations which transform $\epsilon(\boldsymbol{r})$ into itself.

6.6 Consequences of Lattice Periodicity

Here we want to illustrate the consequences of lattice periodicity to the solution of Schrödinger equation or Maxwell's equation. We consider crystals which have d-dimensional translational symmetry and can be represented as a set of points sitting on a Bravais-lattice (for convenience we do not take

into account a possible arrangement of basis atoms around the lattice points
here). We want to denote the translation vectors which form the Bravais
lattice $\{\mathbf{T}\}$ as

$$\mathbf{T} = \sum_{i=1}^{d} m_i \mathbf{a}_i, \qquad m_i = 0, \pm 1, \pm 2 \ldots \tag{6.24}$$

with $\mathbf{a}_i$ being a basic lattice vector. Associated with the set $\{\mathbf{T}\}$ is a set of
symmetry transformations (translations), which leave the crystal invariant.
It can easily established that this set forms a infinite, discrete group, denoted
here as $\mathcal{T}$.

Associated with every lattice point might be a set of symmetry operations
consisting of rotations, mirror reflections and so on, which leave the crystal
and the point they are applied to also invariant. The group formed from this
set is called a point group $\mathcal{G}_0$.

Let us introduce a new symbols to state the considerations made so far
more clearly. The Seitz-operator $\{\mathbf{R}|\mathbf{t}\}$ takes some vector $\mathbf{r}$ to a new vector
$\mathbf{r}'$ by first rotating and then translating it

$$\mathbf{r}' = \{\mathbf{R}|\mathbf{t}\}\mathbf{r} = \mathbf{R}\mathbf{r} + \mathbf{t}. \tag{6.25}$$

Here $\mathbf{R}$ denotes a rotation matrix and $\mathbf{t}$ is a vector. Then the translation
group $\mathcal{T}$ can be characterised by the elements $\{\mathbf{E}|\mathbf{T}\}$ ($\mathbf{E}$ is the unity matrix)
and the elements of the point group elements as $\{\mathbf{R}|\mathbf{0}\}$. The space group
$\mathcal{G}$ of a crystal is defined as a group of operations $\{\mathbf{R}|\mathbf{t}\}$ which contains
as a subgroup the set of all pure primitive translations of a lattice, $\mathcal{T}$, but
which contains no other pure translations. Here we want to consider only
symmorphic space groups whose symmetry operations consists of a rotation
followed by a *primitive* translation $\mathbf{T}$ (this excludes the case of having for
examples glide-planes and screw axis as symmetry elements of the crystal,
where fractions of $\mathbf{T}$, $\mathbf{t} = \mathbf{T}/n$, $n \in \mathbb{N}$ are involved).

In what follows we want first use the properties of $\mathcal{T}$ to arrive at the Bloch-
theorem which makes a statement about the form of the wavefunction in
translational invariant systems. As stated above $\mathcal{T}$ is an infinite group behind
of which the construct of an infinite crystal lies. To proceed we form a *finite*
group[3] from $\mathcal{T}$ by imposing periodic boundary conditions in d dimensions.
We consider a building block of the crystal containing $N_1 \times \cdots \times N_d = N^d$
primitive cells.[4] This imposes the following condition on the wavefunction of
the crystal

---

[3] This is done for convenience, because we then can use all statements for finite
groups.

[4] This is an approximation which is not severe if we disregard surface effects.
Note, that whereas imposing periodic boundary conditions in one dimension can
be thought of forming a ring out of a chain of lattice points, in three dimensions
this is topological not possible.

$$\psi(\mathbf{r}) = \psi(\{\mathbf{E}|N\mathbf{a_i}\}\mathbf{r}) \qquad i = 1, \ldots, d. \tag{6.26}$$

which can be mapped to a condition for function operators

$$\hat{P}(\{\mathbf{E}|N\mathbf{a_i}\}) = \hat{P}(\{\mathbf{E}|\mathbf{a_i}\})^N = \hat{P}(\{\mathbf{E}|0\}) \qquad i = 1, \ldots, d. \tag{6.27}$$

Therefore using (6.26) is equivalent to working with the finite group of function operators $P(\{\mathbf{E}| \sum_i^d m_i\mathbf{a_i}\})$ where $0 \le m_i \le N - 1$ with dimension N^d. As two translations commute this group is abelian, *i.e.* all the irreducible representations are one dimensional. Using (6.27) one can work out a matrix representations for the IRs. Consider a one-dimensional IR Γ of $\mathcal{T}$, and suppose that $\Gamma(\{\mathbf{E}|\mathbf{a_i}\})$ is the representation belonging to a translation by a basic vector $\mathbf{a_i}$.[5] Then from (6.27) it follows immediately that

$$(\Gamma(\{\mathbf{E}|\mathbf{a_i}\}))^N = 1 \tag{6.28}$$

which leads to the following condition for the matrix elements for the IRs

$$\Gamma(\{\mathbf{E}|\mathbf{a_i}\}) = e^{-2\pi i \frac{p_i}{N}}, \qquad p_i \in \mathbb{N}. \tag{6.29}$$

By (6.28) only N values for p_i are allowed and they may be taken to be $p_i = 1, \ldots, N - 1$. If one now defines the basic vectors $\mathbf{b_j}$ of the reciprocal lattice by

$$\mathbf{a_i} \cdot \mathbf{b_j} = 2\pi \delta_{ij}, \qquad i, j = 1, \ldots, d \tag{6.30}$$

and allowed $\mathbf{k}$-vectors by

$$\mathbf{k} = \sum_{i=1}^{d} k_i \mathbf{b}_i \tag{6.31}$$

with $k_i = p_i/N$ one can express all IRs in terms of

$$\Gamma^{\mathbf{k}}(\{\mathbf{E}|\mathbf{T}\}) = e^{-i\mathbf{k}\cdot\mathbf{T}} \tag{6.32}$$

numerated by N^d allowed $\mathbf{k}$-vectors. The character table is very simple and shown in Table 6.6.[6] The basis functions for the IR are then labelled by $\mathbf{k}$ and their transformation properties are given by

$$\hat{P}(\{\mathbf{E}|\mathbf{T}\})\phi_{\mathbf{k}}(\mathbf{r}) = e^{-i\mathbf{k}\cdot\mathbf{T}}\phi_{\mathbf{k}} = \phi_{\mathbf{k}}(\{\mathbf{E}|\mathbf{T}\}^{-1}\mathbf{r}) = \phi_{\mathbf{k}}(\mathbf{r} - \mathbf{T}) \tag{6.33}$$

This leads to Bloch's theorem which states that the basis function of the IRs of the translation group $\mathcal{T}$ can be chosen as Bloch functions

---

[5] From now on we denote the *finite* group of translations as $\mathcal{T}$. No confusion with the infinite group mentioned before shall arise.

[6] As the IRs are one dimensional we obtain the character table immediately.

	$\mathbf{T_1}$	$\mathbf{T_2}$	...	$\mathbf{T}_{Nd}$
$\Gamma^{\mathbf{k}_1}$	$e^{-i\mathbf{k}_1\cdot\mathbf{T}_1}$	$e^{-i\mathbf{k}_1\cdot\mathbf{T}_2}$	...	$e^{-i\mathbf{k}_1\cdot\mathbf{T}_N}$
...				
$\Gamma^{\mathbf{k}_{Nd}}$	$e^{-i\mathbf{k}_{Nd}\cdot\mathbf{T}_1}$	$e^{-i\mathbf{k}_{Nd}\cdot\mathbf{T}_2}$	...	$e^{-i\mathbf{k}_{Nd}\cdot\mathbf{T}_{Nd}}$

$$\phi_{\mathbf{k}}(\mathbf{r}) = e^{i\theta}e^{-i\mathbf{k}\cdot\mathbf{T}}u_{\mathbf{k}}(\mathbf{r}), \qquad (6.34)$$

where $u_{\mathbf{k}}(\mathbf{r}) = u_{\mathbf{k}}(\mathbf{r}-\mathbf{T})$ has the periodicity of the lattice. The phase factor $e^{i\theta}$, $\theta \in \mathbb{R}$, is omitted in the following considerations. The label $\mathbf{k}$ for the wave function is called quantum number and in the limit $N \to \infty$, $\phi_{\mathbf{k}}$ is a smooth function of the wave vector $\mathbf{k}$.[7]

So far we have seen that the translational symmetry imposed certain conditions on the functional form of the wave-function. If now the space group $\mathcal{G}$ contains apart from the translational part point group symmetries we can find further IRs of $\mathcal{G}$ which are not necessarily one dimensional. Important here is the fact that the basis functions of these IRs can be constructed from the Bloch-functions (6.34) alone (this is not proven here). Before we can do this, we have to introduce some new concepts.

Definition: The point group $\mathcal{G}_0(\mathbf{k})$ of the wave-vector $\mathbf{k}$ is the subgroup of the point group $\mathcal{G}_0$ of the space group $\mathcal{G}$ that consists of all rotations $\{\mathbf{R}|\mathbf{0}\}$ that rotate $\mathbf{k}$ into itself or an equivalent vector $\mathbf{k}+\mathbf{G}$ where $\mathbf{G}$ is a vector of the reciprocal lattice given by: $\mathbf{G} = \sum_i n_i \mathbf{b}_i$, $n_i = 0, \pm 1, \pm 2, \ldots, \quad i = 1, \ldots, d$.

Given a vector $\mathbf{k}' = \mathbf{k} + \mathbf{G}$ one realises that a IR $\Gamma^{\mathbf{k}}$ of $\mathcal{T}$ could also be labelled by $\mathbf{k}'$ because of $e^{i(\mathbf{k}+\mathbf{G})\cdot\mathbf{T}} = e^{i\mathbf{k}\cdot\mathbf{T}}$; $\mathbf{k}$ and $\mathbf{k}'$ are said to be equivalent. This brings us to the concept of the Brillouinzone (BZ). The first BZ consists of the points of $\mathbf{k}$-space which lie closer to $\mathbf{k} = \mathbf{0}$ than any other reciprocal lattice points. The boundaries of the first BZ are given by the planes consisting of the points satisfying:

$$\mathbf{k} \cdot \mathbf{G} = \frac{1}{2}|\mathbf{G}|^2 \qquad (6.35)$$

where here the $\mathbf{G}$'s are the nearer reciprocal lattice points. As an example in Fig. 6.2 the first BZ of an *fcc*-lattice is shown. In the BZ a high symmetry point is defined to have more members in $\mathcal{G}_0(\mathbf{k})$ than any neighbouring point $\mathbf{k}'$ in the BZ. Neighbouring points having the same $\mathcal{G}_0(\mathbf{k})$ are defined to form high symmetry lines and planes, respectively.

Working out the character table for a space group would be tedious, because of the vast number of IR that $\mathcal{G}$ possesses. However a fundamental theorem for symmorphic space groups exists (see *e.g.* [6.1]) which implies that any IR of $\mathcal{G}$ can be labelled by two quantities, namely the allowed $\mathbf{k}$-vector and a label p specifying the IR of $\mathcal{G}_0(\mathbf{k})$. Therefore one can label the

---

[7] If the periodicity is lost in one spatial direction, say the z direction perpendicular to a surface of a crystal, then k_z cannot be used as quantum number to label the states.

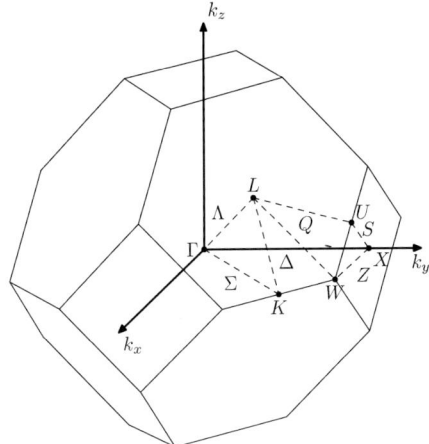

Fig. 6.2. Brillouin zone of the *fcc* structure. Symmetry points and symmetry lines are indicated.

IR's of $\mathcal{G}$ by $\Gamma^{\mathbf{k}p}$. Further basis functions transforming as the ith row of the pth IR of $\mathcal{G}_0(\mathbf{k})$ are labelled by $\phi^p_{\mathbf{k}i}(\mathbf{r})$.

The following matrix element theorem for Bloch-functions (see *e.g.* [6.1]) allows for considerable reduction of eigenvalue problems surfacing in the solution of the Schrödinger equation for crystals (see section 6.7). Given two Bloch functions $\phi^p_{\mathbf{k}i}$ and $\phi^{p'}_{\mathbf{k}'j}$ it states that the matrix elements $(\phi^p_{\mathbf{k}i}, \phi^{p'}_{\mathbf{k}'j})$ and $(\phi^p_{\mathbf{k}i}, \hat{H}\phi^{p'}_{\mathbf{k}'j})$ are both zero unless $\mathbf{k} = \mathbf{k}'$, $i = j$ and $p = p'$. Further $(\phi^p_{\mathbf{k}i}, \phi^p_{\mathbf{k}i})$ and $(\phi^p_{\mathbf{k}i}, \hat{H}\phi^p_{\mathbf{k}i})$ are independent of i.

6.7 Electronic Band Structure

We want to solve the following equation for an electron in an effective potential, which has lattice periodicity $V(\mathbf{r}) = V(\mathbf{r} + \mathbf{R})$.

$$\hat{H}\psi_{\mathbf{k}\nu}(\mathbf{r}) = E_\nu(\mathbf{k})\psi_{\mathbf{k}\nu}(\mathbf{r}) \tag{6.36}$$

The crystal wavefunction $\psi_{\mathbf{k}\nu}(\mathbf{r})$ is expanded with respect to a set of Bloch functions $\phi^n_{\mathbf{k}}(\mathbf{r})$.

$$\psi_{\mathbf{k}\nu}(\mathbf{r}) = \sum_n c^\nu_n(\mathbf{k})\phi^n_{\mathbf{k}}(\mathbf{r}) \tag{6.37}$$

Using this ansatz (6.36) can be transformed into an eigenvalue problem.

$$0 = \sum_n c^\nu_n(\mathbf{k}) \left[\hat{H}\phi^n_{\mathbf{k}}(\mathbf{r}) - E_\nu(\mathbf{k})\phi^n_{\mathbf{k}}(\mathbf{r}) \right] \tag{6.38}$$

$$0 = \sum_n c^\nu_n(\mathbf{k}) \left[\left(\phi^m_{\mathbf{k}}, \hat{H}\phi^n_{\mathbf{k}} \right) - E_\nu(\mathbf{k}) \left(\phi^m_{\mathbf{k}}, \phi^n_{\mathbf{k}} \right) \right] \tag{6.39}$$

A nontrivial solution requires of (6.39) requires:

$$0 = \det \left| \left(\phi_{\mathbf{k}}^m, \hat{H} \phi_{\mathbf{k}}^n \right) - E_\nu(\mathbf{k}) \left(\phi_{\mathbf{k}}^m, \phi_{\mathbf{k}}^n \right) \right|. \tag{6.40}$$

The dimension of the eigenvalue problem is given by the number of basis functions used in ansatz (6.37). If complex crystal structures have to be considered, the number of basis functions can be large. The problem can be simplified drastically at certain $\mathbf{k}$ points. This will be discussed later. If we plot the eigenvalues $E_\nu(\mathbf{k})$ along special lines in the Brillouin zone, we will get the so called band structure. The solutions are labelled with respect to the fundamental theorem by the allowed $\mathbf{k}$ vector and a label p specifying the IR of $\mathcal{G}_0(\mathbf{k})$ as it can be seen in Fig. 6.3.

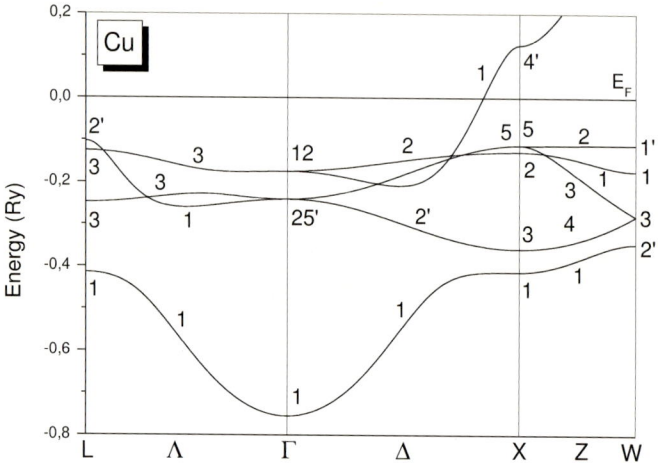

Fig. 6.3. Electronic band structure of Cu along the symmetry lines L-Γ-X-W

6.7.1 Compatibility Relations

The band structure of fcc Copper in Fig. 6.3 is shown between the high symmetry points L, Γ, X and W. The lines between these symmetry points are usually labelled with Λ, Δ and Z. The question is, why are some bands crossing and others not or why are degeneracies at symmetry points are released along the symmetry lines? The reason for this are the different symmetry groups $\mathcal{G}_0(\mathbf{k})$ at the different $\mathbf{k}$-points. At Γ ($\mathbf{k} = (0,0,0)$) there is the highest symmetry. For the fcc structure the pointgroup $\mathcal{G}_0(\Gamma)$ is O_h. The other $\mathbf{k}$-points have lower symmetries, but these groups are all subgroups of O_h. Let us concentrate on the representation Γ_{12} at the Γ point. We can read from the character table (Table 6.1) that the Γ_{12} representation is twodimensional.

Therefore the corresponding energy value is twofold degenerate. The lower symmetry along Δ causes the split into two energy bands. The irreducible representations, corresponding to those bands, have to be compatible with the representation at Γ. This can be calculated with formula 6.6.

Table 6.1. Character table of the point group O_h

	E	$8C_3$	$3C_2$	$6C_4$	$6C_2'$	i	$8S_6$	$3\sigma_h$	$6S_4$	$6\sigma_d$
A_{1g} Γ_1	1	1	1	1	1	1	1	1	1	1
A_{2g} Γ_2	1	1	1	-1	-1	1	1	1	-1	-1
E_g Γ_{12}	2	-1	2	0	0	2	-1	2	0	0
T_{1g} $\Gamma_{15'}$	3	0	-1	1	-1	3	0	-1	1	-1
T_{2g} $\Gamma_{25'}$	3	0	-1	-1	1	3	0	-1	-1	1
A_{1u} $\Gamma_{1'}$	1	1	1	1	1	-1	-1	-1	-1	-1
A_{2u} $\Gamma_{2'}$	1	1	1	-1	-1	-1	-1	-1	1	1
E_u $\Gamma_{12'}$	2	-1	2	0	0	-2	1	-2	0	0
T_{1u} Γ_{15}	3	0	-1	1	-1	-3	0	1	-1	1
T_{2u} Γ_{25}	3	0	-1	-1	1	-3	0	1	1	-1

The use of group theory proceeds as follows:
Let $\mathcal{G}_0(\Gamma)$ and $\mathcal{G}_0(\Delta) = C_{4v}$ be the pointgroup $\mathcal{G}_0(k)$ for k at Γ and Δ. Then $\mathcal{G}_0(\Delta)$ is a subgroup of $\mathcal{G}_0(\Gamma)$ and so the irreducible representations (IRs) of $\mathcal{G}_0(\Gamma)$ are in general reducible representations of $\mathcal{G}_0(\Delta)$. The number n_p, how often an irreducible representation $\Gamma^{p\Delta}$ of $\mathcal{G}_0(\Delta)$ is contained in the reduction of the reducible representation Γ^Γ, wich is irreducible with respect to $\mathcal{G}_0(\Gamma)$ is given by

$$n_p = \frac{1}{g} \sum_{T \in \mathcal{G}_0(\Delta)} \chi^{p\Delta}(T)\, \chi^\Gamma(T)^* \tag{6.41}$$

where g is the order of the group $\mathcal{G}_0(\Delta)$. The evaluation of 6.41 for Γ_{12} along Δ shows, using the character tables 6.1 and 6.4, that the representation Γ_{12} can be written as direct sum of Δ_1 and Δ_2

$$\Gamma_{12} = \Delta_1 \oplus \Delta_2. \tag{6.42}$$

because only the numbers of n_{Δ_1} and n_{Δ_2} are one, the other are all zero. Working out all the relations for the different symmetry points and lines, we get the so called *compatibility relations*, given in Table 6.2.

It can be seen from the band structure, that bands of the same symmetry never cross. This can be understood by the following argumentation. Lets assume two bands corresponding to the same onedimensional irreducible representation will cross. At the crossing point the degeneracy is twofold corresponding to a twodimensional irreducible representation. This is a contradiction.

Table 6.2. Compatibility relations between the high symmetry points Γ, X, L and W and the symmetry axes Λ, Δ and Z for the face-centered cubic space group O_h^5. (see picture 6.3)

Γ_1	Γ_2	Γ_{12}	Γ_{15}	Γ_{25}	$\Gamma_{1'}$	$\Gamma_{2'}$	$\Gamma_{12'}$	$\Gamma_{15'}$	$\Gamma_{25'}$
Δ_1	Δ_2	$\Delta_1\ \Delta_2$	$\Delta_1\ \Delta_5$	$\Delta_2\ \Delta_5$	$\Delta_{1'}$	$\Delta_{2'}$	$\Delta_{1'}\Delta_{2'}$	$\Delta_{1'}\ \Delta_5$	$\Delta_{2'}\ \Delta_5$
Λ_1	Λ_2	Λ_3	$\Lambda_1\ \Lambda_3$	$\Lambda_2\Lambda_3$	Λ_2	Λ_1	Λ_3	$\Lambda_2\Lambda_3$	$\Lambda_1\Lambda_3$

X_1	X_2	X_3	X_4	X_5	$X_{1'}$	$X_{2'}$	$X_{3'}$	$X_{4'}$	$X_{5'}$
Δ_1	Δ_2	$\Delta_{2'}$	$\Delta_{1'}$	Δ_5	$\Delta_{1'}$	$\Delta_{2'}$	Δ_2	Δ_1	Δ_5
Z_1	Z_1	Z_4	Z_4	Z_2Z_3	Z_2	Z_2	Z_3	Z_3	Z_1Z_4

W_1	W_2	$W_{1'}$	$W_{2'}$	W_3
Z_1	Z_2	Z_2	Z_1	Z_3Z_4

L_1	L_2	L_3	$L_{1'}$	$L_{2'}$	$L_{3'}$
Λ_1	Λ_2	Λ_3	Λ_2	Λ_1	Λ_3

6.7.2 Symmetry-Adapted Basis Functions

Making use of the symmetry in the solution of the eigenvalue problem (6.40) is now quite simple. Suppose the Schrödinger equation is $\hat{H}\psi(\mathbf{r}) = E\psi(\mathbf{r})$. The group of the Schrödinger equation is $\mathcal{G}$. The function $\psi(\mathbf{r})$ can be expressed as

$$\psi(\mathbf{r}) = \sum_{i=1}^{\infty}\sum_{p}\sum_{m} C_{ip}^{m}\phi_{im}^{p}(\mathbf{r}) \tag{6.43}$$

The transformation to the eigenvalue problem leads therefore to:

$$det\left|\left(\phi_{jn}^{q}, \hat{H}\phi_{im}^{p}\right) - E\left(\phi_{jn}^{q}, \phi_{im}^{p}\right)\right| = 0 \tag{6.44}$$

Corresponding to the matrix element theorems the determinant takes a block form, i.e. the large eigenvalue problem splits in a set of smaller ones. To use such properties, we have to construct symmetry-adapted basis functions. Projection operators can be used to generate the symmetry-adapted basis functions.

Symmetry-adapted plane waves. A plane wave

$$\phi_{\mathbf{k}m}(\mathbf{r}) = e^{i(\mathbf{k}+\mathbf{K}_m)\cdot\mathbf{r}} \tag{6.45}$$

is a Bloch function. The vector $\mathbf{K}_m$ is a reciprocal lattice vector. Therefore the crystal wave function can be expanded in a set of such plane waves. An expansion of such a form converges normally too slowly. A better choice at the point $\mathbf{k}$ of the Brillouin zone will be a function that transforms according to the sth row of the irreducible representation Γ^p of $\mathcal{G}_0(\mathbf{k})$. Such a function is obtained by operating on $\phi_{\mathbf{k}m}(\mathbf{r})$ with the projection operator $\mathcal{P}_{ss}^{p}$. (cf. 6.13). The application of $\hat{P}(T)$ to the plane wave gives

$$\hat{P}(T)\phi_{\mathbf{k}m}(\mathbf{r}) = e^{i\mathbf{R}(\mathbf{k}+\mathbf{K}_m)\cdot\mathbf{r}} \qquad (6.46)$$

whereas $\mathbf{R}$ denotes the rotation matrix connected with the symmetry element T. Because $T = \{\mathbf{R} \mid 0\} \in \mathcal{G}_0(\mathbf{k})$, $\mathbf{R} \cdot \mathbf{k}$ is equivalent to $\mathbf{k}$. Because $\mathbf{R} \cdot \mathbf{K}_m$ is also a lattice vector of the reciprocal lattice, there exists a lattice vector $\mathbf{K}_n$ such, that

$$\mathbf{R}(\mathbf{k} + \mathbf{K}_m) = \mathbf{k} + \mathbf{K}_n \quad , \quad \mid \mathbf{R}(\mathbf{k} + \mathbf{K}_m) \mid = \mid \mathbf{k} + \mathbf{K}_m \mid \qquad (6.47)$$

The second equation holds because of the orthogonality of $\mathbf{R}$. Let us consider an example for the face centered cubic space group O_h^5. The point group of the point L ($\mathbf{k_L} = \frac{\pi}{a}(1,1,1)$) is D$_{3d}$. If we use the plane wave $\phi_{\mathbf{k_L}0}(\mathbf{r}) = e^{i\mathbf{k_L}\cdot\mathbf{r}}$ and apply the character projection operator to all irreducible representations of D$_{3d}$ we get:

$$L_1 : \frac{1}{2}\left(e^{-i\pi(x+y+z)} + e^{-i\pi(x+y+z)}\right)$$

$$L_{2'} : \frac{1}{2}\left(-e^{-i\pi(x+y+z)} + e^{-i\pi(x+y+z)}\right)$$

$$L_3 : \frac{1}{4}\left(e^{-i\pi(3x+y+z)} - e^{-i\pi(x+3y+z)} + e^{i\pi(3x+y+z)} - e^{i\pi(x+3y+z)}\right)$$

$$: \frac{1}{12}\left(e^{-i\pi(3x+y+z)} + e^{-i\pi(x+3y+z)} + e^{i\pi(3x+y+z)} + e^{i\pi(x+3y+z)}\right.$$
$$\left. -2e^{-i\pi(x+y+3z)} - 2e^{i\pi(x+y+3z)}\right)$$

To all the other IRs no symmetrized basis function can be projected out. One has to choose reciprocal lattice vectors different from zero to project basis functions corresponding to the other IRs out ouf a plane wave.

Symmetry-adapted spherical harmonics. The spherical harmonics $Y_{l,m}(\theta,\phi)$ are basis functions of irreducible representations of dimension $(2l + 1)$ of the rotation group $O(3)$. The group C_{4v} is a subgroup of $O(3)$, therefore the representations of $O(3)$ will be reducible with respect to C_{4v}, in general. On the other hand we can construct linear combinations of spherical harmonics, which are basis function of the IRs of C_{4v}.

It is obvious, that all the spherical harmonics to $m = 0$ are basis function of Γ^1, because they are independent of the azimuthal angle ϕ. For higher angular momentum $l \geq 4$ we can built linear combinations which transform like Γ^1 from the functions $Y_{l\pm4}$ because they correspond to the fourfold rotation axis. *Lattice Harmonics* in general are symmetry-adapted spherical harmonics which are given as linear combinations of those. Lattice harmonics are tabulated in the literature [6.26, 6.27] or can be generated automatically using our projection operator technique.

Table 6.3. Linear combinations of spherical harmonics as basis functions of the point group C_{4v}. ($S^+_{lm} = (Y_{lm} + Y_{l,-m})/\sqrt{2}$, $S^-_{lm} = (Y_{lm} - Y_{l,-m})/(\sqrt{2}i)$

$l =$	0	1	2	3	4	5
$\Gamma^1(A_1)$	Y_{00}	Y_{10}	Y_{20}	Y_{30}	$Y_{40}\ S^+_{44}$	$\cdots$
$\Gamma^2(A_2)$	–	–	–	–	S^-_{44}	$\cdots$
$\Gamma^3(B_1)$	–	–	S^+_{22}	S^+_{32}	S^+_{42}	$\cdots$
$\Gamma^4(B_2)$	–	–	S^-_{22}	S^-_{32}	S^-_{42}	$\cdots$
$\Gamma^5(E)$	–	S^+_{11}	S^+_{21}	$S^+_{31}\ S^+_{33}$	$S^+_{41}\ S^+_{43}$	$\cdots$
	–	S^-_{11}	S^-_{21}	$S^-_{31}\ S^-_{33}$	$S^-_{41}\ S^-_{43}$	$\cdots$

6.8 Discussion of Photonic Band Structures

To calculate the photonic band structures, we will use the program developed by Johnson *et al.* [6.28] The fully-vectorial eigenmodes of Maxwell's equations with periodic boundary conditions are computed by preconditioned conjugate-gradient minimization of the block Rayleigh quotient in a plane wave basis in this package. The program itself calculates the bands, but will not directly assign the irreducible representation which belongs to a certain band at a special $\boldsymbol{k}$. First we will demonstrate how this can be done easily by an investigation of the fields.

6.8.1 Assignment of the IRs to the Photonic Band Structure

As an example we will discuss a simple two-dimensional photonic crystal. Therefore we have to deal with scalar fields which are solutions of (6.22). Our example consists of circular air-pores in a dielectric medium of $\epsilon = 2.1$ arranged in a square lattice. Fig. 6.4 shows the first Brillouin zone of the square lattice. The point groups $\mathcal{G}_0(\boldsymbol{k})$ for the points Γ, X and Δ are C_{4v}, C_{2v} and C_{1h}. The character tables are given in Table 6.4.

At Γ we have a higher symmetry than along the line Δ. This implicates that the IRs of the group C_{4v} will be reducible representations of the group C_{1h}. From (6.6) we know how to calculate how often an IR occurs in a representation. The result of such considerations are the so called *compatibility relations*. In our case we find easily the following: The group C_{4v} contains only one two-dimensional representation. Therefore the two-fold degenerate eigenvalues at Γ have to be labeled E. We can calculate the compatibility of E with the IRS of C_{1h} using (6.6) and get: $E = A \oplus B$, i.e. E is a direct sum of the IRs A and B of C_{1h}.

Figure 6.5 presents the band structure for TM polarization, respectively. The irreducible representations with respect to the groups $\mathcal{G}_0(\boldsymbol{k})$ at the differ-

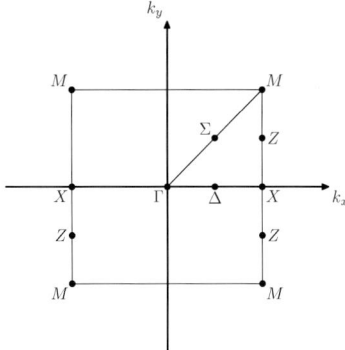

Fig. 6.4. Brillouin zone of the square lattice.

Table 6.4. Character tables of the point groups C_{4v}, C_{2v} and C_{1h}

C_{4v}		E	$2C_4$	C_2	$2\sigma_v$	$2\sigma_d$
A_1	Δ_1	1	1	1	1	1
A_2	$\Delta_{1'}$	1	1	1	-1	-1
B_1	Δ_2	1	-1	1	1	-1
B_2	$\Delta_{2'}$	1	-1	1	-1	1
E	Δ_5	2	0	-2	0	0

C_{2v}	E	C_2	σ_y	σ_x
A_1	1	1	1	1
A_2	1	1	-1	-1
B_1	1	-1	1	-1
B_2	1	-1	-1	1

C_{1h}	E	σ
A	1	1
B	1	-1

ent symmetry points and lines are already indicated. It is not possible to find all IRs without doubt at all symmetry points and lines indicated in Fig. 6.5 only by the inspection of the band structure. In the TM-mode picture of Fig. 6.5 three special areas are indicated. **1** and **3** show, that bands of the same symmetry will not cross each other. The reason is of course the same like in the electronic case. A is a one-dimensional representation of C_{1h}. If two A representations would cross along Σ, the dimension of the representation would have to bee two-dimensional in the crossing point. This is a contradiction and therefore crossing of bands of the same symmetry is forbidden. In **2** the representations are not indicated. We will dicuss now, how to get quickly information about the symmetry. First we calculate the fields. In case of TM-polarization this will be the z-component of the E-field at the point Γ of the Brillouin zone. Now we apply the character projection operator, defined in (6.14) to the fields pixel by pixel with respect to all IRs of the corresponding group $\mathcal{G}_0(\Gamma) = C_{4v}$. If we get the same figure after transformation, we have found the correct IR. Figure 6.6 shows the result of such an analysis for the bands 2-5 at the Γ point and TM-polarization for our example.

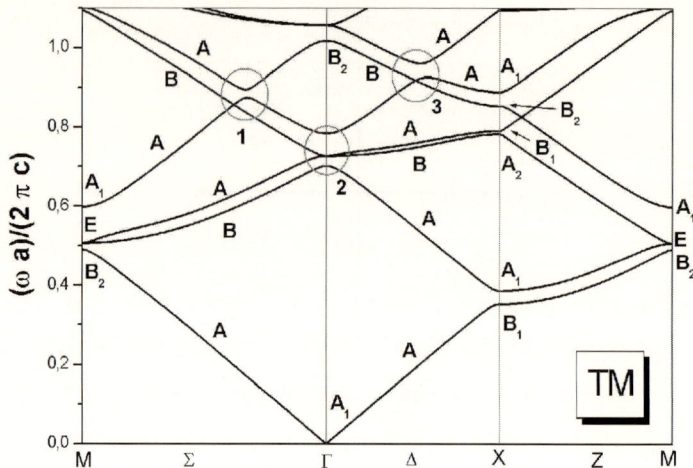

Fig. 6.5. Band structure of a 2D photonic crystal. Circular air-pores are arranged in a square lattice. The filling factor of the structure is f=0.5. The dielectric constant of the material is $\epsilon = 2.1$

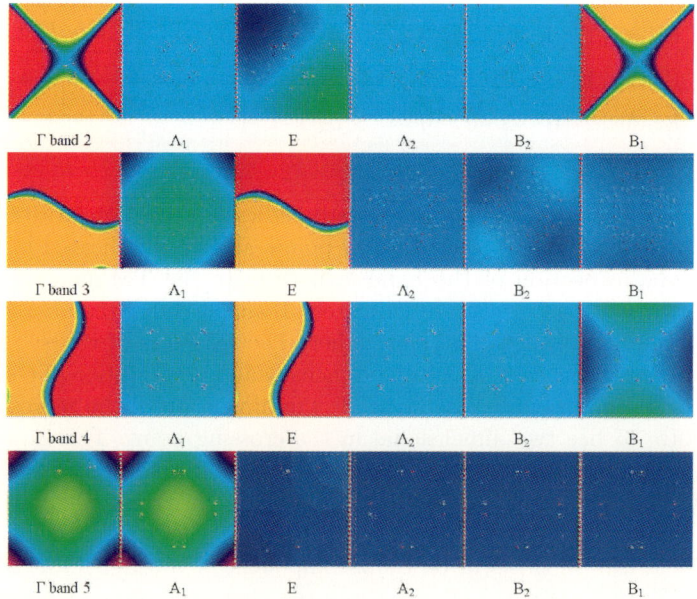

Fig. 6.6. z-component of the E-field at point Γ for bands 2 to 5 of the band structure of Fig. 6.5 for TM-polarization

Another possibility to investigate the symmetry properties in our example is the discussion of symmetrized plane waves. An extensive study of the symmetrized plane waves was given by Luehrmann. [6.29] To solve (6.22) we expand the solution in terms of plane waves:

$$E_z(\boldsymbol{r}_{||}) = \sum_{\boldsymbol{G}_{||}} C_{\boldsymbol{k}_{||}}(\boldsymbol{G}_{||}) e^{i(\boldsymbol{k}_{||}+\boldsymbol{G}_{||})\cdot\boldsymbol{r}_{||}} \tag{6.48}$$

The projection operator technique can be used to construct symmetrized plane waves. Basis functions constructed from a plane wave

$$\phi_{\boldsymbol{k}_{||},\boldsymbol{G}_{||}}(\boldsymbol{r}_{||}) = e^{i(\boldsymbol{k}_{||}+\boldsymbol{G}_{||})\cdot\boldsymbol{r}_{||}} \tag{6.49}$$

that transform like the sth row of the irreducible representation Γ^p of $\mathcal{G}_0(\boldsymbol{k})$ are obtained by simply operating on $\phi_{\boldsymbol{k}_{||},\boldsymbol{G}_{||}}(\boldsymbol{r}_{||})$ with the projection operator $\mathcal{P}_{ss}^p$. If we want to construct symmetrized plane waves at the Γ point ($\boldsymbol{k}_{||} = 0$) from the reciprocal lattice vectors with $|\boldsymbol{G}_{||}| = 2\pi/a$ we get the following symmetrized linear combinations with respect to IRs of C_{4v} belonging to A_1 and A_2 and E:

$$A_1 : \frac{1}{4}\left(e^{-2\pi x/a} + e^{-2\pi x/a} + e^{-2\pi y/a} + e^{-2\pi y/a}\right)$$

$$A_2 : \frac{1}{4}\left(e^{-2\pi x/a} + e^{-2\pi x/a} - e^{-2\pi y/a} - e^{-2\pi y/a}\right)$$

$$E : \frac{1}{2}\left(-e^{-2\pi x/a} + e^{-2\pi x/a}\right)$$

$$E : \frac{1}{2}\left(-e^{-2\pi y/a} + e^{-2\pi y/a}\right)$$

Symmetrized linear combinations of plane waves to the IRs B_1 and B_2 have to be constructed from different reciprocal lattice vectors.

The *Mathematica* package allows to calculate the photonic band structure for two-dimensional photonic crystals. We get the IRs for a given eigenfrequency $\omega(\boldsymbol{k})$ by comparison of the signs of the plane wave components in the series expansion (6.48) with the symmetrized plane waves which belong to $\mathcal{G}_0(\boldsymbol{k})$.

The methods, explained in this chapter can be extended to the three-dimensional case in a straightforward manner.

6.9 Conclusions

In summary, we have introduced the group theoretical treatment of photonic band structures. We have demonstrated that computer algebra systems like *Mathematica* can be used to investigate in an effective way the symmetry properties of two- and three-dimensional photonic band structures. The symmetry properties of the fields can be investigated directly by the projection operator technique. The properties of symmetrized basis functions like vector spherical harmonics can be considered easily. We believe that such tools will be helpful for detailed analysis of photonic band structure calculations or the interpretation of experimental results.

References

[6.1] J.F. Cornwell, Group Theory and Electronic Energy Bands in Solids, Selected Topics in Solid State Physics Vol. 10 (North Holland, Amsterdam, 1969).

[6.2] J.F. Cornwell, Group Theory in Physics, Techniques in Physics Vol.7 (Academic Press, London 1984)

[6.3] W. Ludwig and C. Falter, Symmetry in Physics, Springer Series in Solid State Sciences 64 (Springer, Berlin, 1988).

[6.4] B.S. Tsukerblat, Group Theory in Chemistry and Spectroscopy - A simple guide to advanced usage (Academic Press, London, 1994).

[6.5] J.D. Joannopoulos, R.D. Meade, J.N. Winn, Photonic Crystals (Princeton University Press, Princeton 1995).

[6.6] K. Sakoda, Phys. Rev. B **52**, 7982 (1995).

[6.7] K. Sakoda, Phys. Rev. B **55**, 15345 (1997).

[6.8] T. Ochiai and K. Sakoda, Phys. Rev. B **64**, 045108 (2001).

[6.9] K. Sakoda, Optical Properties of Photonic Crystals, Springer Series in Optical Sciences Vol.80 (Springer, Berlin, 2001).

[6.10] K. Ohtaka and Y. Tanabe, J. Phys. Soc. Japan **65**, 2670 (1996).

[6.11] Computeralgebra in Deutschland, Bestandsaufnahme, Möglichkeiten, Perspektiven, Fachgruppe Computeralgebra der GI, DMV und GAMM, Passau und Heidelberg 1993

[6.12] K. Shirai, Theories of point group in solid state physics, package available from the *Mathematica* resource library http://library.wolfram.com/database/.

[6.13] The package can be found at http://www.physik.uni-halle.de/Heraeus/Summer/f_main.htm. The software is still under development. Critical comments and hints are welcome.

[6.14] C.F. Bohren, D.R. Huffman, Absorption and Scattering of Light by Small Particles, (John Wiley & Sons, New York).

[6.15] G.B. Arfken, H.J. Weber, Mathematical Methods for Physicists (Academic Press, San Diego 1995)

[6.16] J.A. Stratton, Electromagnetic Theory (McGraw-Hill, New York 1941).

[6.17] R.F. Egorov, B.I. Reser, V.P. Shirkovskii, phys. stat. sol. **26**, 391 (1968).

[6.18] E. Lidorikis, M.M. Sigalas, E.N. Economou, and C.M. Soukoulis, Phys. Rev. Lett. **81**, 1405 (1998).

[6.19] J.P. Albert, C. Jouanin, D. Cassagne, and D. Monge, Opt. and Quantum Electronics **34**, 251 (2002).

[6.20] M. Le Vassor d'Yerville, D. Monge, D. Cassange, and J.P. Albert, Opt. and Quantum Electronics **34**, 445 (2002).

[6.21] J.P. Albert, C. Jouanin, D. Cassagne, and D. Bertho, Phys. Rev. B **61**, 4381 (2000).

[6.22] M. Bayindir, B. Temelkuran, and E. Ozbay, Phys. Rev. B **61**, R11855 (2000).

[6.23] M. Bayindir, B. Temelkuran, and E. Ozbay, Phys. Rev. Lett. **84**, 2140 (2000).

[6.24] M. Bayindir and E. Ozbay, Phys. Rev. B **62**, R2247 (2000).

[6.25] S. Wolfram, Das *Mathematica*-Buch, Die offizielle Dokumentation (Addison-Wesley-Longman, Bonn 1997).

[6.26] S.L. Altmann and A.P. Cracknell, Rev. Mod. Phys. **37**, 19 (1965).

[6.27] S.L. Altmann and C.J. Bradley, Rev. Mod. Phys. **37**, 33 (1965).

[6.28] S.G. Johnson and J.D. Joannopoulos, Optics Express **8**, 173 (2001).

[6.29] A.W. Luehrmann, Adv. in Physics, **17**, 3 (1968).

[6.30] W.M. Robertson, G. Arjavalingam, R.D. Meade, K.D. Brommer, A.M. Rappe, and J.D. Joannopoulos, Phys. Rev. Lett. **68**, 2023 (1992).

[6.31] W.M. Robertson, G. Arjavalingam, R.D. Meade, K.D. Brommer, A.M. Rappe, and J.D. Joannopoulos, J. Opt. Soc. Am B **10**, 322 (1993).

[6.32] F. López-Tejeira, T. Ochiai, K. Sakoda, and J. Sánchez-Dehesa, Phys. Rev. B **65**, 195110 (2002).

Part II

Simulation from Nanoscopic Systems to Macroscopic Materials

7 From the Cluster to the Liquid: Ab-Initio Calculations on Realistic Systems Based on First-Principles Molecular Dynamics

C. Massobrio[1], M. Celino[2], Y. Pouillon[1,2], and I.M.L. Billas[3]

[1] Institut de Physique et de Chimie des Matériaux de Strasbourg, 23 rue du Loess, BP 43, 67034 Strasbourg Cedex 2, France

[2] Ente per le Nuove Tecnologie, l'Energia e l'Ambiente, C.R. Casaccia, CP 2400, 00100 Roma, Italy and Istituto Nazionale per la Fisica della Materia, Unità di Ricerca Roma1, Italy

[3] Département de Biologie et de Génomique Structurales, Institut de Génetique et de Biologie Moléculaire et Cellulaire, 1 rue Laurent Fries, BP 10142, 67404 Illkirch, France

Abstract. In both clusters and disordered systems the determination of structural properties often relies on qualitative interpretations of experimental data. First-principles molecular dynamics provides a reliable atomic-scale tool to optimize geometries and follow the dynamical evolution at different temperatures. We present three examples of application of first-principles molecular dynamics to the study of finite systems and disordered, bulk networks. In the first case, devoted to the copper oxide clusters CuO_2 and CuO_6, the account of temperature effects and a careful search of all isomer allows to complement effectively photoelectron spectroscopy data. In the second example, we analyze the behavior of the C_{60} fullerene when one or two silicon atoms are inserted in the cage to replace carbon atoms. Silicon atoms correspond to chemically reactive sites of the fullerenes, giving rise to local structural distortions. Then, we describe the determination of the structure for liquid $SiSe_2$ at thermal equilibrium. The microscopic origins of intermediate range order are rationalized in terms of network connectivity and specific features appearing in the structure factors. Overall, first-principles molecular dynamics appears as a convincing method to corroborate experimental work and make reliable predictions based on well-established electronic structure techniques.

7.1 Introduction

Modern materials science is based on the precise knowledge of the interplay between different kinds of properties, such as structural, electronic, magnetic and thermodynamical. The ultimate goal is to control the fabrication of new compounds of specific characteristics and suitable sizes, from the nanoscale dimensions to the massive, macroscopic quantities. Even though several viewpoints can be adopted, mostly reflecting approaches to materials science which stem either from physics, chemistry or even biology, a necessary prerequisite to any investigation in the area of condensed matter is the

C. Massobrio, M. Celino, Y. Pouillon, and I.M.L. Billas, From the Cluster to the Liquid: Ab-Initio Calculations on Realistic Systems, Lect. Notes Phys. **642**, 129–157 (2004)
http://www.springerlink.com/ © Springer-Verlag Berlin Heidelberg 2004

determination of the atomic configurations. The atomic structure is the very first information one needs to label a system according to its arrangement in space. Structural information is of paramount importance to seek first hints on the nature of electronic bonding, identify the defects and understand their role in determining macroscopic properties. A more ambitious challenge is to follow the mechanism of atomic movement, which amounts to monitoring the changes occurring to a configuration as a result of mechanically or thermally activated displacements. Diffraction and microscopy techniques are of invaluable help in the case of periodic, ordered systems, for which the data can be readily converted to real space positions. Although these procedures are not necessarily straightforward to implement, their feasibility can lead to high quality structural determinations.

The situation becomes more delicate for two classes of systems: clusters, i.e. collection of atoms not replicating in space along any directions, and disordered networks (liquids and amorphous solids) characterized by finite-range correlations among atomic positions. Cluster science owes the identification of geometries, giving precious input to corroborate the longstanding quest for size-dependent structural properties, to two experimental techniques: mass spectrometry and photoelectron spectroscopy [7.1]. In mass spectrometry, detection rests on the possibility of selecting clusters according to the number of constitutive atoms. Enhanced stability for specific sizes is ascribed to optimized bonding and/or closing shell electronic effects. Photoelectron spectroscopy is based on the collection of kinetic energies of electrons ejected via incoming photons, leading to a map of electronic states up to the Fermi level [7.2]. In both approaches the assignement of geometries to the spectral features results from educated guesses and do not provide any information on the interatomic distances. Moreover, different isomers of the same cluster can hardly be distinguished, making the interpretation far too qualitative. Despite the success of these techniques, it has become customary to end successful experimental reports on clusters by calling for atomic scale calculations expected to confirm or disprove the analysis drawn from the measurements.

Structural properties of disordered systems are in principle accessible via x-ray or neutron diffraction. Collection of data in reciprocal space and their conversion in real space have profited of advances in the decomposition of the structure factors in partial contributions, allowing a quantification of chemical order [7.3, 7.4]. However, despite these accurate interpretations of pair distribution functions, relevant controversies exist on the number of defects and their influence on the establishement of short and intermediate range order. Moreover, while the evolution of diffraction properties with temperature and the implications on the structure can be analyzed, it remains hard to picture atomic scale movements, responsible of diffusive motions. As a consequence, time trajectories obtained via molecular dynamics are often explicitly invoked as a necessary complement to go beyond the structural information contained in diffraction data for disordered systems.

Nowadays computer simulation at the atomic scale is an affordable and reliable tool to tackle structural determination, providing access to both optimized minimum energy structures and to their time evolution at finite temperatures [7.5]. The predictive power of molecular dynamics is crucially dependent on the accuracy of the total energy and interatomic forces. Implementations can be divided into two main classes. The first goes under the name of *classical* molecular dynamics and covers all expressions for the total energy depending on the ionic coordinates only, with no explicit account of the electronic structure. In these schemes the nature of bonding is taken into account through the choice of appropriate analytical functions for the atomic interactions, called effective potentials. Parameters fitting is usually performed on experimental data, even though first-principles results are also increasingly employed. Well-known examples are the rigid ion or shell model Born-Mayer Coulomb potentials for ionic solids [7.6], the many-body embedded-atom potential for metals [7.7] or the three-body angular potentials for covalent systems like silicon [7.8]. These tools allow easy access to large system sizes and are especially reliable when used for configurations close to those entering the fit of the parameters. They appear to fail for intermediate situations of bonding, as those encountered when the variations in the coordination numbers are driven by temperature-induced phase transformations. Also, drastic reductions in the number of neighbors with respect to bulk, as those occurring in cluster and surface configurations, cannot be handled properly. This shortcoming is more severe in clusters, which have binding properties often strongly varying with size. Interatomic potentials are largely appealing for the description of phenomena irrespective of the specific system involved. This is more the case in physics and in materials science than in chemistry. Potentials meet a great success in areas where the spatial scale of the process is a prior issue before bonding accuracy, for instance in investigations of extended inhomogeneous defects.

To the second class of molecular dynamics simulations belongs those applications which combine a determination of the electronic structure for a given atomic configuration to the evolution of the latter under the action of interatomic forces. In turn, these forces are those resulting from a potential energy depending explicitly on both ionic and electronic degrees of freedom. According to the original proposal, first-principles molecular dynamics is based on the self-consistent evolution of the ions, which interact through forces derived within density functional theory [7.9]. Despite the considerable increase in the computational effort, structural and electronic properties can be equally treated at the same level of reliability. Thus, the bonds adjust simultaneously to the changes in the atomic configurations, by allowing average values of properties to be taken along the temporal trajectories, with the electronic structure fully determined at each single step.

In this work we describe applications of the first-principles molecular dynamics technique to the determination of atomic configurations in three specific cases. The first two deal with isolated clusters in two different size ranges, small copper oxide clusters (CuO_m, m=2,6) and Si-doped fullerenes ($C_{59}Si$ and $C_{58}Si_2$). In these two examples structural optmization is the key strategy to identify isomers and characterize them. Copper oxides are involved in an extremely wide range of fields, going from pyrotechnics to biochemistry, through catalysis, corrosion and coloring of ceramics. From a more fundamental viewpoint the evolution of the Cu-O bond in these oxides is reflected by the variations of their physico-chemical properties. These variations follow the changes in composition and nature of the atomic or molecular components. However, whereas the Cu-O bond has been firmly established as predominantly ionic in solid compounds like Cu_2O [7.10], only few information are available for other compounds, particularly for nanostructures, to make a clear statement about its nature.

Substitutionally doped heterofullerenes with silicon atoms are of great interest in many different respects. Although Si belongs to the same group of C, its chemical behavior is very different. The chemistry of carbon is characterized by the very flexible bonding character of this element, which has the ability to form single, double and triple bonds with itself and other atoms. On the other hand, silicon prefers to form multidirectional single bonds sp^3 [7.11]. A crucial difference between Si and C is the large number of Si core electrons, making more difficult for two silicon atoms to form double and triple bonds. It appears that the stability and bonding of systems obtained via Si substitution in a carbon environment with a strong sp^2 character deserve accurate theoretical studies.

In the third example we deal with a disordered network-forming material, liquid $SiSe_2$. Statistical averages over time trajectories reveal the morphology of the network as well as fundamental correlations between topological and chemical order. The main issue is to provide hints on the microscopic origin of intermediate range order which establishes on distances much larger than nearest-neighbor interactions. Within the family of AX_2 glasses and liquids (A= Si, Ge, X=0, Se, S), effective potentials are unable to discriminate between different network structures [7.12]. Their strong ionic character, due to the predominant Coulomb interaction inherent in the interatomic potential, makes all systems equally composed of undefected tetrahedra. First-principles calculations allow to improve upon this description via a quantified description of the relevant structural units and defects. The paper is organized as follows. Sect. 7.2 is devoted to the theoretical methods. We briefly review the basic ideas of the first-principles molecular dynamics technique. Then, we give more specific technical details on the three sets of simulations. Results are contained in Sect. 7.3 which is divided in three parts, devoted to copper oxide clusters, silicon doped fullerenes and chalcogenide disordered materials respectively. The paper concludes with Sect. 7.4.

7.2 Theoretical Methods

7.2.1 First-Principles Molecular Dynamics

The basic concepts underlying the first-principles molecular dynamics method (FPMD), according to the original formulation worked out by R. Car and M. "Parrinello [7.9]", can be summarized as follows. By assuming that the electronic ground state for a fixed set of ionic coordinates is available, the quantum mechanical force could in principle be calculated by means of the Hellmann-Feynman theorem [7.13]. This states that in the electronic ground state the forces on each ion are given by (minus) the partial derivative of the total energy with respect to the ionic coordinates. Implementing this recipe amounts to performing four steps for the solutions of the equations of motions, namely *a)* calculate the Hellmann-Feynman forces, *b)* move the ions according to the laws of Newton, *c)* solve the Kohn-Sham equations [7.14] and *d)* iterate the procedure in time. The electrons will follow the ions by lying exactly, at each time step, on the Born-Oppenheimer (BO) surface. Calculations of this kind, currently pursued by some groups, are termed Born-Oppenheimer molecular dynamics simulations [7.5]. The computational cost of this approach might be prohibitive since each new set of ionic coordinates requires a full search of the groud state for the electrons. This difficulty can be overcome by introducing a fictitious dynamical system associated with the physical one. The fictitious system is devised in a such a way that the trajectory generated by its dynamics reproduces very closely that of the physical system. Accordingly, the method is based on the evolution of a combined classical and quantum degrees of freedom defined by the Lagrangian

$$\mathcal{L} = \sum_i^{occ} \int d\mathbf{r}\, \mu_i \left| \dot{\psi}_i(\mathbf{r}) \right|^2 + \frac{1}{2} \sum_I M_I \dot{\mathbf{R}}_I^2 - E\left[\{\psi_i\}, \mathbf{R}_I \right]$$
$$+ \sum_{ij} \Lambda_{ij} \left(\int dr\, \psi_i^* \psi_j - \delta_{ij} \right) \tag{7.1}$$

where μ_i are arbitrary parameters which play the role of generalized masses for the electronic degrees of freedom. The first and second terms in the Lagrangian are the electronic K_e and ionic K_I kinetic energies respectively, while E is the Kohn-Sham potential energy. The Lagrange multipliers Λ_{ij} are used to impose the orthogonality condition on the ψ_i and are simply holonomic constraints. The corresponding equations of motion are

$$\mu \ddot{\psi}_i = -\frac{\delta E}{\delta \psi_i^*} + \sum_j \Lambda_{ij} \psi_j \tag{7.2}$$

$$M_I \ddot{\mathbf{R}}_I = -\frac{\partial E}{\partial \mathbf{R}_I} \tag{7.3}$$

A crucial point concerning the dynamical evolution expressed in (7.1) and (7.3) is the accuracy by which the trajectories derived from the extended Lagrangian are close to those defined in the BO surface, as required by a derivation of interatomic forces from first principles. For finite systems as well as bulk insulators and semiconductors the BO surface is followed accurately, provided a very small μ is chosen with respect to the value of the physical mass M. The closing of a gap causes instabilities in the equations of motion, due to the transfer of energy from the ions to the electrons. In some cases, this shortcoming can be avoided by using dynamical thermostats, so as to control separately the temperature associated to the ionic motion and the fictitious electronic motion [7.15, 7.16].

7.2.2 Details of Calculations

All calculations reported here are based on density functional theory and employ a plane-wave basis set with periodic boundary conditions. For clusters, this means that the system sizes have to be chosen much larger than the estimated spatial extension of the optimized structures to eliminate spurious interactions with periodically repeated images. In our calculations of fullerenes the edge of the face-centered cubic cell is equal to 21.17 Å. In the case of copper oxide clusters [7.17, 7.18], a cubic cell of side 13.2 Å proved sufficiently large. The cell dimensions of disordered sytems have to be compatible with investigations of intermediate range order distances (typically in between 5 and 15 Å), which cause the appearance of a first sharp diffraction peak in the total and in some of the partial structure factors. To fulfill this condition our simulations on SiSe$_2$ were performed at constant volume on systems consisting of 120 atoms (40 Si and 80 Se) [7.19, 7.20]. The edge of the cell is of about 16 Å. This size is sufficiently large to cover the region of wavevectors in which the FSDP occurs, $k_{FSDP}=1$ Å^{-1}. For the exchange-correlation part of density functional Kohn-Sham total energy various generalized gradient approximations are employed. SiSe$_2$ and CuO$_n$ clusters are modeled with the GGA introduced by Perdew and Wang [7.21]. In the case of silicon doped fullerenes the GGA is after Becke [7.22] for the exchange energy and Lee, Yang and Parr [7.23] for the correlation energy. Pseudopotentials are of norm-conserving type for SiSe$_2$ and fullerenes studies (with energy cutoffs E$_{cut}$ equal to 20 and 40 Ry respectively), while the ultrasoft scheme, due to Vanderbilt [7.24], has been preferred in the case of copper oxides to reduce the computational costs. The implementation of first-principles molecular dynamics within this framework has been extensively presented in [7.25], from which the equations of motion for the norm-conserving case can also be easily recovered. Two cutoffs energies are needed with ultrasoft pseudoptentials (UP), E$_{cut}^{UP}$ for the electronic orbitals and E$_{cut}^{\rho}$ for the augmented electronic density. These values are 20 Ry and 150 Ry, respectively. This scheme is more convenient than the norm-conserving one provided E$_{cut}^{UP} < 2^{2/3}$ E$_{cut}$, a condition largely satisfied in view of the norm-conserving energy cutoff currently employed to obtain converged properties for Cu and O systems (~ 90 Ry). Different optimization approaches have been adopted for clusters. Fullerene structures are relaxed by minimizing the forces on all atoms using the method of direct inversion in the iterative subspace (DIIS) [7.26–7.28]. The configurational ground state of copper oxide clusters is reached via the damped molecular dynamics scheme [7.29]. First-principles molecular dynamics has been carried out for liquid $SiSe_2$. Temperature control is implemented for both ionic and electronic degrees of freedom by using Nosé-Hoover thermostats [7.15, 7.16]. Energy conservation is optimal for fictitious electron masses μ in between 3000 and 5000 a.u. and time steps not larger than $\delta t = 0.5$ fs to integrate the equations of motion. The simulations on copper oxide clusters and liquid networks were performed using the computer program described in [7.25]. This code is highly performant in vectorial mode and can be straightforwardly adapted to norm-conserving calculations. Fullerenes have been studied by using the parallel CPMD program [7.30].

7.3 Selected Applications to Clusters and Disordered Systems

7.3.1 CuO_n Clusters

The search of constitutive building blocks in increasingly larger Cu_nO_m clusters and the existence of unexplained features in the photoelectron spectra of CuO_m species are strong motivations for a thorough structural study of the CuO_m(m=2-6) series. Both neutral and negatively charged systems have been considered. We report here results for the smallest (CuO_2) and the largest (CuO_6) member of the group [7.17, 7.18]. Among the negatively charged molecules (see Fig. 7.1), the linear $OCuO^-$ is the most stable one (Table 7.1), particularly in the S=1 spin state. This isomer is also the dominating species in the photoelectron spectra of [7.31] and it lies at high (4 eV) binding energies. At lower binding energies (1.5 eV) a minority species was taken to correspond to the CuO_2^- complex, but no conclusive identification was possible. Our results (see Table 7.1) show that the bent isomer is energetically favored in all calculations. However, the energy difference between the bent and the side-on isomers, which is 0.28 eV for S=0, reduces to 0.12 eV for the S=1 ground state. Very small is also the energy separation between the S=0 and the S=1 optimized configurations of the bent. These small energy differences are comparable to those involved in temperature effects typical of supersonic expansion techniques [7.1]. To better control these factors, we have performed first-principles molecular dynamics simulations by taking as initial configurations both the negatively charged bent and side-on isomers in the S=0 and S=1 spin states. Accordingly, four distinct temporal evolutions have been followed with the ionic temperature lying between 950 K and 1050 K over independent trajectories of 2 ps each. In Fig. 7.2, we report the average bonding distances and angles of the anionic bent and side-on isomers. For S=0, the CuOO⁻ bent is found to be dynamically stable on the time scale of our simulations, while the CuO_2^- side-on fragments readily (less than 0.5 ps) and yields a Cu atom and a O_2 dimer. On the other hand, in the S=1 case,

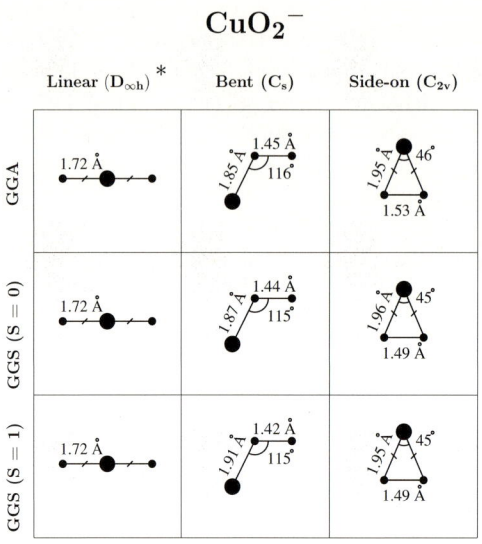

Fig. 7.1. Calculated equilibrium geometries of CuO_2^- molecules. The ground state geometry is the linear structure (indicated by a star), for the spin-unpolarized and the two spin-polarized calculations (S=0 and S=1).

Table 7.1. Bond lengths (given in Å), bond angle and relative energy ΔE_b (in eV) with respect to the most stable structures for the three isomers OCuO⁻, CuOO⁻ and CuO_2^- side-on. ΔE_b is obtained by considering the binding energies of the different isomers calculated within the same theoretical framework. Accordingly, $\Delta E_b = 0$ (given in bold) corresponds to the configurational ground state.

	d_{O-O}	d_{Cu-O}	Bond Angle	ΔE_b
OCuO⁻				
DFT-GGS, S=0	3.44	1.72	180°	0.38
DFT-GGS, S=1	3.44	1.72	180°	**0**
CuOO⁻				
DFT-GGS, S=0	1.44	1.87	115°	1.95
DFT-GGS, S=1	1.42	1.91	115°	1.92
CuO_2^- side-on				
DFT-GGS, S=0	1.49	1.96	45°	2.23
DFT-GGS, S=1	1.49	1.95	45°	2.04

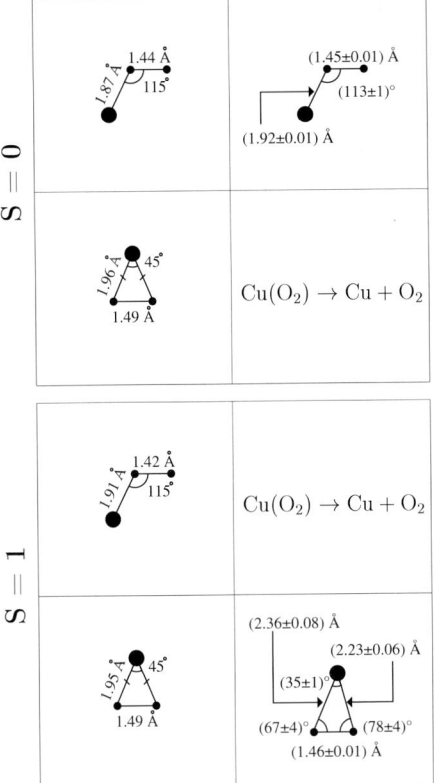

Fig. 7.2. Initial configurations (at time t=0, on the left) and average configurations (on the right) over a time trajectory of 2 ps for the negatively charged bent and side-on isomers in the spin configurations S=0 and S=1. Error bars are the standard deviation of the mean over subtrajectories of 0.05 ps.

the bent $CuOO^-$ is unstable, while the side-on remains bound, despite a shift of the Cu atom to larger interatomic Cu-O distances. These behaviors show that both the bent and the side-on, in different spin states (S=0 for the bent and S=1 for the side-on) can be found in the low binding energy portion of the photoelectron spectra reported in [7.31].

In larger clusters as CuO_6, two constitutive subunits are found to compete: those formed by assembling CuO_2 side-on and bent geometries and those containing a planar $Cu(O_3)$ ring. The specific arrangement of three oxygen atoms forming a loop with a copper atom is termed ozonide. $Cu(O_3)$ subunits cannot be discarded a priori when proposing structural models for the larger clusters, since their presence have been detected by photoelectron spectroscopy, although at very low intensities citewu1. The isomer search for CuO_6 has been carried out by selecting 20 initial sets of positions randomly distributed in space for the neutral case in the S=1/2 spin state. The six

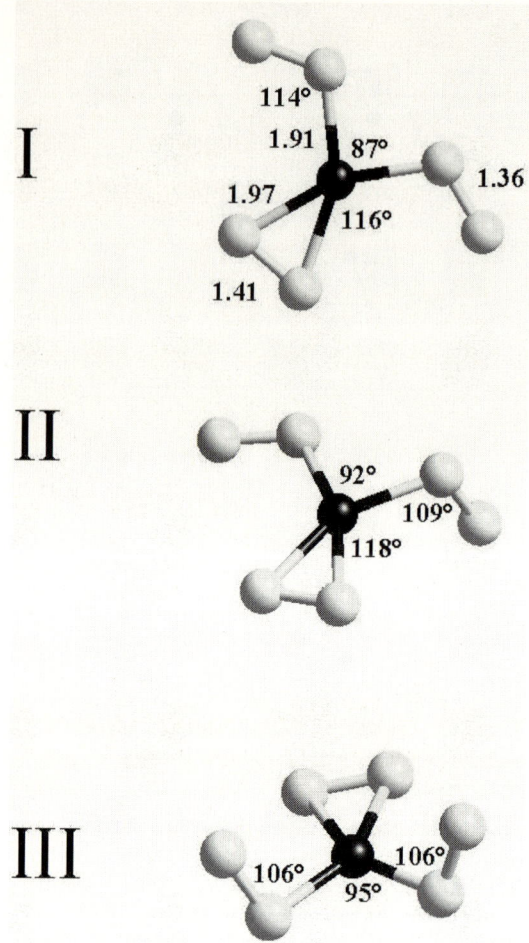

Fig. 7.3. Calculated equilibrium geometries of CuO$_6$ clusters in the S=1/2 spin state. The isomers shown are characterized by the presence of Cu(O$_2$) (side-on) and OOCu (bent) units. Distances are in Å.

most stable isomers are shown in Fig. 7.3 and Fig. 7.4, with their relative binding energy given in Table 7.2. The isomers shown in Fig. 7.3 are made of one side-on Cu(O$_2$) and two OOCu bent units sharing a common Cu atom. In the lowest-energy structure (isomer **I**) all atoms lie on the same plane. The O−O bonds which are part of the OOCu motifs can also be found rotated by 90° with respect to the Cu(O$_2$) plane. This is the case of isomer **II** and **III**, where one and two of these O−O bonds are involved in such rotation, respectively. Accordingly, the difference in binding energy between isomer **I** and isomer **III** (0.01 eV/atom, i.e. 0.07 eV) can simply be associated with the rotation of the O−O bonds out of the Cu(O$_2$) plane. As shown

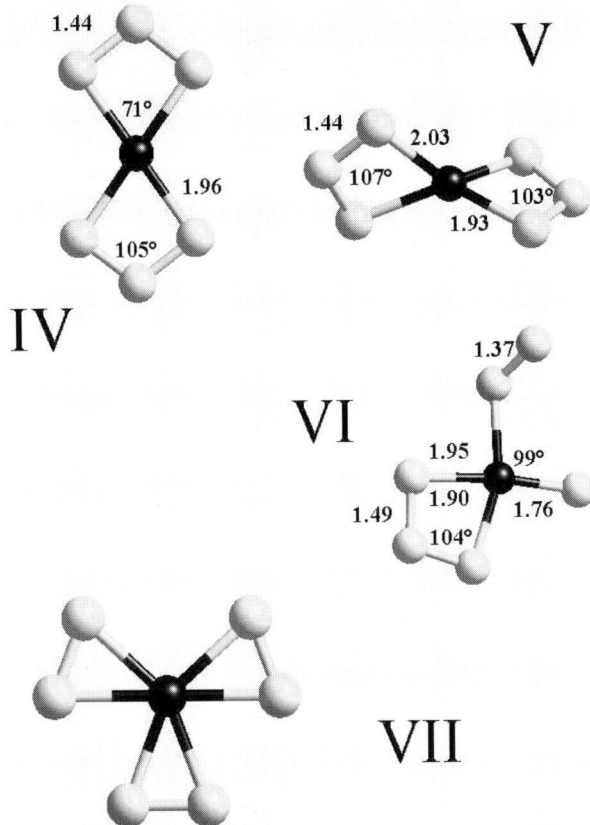

Fig. 7.4. Calculated equilibrium geometries of CuO$_6$ clusters in the S=1/2 spin state. The isomers shown are characterized by the presence of ozonide Cu(O$_3$) units. Isomer **VII** was proposed in [7.32] to rationalize the photoelectron spectra but is actually unstable. Distances are in Å.

Table 7.2. Binding energies differences ΔE_b (given in eV/atom) with respect to the lowest energy isomer for CuO$_6$ in the spin state S=1/2.

	Isomer	ΔE_b
CuO$_6$, S=1/2		
	I	0.01
	II	0.01
	III	0.02
	IV	0.00
	V	0.05
	VI	0.17

in Fig. 7.4, isomers with one or two ozonide $Cu(O_3)$ units are also found. Isomer **IV** is planar, while in isomer **V** the two $Cu(O_3)$ units are unchanged but lie in perpendicular planes. Transition to this new configuration yields higher binding energies by an amount ranging 0.05 ev/atom for the neutral CuO_6 with S=1/2. The presence of isomer **VI** allows to estimate the energetic cost associated with breaking one O−O bond within the OCuOO ring. The planar character of the isomer is lost, and the surviving O−O bond is tilted out of the original plane, while the corresponding OCuO angle changes from 71° to 99°. The energy difference between isomer **IV** and isomer **VI** amounts to 0.17 eV/atom. The planar double-ozonide $Cu(O_3)$ and the three isomers containing a $Cu(O_2)$ unit are found in a narrow interval of energies (0.02 eV/at at most), with a slight preference for the ozonide. Similar indications have been collected by extending our investigation to negatively charged clusters [7.18]. These findings do not agree with the qualitative predictions of [7.32], where the presence of $Cu(O_3)$ units was ruled out in the interpretation of the photoelectron spectra of CuO_6^-. Focusing on the isomers made of $Cu(O_2)$ units, we tested the stability of the planar geometry (isomer **VII**) proposed by Wu, Desai and Wang for CuO_6 via an additional set of calculations for the neutral (S=1/2, S=3/2) and anionic (S=0, S=1) CuO_6. This structure, depicted in Fig. 7.4, is made of three $Cu(O_2)$ units and was selected as the starting configuration. However, it turned out to be unstable and readily took the shape of either isomer **I**, **II** or **III**. These results point out that the information available on the structure of clusters is quite often limited to educated guesses proposed in the framework of experimental analysis of data. Here we have shown that first-principles structural optimization of structures considerably enriches the interpretation of the photoelectron spectra, either discarding unstable geometries or allowing critical comparison of the binding energies involved. It remains to be kept in mind that the theoretical study of photoelectron spectra on the basis of one-electron, self-consistent ground state calculations have intrinsic limitations and can only be considered as a first step toward a more complete description involving electronic excitations.

7.3.2 Si-Doped Heterofullerenes $C_{59}Si$ and $C_{58}Si_2$

Since all sites in C_{60} are equivalent, substitution of one C by one Si atom leads to a single isomer, the heterofullerene $C_{59}Si$. The system reacts to this replacement by adopting a less strained configuration with an energy gain of about 8 eV. The binding energy of the relaxed structure of $C_{59}Si$ shown in Fig. 7.5 is 0.12 eV/atom smaller than the binding energy of C_{60}. The Si−C bond lengths are much longer than the corresponding C–C bonds of C_{60}, as shown in Table 7.3, but their conjugation pattern is preserved. The C–C single bonds of $C_{59}Si$ which are located in the vicinity of the dopant atom exhibit the largest deviations (up to 3 %) with respect to the C–C single bond lengths found in C_{60}. We recall that in fullerenes a 5−6' bond is defined as the

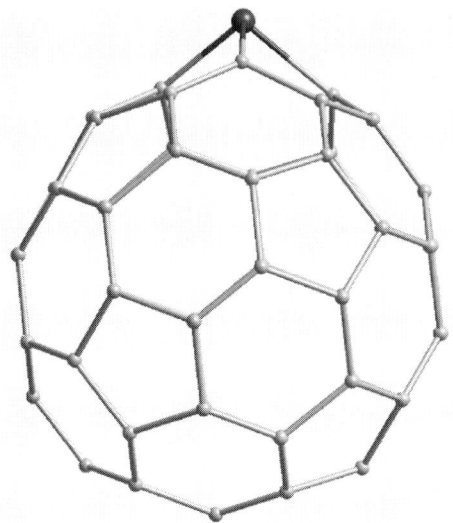

Fig. 7.5. Ball and stick representation of the structure of $C_{59}Si$. (The large dark atom is silicon, the smaller light grey atoms are carbon atoms.)

Table 7.3. Bond lengths of $C_{59}Si$ and C_{60}. For the C-C bond lengths, the smallest and the largest value found are indicated. The distances are expressed in $\overset{\circ}{A}$.

	$C_{59}Si$	C_{60}	C_{60} (Ref. [7.33])
Si-C 5−6' bonds	1.90		
Si-C 6−6' bonds	1.85		
C-C 5−6' bonds	1.46-1.50	1.46	1.458(6)
C-C 6−6' bonds	1.40-1.43	1.41	1.401(10)

bond between one pentagon and one hexagon and a 6−6' bond as the bond between two hexagons. In addition to this geometrical definition, used here for the Si-C bonds, one can also refer to single and double bonds in a chemical sense. The magnitude of the deformations induced by Si in-cage substitution is much larger than in the case of nitrogen and boron substitutional doping, the elongation of the B–C and N–C bonds with respect to the C–C bonds not exceeding a few percent [7.34, 7.35]. While B and N atoms have size comparable to that of C, the Si atom is much larger and requires more room to fit into the fullerene network.

Substitutional doping with more than one heteroatom results in a number of different isomers which correspond to all possible arrangements of the dopant atoms in the fullerene cage. In the case of $C_{58}Si_2$ we consider nine different isomers whose equilibrium structures are shown in Fig. 7.6. We include the isomers for which the two Si atoms are *a)* first nearest neighbors (*ortho-isomers:* **1** and **2**), *b)* second nearest neighbors (*meta-isomers:* **3** and

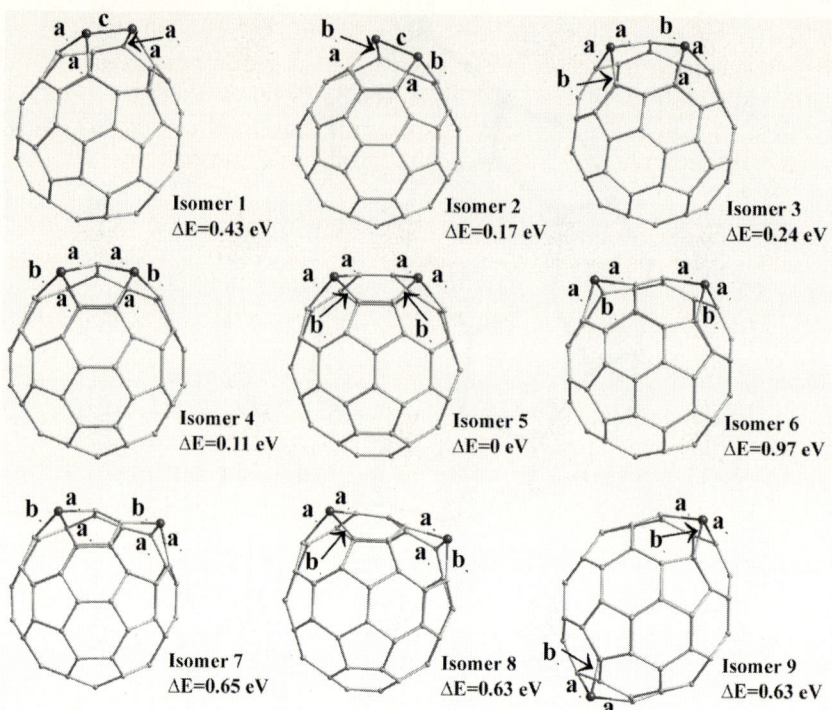

Fig. 7.6. Ball and stick representations of the structures of the various isomers of
$C_{58}Si_2$ considered in this work which are specified by a number, as described in the
text; a_1, a_2, a_3, a_4, b_1, b_2 denote the Si-C bonds and c corresponds to Si-Si bonds.
The lengths of these bonds are given in Table 7.2. The large dark atoms are the
silicon atoms and the smaller light gray atoms are carbon atoms. We also indicate
the total energies for the various isomers, relative to the energy of isomer **5**.

4), *c)* third nearest neighbors (**5**, **6** and **7**), *d)* one of the isomers for which
the two Si atoms are fourth nearest neighbors (**8**) and *e)* the isomer corresponding to the two Si atoms lying at diametrically opposite sites on the cage
(ninth nearest neighbor position) (**9**). The ortho-isomers **1** and **2** show a bond
between the two silicon atoms. In isomer **2**, the Si–Si bond length is slightly
smaller (2.30 Å) than the experimental values measured for bulk Si (2.346
Å). The conjugation pattern of the Si–Si bond in **2** is that of a single bond,
while for isomer **1** the conjugation pattern is that of a Si–Si double bond.
Correspondingly, the Si–Si bond of **1** takes a value (2.23 Å) between the experimental one of complexes with a Si–Si double bond (2.14−2.16 Å) [7.36]
and the bond length of silicon dimer Si_2 (2.246 Å) [7.37].

The energies of the $C_{58}Si_2$ isomers (see Fig. 7.6) span an interval of 0.97 eV
and the energy differences between next lying configurations vary in from
0.1 eV to 0.35 eV. The energetically lowest isomer is **5**, carrying two Si
on the same hexagon at para-sites as shown in Fig. 7.6. Its binding energy

is 0.14 eV/atom smaller than the binding energy of C_{60}. The two lowest energy isomers (**5** and **4**) are separated by 0.11 eV. The next isomer in this energy scale is the ortho-isomer **2** which is 0.06 eV less stable than **4** and lies at 0.07 eV below **3**. The other ortho-isomer **1** is located at a higher energy and is well separated (by about 0.2 eV) from the energetically very close isomers **9**, **8** and **7**. At the top of this energy scale, at ~0.32 eV above **7**, we located isomer **6**. The most favorable configuration found in the lowest energy $C_{58}Si_2$ isomer, **5**, is neither the one where the two Si atoms are at the largest distance on the cage, nor the one where the two Si atoms are nearest neighbors. This differs from the results of experimental and computational studies of $C_{58}B_2$. In that case it was found that the two B atoms tend to be well separated from each other in the fullerene network [7.38, 7.39].

Since Si and C have the same number of valence electrons, doping of C_{60} with silicon does not change the net occupancy of the energy levels in contrast to B and N in–cage doping [7.34]. In the case of $C_{59}Si$, the degeneracy of both HOMO and LUMO of C_{60} is partly removed by the doping, as shown in Fig. 7.7. The multiplet which derives from the C_{60} HOMO consists of four closely spaced levels plus one which lies ~ 0.05 eV higher in energy. The splitting of the C_{60} LUMO is more significant: one level is shifted to lower energy with respect to the other two levels by ~ 0.4 eV. By projecting

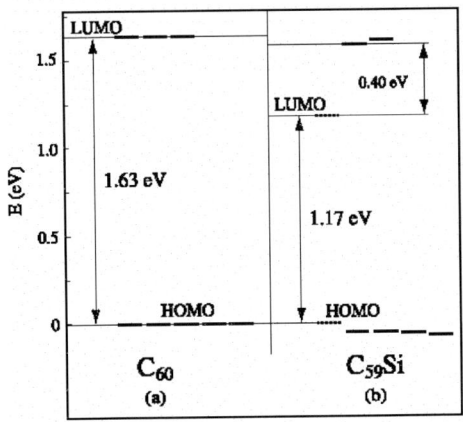

Fig. 7.7. Kohn-Sham energy levels in the HOMO-LUMO region of C_{60}. (a) C_{60}; (b) $C_{59}Si$. $E = 0$ is arbitrarily fixed at the uppermost level (HOMO) for both cases. Solid and dashed lines in (b) indicate levels with a dominant carbon-like and silicon-like character, respectively. Also indicated are the values of the HOMO-LUMO energy gaps for C_{60} and $C_{59}Si$.

the KS HOMO orbital of $C_{59}Si$ on silicon pseudo-atomic orbitals centered on the dopant atom, we find that the Si 3s and 3p pseudo-atomic orbitals account for a large fraction (more than 57 %) of the $C_{59}Si$ HOMO. The $C_{59}Si$ HOMO–LUMO energy gap amounts to 1.17 eV, a value almost 0.5 eV smaller than in the C_{60} case. The HOMO–1 and LUMO+1 have a strong carbon-like character and they are separated by an energy difference of 1.62 eV, a value very close to the HOMO–LUMO gap of C_{60}. This shows that the changes in the energy levels from C_{60} to $C_{59}Si$ mainly concern the orbitals close to the gap, the remaining orbitals being barely affected. Therefore Si doping of C_{60} introduces a chemically distinct site in the cage but perturbs only moderately the overall electronic structure of the pure fullerene.

By doping with two silicon atoms, the degeneracy in the electronic levels of C_{60} is removed. In Fig. 7.8 the Kohn-Sham orbitals around the HOMO–LUMO gap are shown for the $C_{58}Si_2$ isomers. The splitting is larger for the states deriving from the C_{60} LUMO than for those deriving from the C_{60} HOMO. The HOMO–LUMO energy gap is found to vary in between 0.44 eV and 1.07 eV depending on the isomer. Low energy isomers do not necessarily have the largest HOMO–LUMO gaps. In particular the most stable structure **5** has an energy gap of 0.63 eV. However its HOMO and LUMO are well separated from the next lying orbitals by as much as 0.29 eV (HOMO and HOMO–1) and 0.68 eV (LUMO and LUMO+1). The degree of localization of the HOMO and the LUMO energy levels on the Si atoms differs from one isomer to the other, but in general it is large for all isomers considered. Among the energy levels derived from the HOMO and LUMO of C_{60} those which have a projection on atomic silicon orbitals larger than 0.2 are indicated

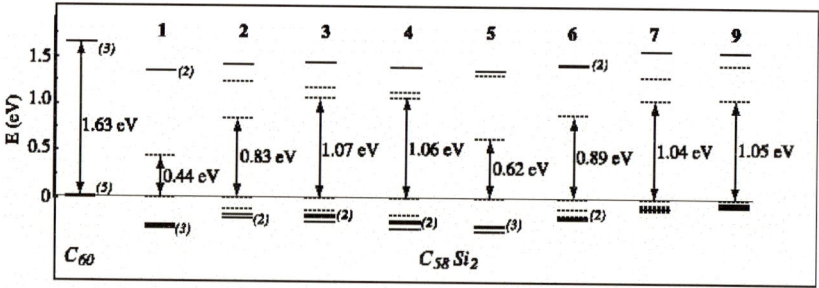

Fig. 7.8. Kohn-Sham energy levels in the HOMO-LUMO region of C_{60} and of various isomers of $C_{58}Si_2$, specified by the corresponding number in the upper part of the figure. $E = 0$ is arbitrarily fixed at the uppermost level (HOMO) for all cases. Solid and dashed lines for $C_{58}Si_2$ isomers indicate levels with a dominant carbon-like and silicon-like character, respectively. Also indicated in the figure are the HOMO-LUMO energy gaps. Numbers in parentheses denote the level degeneracy.

in Fig. 7.8. For isomers **2**, **3**, **4** and **5** the two highest occupied and the two lowest unoccupied Kohn-Sham energy levels are silicon-related orbitals, while for isomer **1** only the HOMO and the LUMO are considerably localized on Si atoms. In the case of the highest energy isomers, **6** and **7**, three and four of the five highest occupied orbitals exhibit an important Si character, respectively. The same can be found in the HOMO, the HOMO−2 and in the two lowest unoccupied energy levels of isomer **9**.

Information concerning bonding in these heterofullerenes can be gained from a Mulliken population analysis. For $C_{59}Si$ the net charges on Si (0.63) and on its three nearest neighbor C atoms (-0.17) are large compared to the values on the C atoms farther apart on the cage (0.03-0.05). For these latter the Mulliken charges are close to the values calculated for C_{60}. Moreover we observe that the charges on the Si atom and on the C nearest neighbors are of opposite sign, indicating a substantial electronic charge transfer from the Si site to the neighboring C sites. The formation of Si–C bonds with strong polar character can be ascribed to the smaller electronegativity of silicon compared to carbon. The case of $C_{58}Si_2$, largely confirms this picture, the Mulliken charge on the Si atoms being equal or slightly larger than the net charge on the Si atom of $C_{59}Si$. Overall, this study provides evidence on the clear correlation between the Kohn–Sham electronic energy levels in the immediate vicinity of the gap (HOMO and LUMO in particular) and the electron localization on Si atoms and, to a lesser extent, on the neighboring C atoms. This goes along with an overall charge depletion from the dopant sites, which is a manifestation of the polar character of the Si–C bonds. Thus the replacement by Si atoms introduces chemically reactive sites in the cage, affecting only locally its electronic structure.

Table 7.4. Population analysis for $C_{59}Si$ and for C_{60} in terms of the Mulliken charge. The Mulliken charge is given for the atom specified in the atom column; C(1) and C(2) represent the carbon atoms involved in the 5-6' and 6-6' Si-C bonds, while C(3) and C(4) are carbon atoms involved in the 5-6' and 6-6' C-C bonds.

Molecule	Atom	Mulliken charge	bond
$C_{59}Si$	Si	0.63	
	C(1)	-0.17	5-6' Si-C
	C(2)	-0.17	6-6' Si-C
	C(3)	0.03-0.05	5-6' C-C
	C(4)	0.03-0.05	6-6' C-C
C_{60}	C(3)	0.03-0.05	5-6' C-C
	C(4)	0.03-0.05	6-6' C-C

7.3.3 Disordered Network-Forming Materials: Liquid SiSe$_2$

Disordered systems are characterized by a residual degree of order which involves nearest-neighbor distances, the short range order (SRO). In certain network-forming liquid and glasses structural organization extends on larger distances and is commonly termed intermediate range order (IRO). The first sharp diffraction peak (FSDP) in the total neutron structure factor is an unambiguous signature of IRO and appears in the case of AX$_2$ oxides and chalcogenides disordered materials (A: Si, Ge; X: O, S, Se) [7.40]. First-principles molecular dynamics is of invaluable help to better understand enhanced forms of order in disordered systems. This issue has been first addressed by focusing on liquid GeSe$_2$ (l-GeSe$_2$) [7.41, 7.42], an interesting case due to the existence of measured partial structure factors [7.43]. It was found that a proper account of the ionic character of bonding brings theory in agreement with experiments [7.41]. In l-GeSe$_2$ regular GeSe$_4$ tetrahedra coexist with Ge$-$Ge and Se$-$Se homopolar bonds as well as miscoordinated Ge and Se atoms. Such a defective structure is consistent with the picture arising from the neutron diffraction data for both the liquid and the glass [7.43, 7.44]. In particular, the calculated total neutron structure factor reproduces the measurements over the entire range of wavevectors.

When modified by the addition of solid electrolytes, amorphous SiSe$_2$ ($a-$SiSe$_2$) transforms to a composite glass with potential applications in the field of batteries, sensors, and displays. This is due to the high mobility of Ag$^+$, Li$^+$, and Na$^+$ ions [7.45–7.48]. The crystalline structure of SiSe$_2$ is characterized by one-dimensional chains (edge-shared (ES) Si(Se$_{1/2}$)$_4$ tetrahedra) making this system suitable for nanowire technologies [7.49] as reinforcing fibers in high-strength and light-weight composite materials. Since local molecular units in the crystalline phase might be retained when this material is quenched, it was conjectured that a-SiSe$_2$ is a good model of edge-sharing glasses. The presence of edge-shared Si(Se$_{1/2}$)$_4$ tetrahedra in a-SiSe$_2$ was first inferred by Tenhover, Hazle, Grasselli and Thompson [7.50, 7.51] on the basis of similarities between the Raman spectra of glassy and crystalline SiSe$_2$. It was later proposed that corner-shared (CS) Si(Se$_{1/2}$)$_4$ tetrahedra cross-link the chains of edge-shared tetrahedra [7.52]. Moreover it has been shown, by combining results of neutron diffraction and EXAFS experiments, that the average coordination numbers of $a-$SiSe$_2$ are close to 4 for Si and to 2 for Se [7.53]. These values are consistent not only with the so-called 8-N rule [7.54], which predicts the number of neighbors for a given atom on the basis of its position in the periodic table, but are also a clear indication that the network of tetrahedra is chemically ordered. As a further evidence, magic angle spinning-nuclear magnetic resonance (MAS-NMR) experiments pointed out the presence of both CS and ES tetrahedra by classifying the Si atoms, according to their belonging to zero (Si(0)), one (Si(1)) or two (Si(2)) fourfold rings [7.55]. Such configurations correspond to Si atoms participating only in CS connections, or in one ES or two ES connections, i.e. one or two fourfold rings. In this latter case the Si atoms form a chain reminiscent of the crystalline structure.

From a theoretical side, the atomic structure of $a-SiSe_2$ has been described by phenomenological and atomic-scale models. Within the first class, it was concluded that the best intepretation of the neutron data is achieved when several kinds of structural units are included, as corner sharing units (<15%), locally parallel chains (<15%) and random edge-sharing tetrahedral chains with a mean length of seven [7.56–7.59]. $SiSe_2$ disordered systems have been also studied through molecular dynamics with effective potentials to describe mechanical fractures in $SiSe_2$ nanowires and their amorphization [7.49, 7.60, 7.61]. Very recently, the Raman spectra of $a-SiSe_2$ has been interpreted on the basis of a first-principles cluster model approach, by focusing on isolated molecular entities which are expected to exist in the real glass [7.62]. These efforts call for the establishment of accurate structural models of $a-SiSe_2$, providing unambiguous information on the identity and the relative importance of the different constitutive units forming the network. With this purpose in mind, we have first developed a reliable first-principles model for liquid $SiSe_2$ ($l-SiSe_2$), that has been employed to produce an amorphous structure [7.63].

By its very definition a liquid does not keep any memory of the initial conditions from which it was created. To comply with this criterion and assess advantages and limitations of our approach, we have performed two MD runs starting from two distinct atomic configurations. The first contains many homopolar bonds (liquid L_1) while the second is highly ordered without homopolar bonds (liquid L_2). In both cases the size of the cubic cell corresponds to the density of the amorphous system at room temperature. For the average results to be reliable, the properties of these two liquids do not have to differ significantly. Our liquids are undercooled, since their equilibrium temperature (T=1000 K) is lower than the melting temperature of crystalline $SiSe_2$ (T=1243 K). The starting configuration for L_1 has been produced by positioning the atoms at random. This procedure ensures the inclusion of several homopolar Si–Si and Se–Se bonds. After 1 ps with no control of the temperature, the system has been heated up to T=4000 K where a run of about 1.5 ps has been performed. Then the system has been cooled and annealed at T=1500 K during 9.7 ps by introducing a dynamical thermostat [7.15, 7.16]. During this first set of simulations, discarded from the calculation of the averages, each atom has moved by more than 7 Å thereby evolving to configurations very different from the initial one. To produce an additional liquid (L_2) we have looked for a starting configuration highly ordered, i.e. with the minimum number of homopolar bonds. This can be achieved by employing the crystalline structure of $SiSe_2$ and adapting it to a cubic cell of volume defined by the density of the amorphous system (L = 31.3 a.u. = 16.585 Å). By adopting the 2x3x2 geometry such approach yields a contraction in both the directions a and b and an expansion along the direction c. The expansion along c is about 40%, more than that used in [7.49] to produce amorphous structures from crystalline nanowires of $SiSe_2$ (13%). After a substantial relaxation, followed by 2 picosends at T=300 K, we have raised the temperature at T=1000 K to produce a second equilibrated liquid. L_2 has been simulated at T=1000 K for a total duration of 15 ps, with statistical averages calculated over the last 6 ps. At equilibrium and for both liquids, the diffusion coefficients are $D_{Si} = 0.67 \cdot 10^{-5}$ cm^2/s and $D_{Se} = 0.95 \cdot 10^{-5}$ cm^2/s. These values are close to those calculated in [7.64] ($D_{Si} = 0.95 \cdot 10^{-5}$ cm^2/s, $D_{Se} = 0.96 \cdot 10^{-5}$ cm^2/s at the temperature of 1293 K) by using classical molecular dynamics.

The calculated total neutron structure factors $S_{tot}(k)$ for both liquids are reported in Fig. 7.9. Despite the overall agreement for values of $k > 2$, some differences are detectable in the low limit of k, ascribed to a non complete full equilibration of liquid L_2. Therefore, our simulations are especially instructive in pointing out that long equilibration runs (> 10 ps) are needed to obtain equivalent intermediate range properties for two liquid samples. A more quantitative insight can be given by the partial pair correlation func-

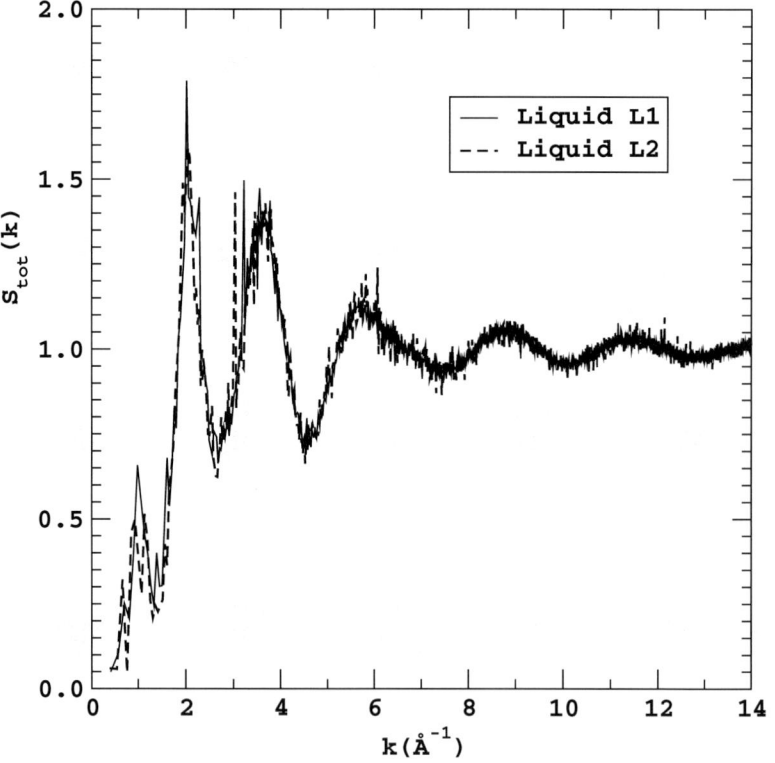

Fig. 7.9. Total neutron structure factor $S_{tot}(k)$ of liquid L_1 (line) and of liquid L_2 (dots).

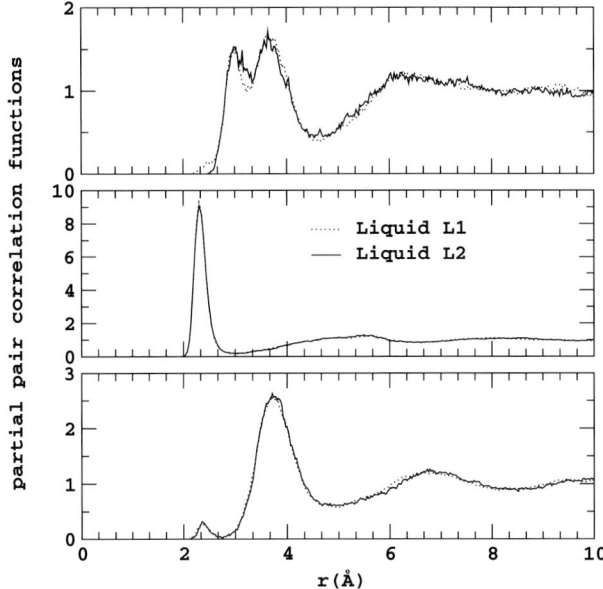

Fig. 7.10. Partial pair correlation functions of the liquids L_1 (dots), L_2 (line) From top to bottom: $g_{SiSi}(r)$, $g_{SiSe}(r)$, $g_{SeSe}(r)$.

tions (see Fig. 7.10) and the coordination numbers $n_{\alpha\beta}$ (see Table 7.5). These quantities refer to the nearest neighbor coordination of each species, $n_{\alpha\beta}$ being the coordination of an atom of species α by an atom of species β. Looking at $g_{Si Si}(r)$ a small tail at short distances in both systems L_1 and L_2 can be found. This tail is confined at $r < 2.6$ Å and is indicative of homopolar bonding, leading to a vanishingly small number for the homopolar coordination $n_{Si Si}$ (0.02 for L_1 and 0.003 for L_2). In both liquids the main peak of $g_{Si Si}(r)$ is at $r=3.0$ Å. This peak is also present in the crystalline structure and corresponds to pairs of Si atoms located on neighboring edge-sharing

Table 7.5. Calculated coordination numbers $n_{\alpha\beta}$ of the partial pair correlation functions $g_{\alpha\beta}(r)$ in liquid $SiSe_2$: $L1$, $L2$. The integration ranges corresponding to the coordination numbers $n_{\alpha\beta}$ are 0−2.6 Å for $g_{Si Si}(r)$, 0−3.0 Å for $g_{Si Se}(r)$ and 0−2.7 Å for $g_{Se Se}(r)$. The corresponding values for the CON and the RCN models are reported for comparison.

	$g_{Si Si}(r)$	$g_{Si Se}(r)$	$g_{Se Si}(r)$	$g_{Se Se}(r)$
$n_{\alpha\beta}$ (L1)	0.02	3.90	1.95	0.13
$n_{\alpha\beta}$ (L2)	0.003	3.92	1.96	0.12
$n_{\alpha\beta}$ (CON)	0	4	2	0
$n_{\alpha\beta}$ (RCN)	2	2	1	1

(ES) tetrahedra. The second main peak is at 3.8 Å and it corresponds to pairs of Si atoms in neighboring CS tetrahedra.

Turning now to the pair correlation function $g_{SiSe}(r)$ (Fig. 7.10), we notice the main peak at the distance of 2.30 Å in both liquids L_1 and L_2. This peak is related to the Si–Se intratetrahedral interactions. Integrating the area under this peak one has n_{SiSe}=3.90 (L_1) and n_{SiSe}=3.92 (L_2). The n_{SeSi} coordination is equal to 1.95 and 1.96 for L_1 and L_2, respectively. In the case of the pair distribution function $g_{SeSe}(r)$, a clear feature is discernible at r =2.35 Å, indicating homopolar bonds. The coordination number n_{SeSe} is similar in the two liquids, 0.12 in L_1 and 0.13 in L_2. The values for n_{SiSi}, n_{SeSi} and n_{SeSe} indicate that l−SiSe$_2$ is chemically ordered, as confirmed by the average coordination numbers $n_{Si} = n_{SiSi} + n_{SiSe}$ and $n_{Se} = n_{SeSe} + n_{SeSi}$, where $n_{SiSe} = 2n_{SeSi}$. One obtains for both liquids n_{Si}=3.92 and $n_{Se} = 2.08$, not differing significantly from those, 4 and 2, expected for a chemically ordered AX$_2$ material. In Tab. 7.5 we also report the coordinations for two model network structures, the random covalent network (RCN) and the chemically ordered network (CON) [7.54]. Both models are compatible with the so-called 8-N rule, which gives the coordination number of an atom as a function of its column (N) in the periodic table. A structure follows the CON model if the number of heteropolar bonds is the largest possible for that given composition. On the other hand, if there is no preference for either homopolar or heteropolar bonds, the RCN model holds. It is worth noticing that, for a given composition, the total coordination numbers for the CON model and the RCN model ($n_\alpha^{tot} = n_{\alpha\beta} + n_{\alpha\alpha}$) are the same.

In order to get an insight into the diffusion mechanism, we turn our attention to snapshots of the atomic motion reported in Fig. 7.11 for L_1. At the beginning, the Si atom 1 is fourfold coordinated with 4 Se atoms and this configuration has been stable for more than 2 ps (Fig. 7.1(a), t=0 ps). Then, (Fig. 7.11(b), t=0.01 ps) the bond between Se atom 2 and Si atom 1 breaks down: the Si atom 1 becomes threefold coordinated. This configuration is not long living because the tetrahedra is restored through a bond formed with the Se atom 3 (Fig. 7.11(c), t=0.02 ps). This new configuration lasts more than 2 ps and is followed by a time interval featuring the Si atom fivefold coordinated (Fig. 7.11(d), t=2.10 ps). In a subsequent step, the Si atom recovers a tetrahedral environment with four Se neighbors (Fig. 7.11(e), t=2.12 ps). After 0.18 ps, the Se atom 3 takes part to a new tetrahedral configuration with a different Si atom (Fig. 7.11(f), t=2.3 ps). This new configuration is no more the same as the starting one, since the Se atom 3 is now part of a completely new tetrahedra. These observations suggest that diffusive events consist of temporary defects created by occasional neighbor exchanges, with the tetrahedra restored in less than 0.1 ps. The Si atoms are fourfold coordinated during most of the time. A typical diffusive event is characterized by the motion of a Se atom from the coordination sphere of a given Si atom to that of an adjacent Si atom. When this occurs, the missing Se atom is im-

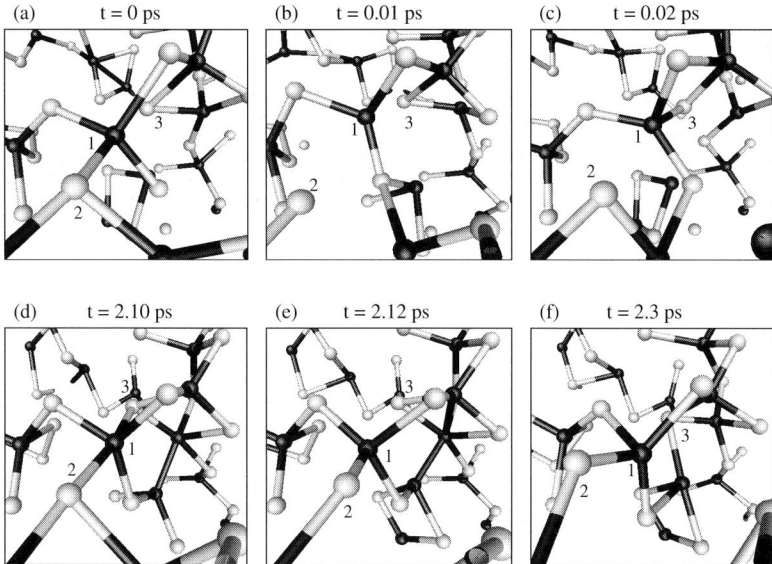

Fig. 7.11. Snapshots of configurations taken from liquid L_1 to elucidate the diffusion mechanism of the Se atoms in l-SiSe$_2$. Black spheres are silicon atoms and white spheres are selenium atoms: atom 1 is a Si atom, atom 2 and 3 are Se atoms.

mediately replaced. This picture is consistent with the presence of a strong bond tetrahedral network even in the liquid phase.

Further information on the structural properties can be gained by studying the partial neutron structure factors [7.65]. To understand the different contributions to the FSDP in S_{tot} we decompose it by using the Faber-Ziman (FZ) partials S_{SiSi}, S_{SiSe} and S_{SeSe}. The FZ partials of both liquids L_1 and L_2 are reported in Fig. 7.12. At values $k \geq 2$ there is a good agreement between the two liquids in all partials. On the other hand, at small values of k, it is possible to notice some differences. In particular in the L_1 S_{SiSi} partial, the FSDP seems to be shifted at smaller values than in L_2. Furthermore, with respect to the S_{SiSe} partial of liquid L_2, a double peak is present in the region of the FSDP, very similar in shape to that observed in the total neutron structure factor shown in Fig. 7.9. We can explicit the decomposition of the total neutron structure factor $S_{\text{tot}}(k)$ in terms of FZ partials

$$S_{\text{tot}}(k) = A\,S_{\text{SiSi}} + B\,S_{\text{SeSe}} + C\,S_{\text{SiSe}} \tag{7.4}$$

with A $= c_{\text{Si}}^2\,b_{\text{Si}}^2/\langle b \rangle^2$, B $= c_{\text{Se}}^2\,b_{\text{Se}}^2/\langle b \rangle^2$, C $= 2\,c_{\text{Si}}\,c_{\text{Se}}\,b_{\text{Si}}\,b_{\text{Se}}/\Delta b^2$ where $\Delta b = b_{\text{Si}} - b_{\text{Se}}$ and $\langle b \rangle = c_{\text{Si}}b_{\text{Si}} + c_{\text{Se}}b_{\text{Se}}$. The values used for the scattering factors are $b_{\text{Si}} = 4.149$ fm and $b_{\text{Se}} = 7.97$ fm. Those of the coefficients A, B, C are $A = 0.0426$, $B = 0.63$, $C = 0.33$, respectively. Despite the well-defined peak in S_{SiSi} at $k \sim 1$ Å, the FSDP in the total structure factor is mostly due to the feature appearing at the FSDP location in the S_{SiSe} structure factor. Indeed,

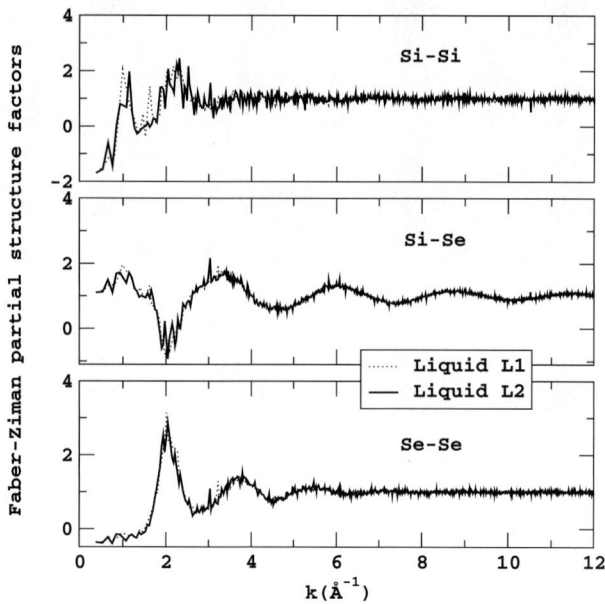

Fig. 7.12. Faber-Ziman partial structure factors of liquid L_1 (dots) and of liquid L_2 (line).

the AS_{SiSi} product in (7.4) accounts for only 12% of the total structure factor at $k \sim 1$ Å.

The concentration-concentration partial structure factor $S_{\mathrm{CC}}(k)$, averaged over the trajectories corresponding to L_1 and L_2 is shown in Fig. 7.13.

$$S_{\mathrm{CC}}(k) = c_{\mathrm{Si}}c_{\mathrm{Se}} \left\{ 1 + c_{\mathrm{Si}}c_{\mathrm{Se}} \left[(S_{\mathrm{SiSi}}(k) - S_{\mathrm{SiSe}}(k)) + (S_{\mathrm{SeSe}}(k) - S_{\mathrm{SiSe}}(k)) \right] \right\} \tag{7.5}$$

We note the small peak at the FSDP location. A peak at a given wave vector in the $S_{\mathrm{CC}}(k)$ stems from the sensitivity of a given atom (Si or Se) to the chemical nature of its neighbors on the length scale associated to that specific value of k. On the contrary, the absence of a peak corresponds to an equivalent tendency to homo- heteropolar neighbors. It was conjectured, as a result of predictions of classical molecular dynamics models, that the absence of FSDP in the $S_{\mathrm{CC}}(k)$ partial structure factor is a general property of binary glasses [7.66, 7.67]. However a FSDP in the S_{CC} structure factor of liquid GeSe$_2$ was experimentally observed [7.68]. In view of the absence of this peak in the corresponding calculated quantity for liquid GeSe$_2$ [7.41], we have analyzed the correlation between the degree of short range chemical order and the possible appearence of the FSDP [7.69]. For a system fully chemically ordered (as amorphous and liquid SiO$_2$ at not too high temperatures) the FSDP in the S_{CC} is absent. The same can be said of liquid GeSe$_2$ modeled

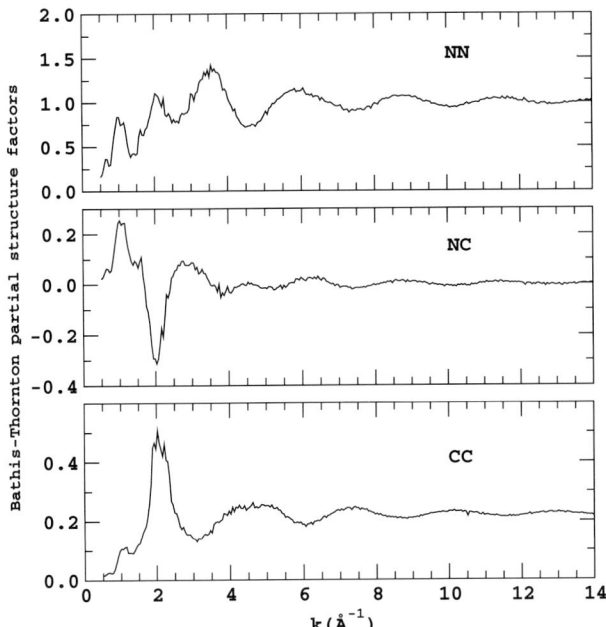

Fig. 7.13. Bathia-Thornton partial structure factors averaged over the trajectories corresponding to liquids L_1 and L_2.

within DFT [7.70]. Here we show that a small feature in the S_{CC} structure factor shows up at the FSDP location in liquid SiSe$_2$, which is less chemically ordered than SiO$_2$ and more chemically ordered than the model liquid for GeSe$_2$ obtained by DFT calculations. This suggests that the degree of short range chemical order is a crucial property able to trigger the occurrence of concentration fluctuations (i.e. the FSDP in the S_{CC} structure factor) on intermediate range distances.

   Our modeling of $l-$SiSe$_2$ has allowed the production of an amorphous phase consistent with experimental evidence [7.71]. Our results provide also an accurate counterpart for network models based on bond-constraint theory [7.72]. The amorphous phase is highly chemically ordered with a majority of Si atoms at the center of tetrahedra in the ES configuration, in good agreement with MAS-NMR experimental data [7.55]. The evaluation of the rings distribution has allowed to distinguish three sets of Si atoms according to their being part of zero (Si(0)), one (Si(1)) or two (Si(2)) fourfold rings. Chain fragments composed by three tetrahedra (triads) have been counted. Si(1)-Si(2)-Si(1) is the most frequent triad with a Si(2) atom in the center. These kind of triads are remindful of the crystalline structure of SiSe$_2$. However, our study has revealed that other types of triads are present and even more abundant, the most frequent being Si(1)-Si(1)-Si(0) and Si(1)-Si(0)-Si(0). Overall, the amount of the various Si(0)- and Si(1)-centered triads indicates that the dominant structural motifs in a-SiSe$_2$ contain both CS and ES tetrahedra. It appears that the count of sequences expected to characterize the network cannot *a priori* be restricted to only one kind (CS or ES) of intertetrahedral connection but it has to include combinations of both. We have focused our analysis on series of three tetrahedra but this conclusion applies equally well to longer chains. Among these, longer chains with sequences of Si(2) silicon atoms in the center are not found.

7.4 Concluding Remarks

Together with a brief outline of the theoretical framework, we have detailed three specific examples of application of the first-principles molecular dynamics approach. Two methodological features deserve to be recalled here. First, the possibility of describing interactions without the use of adjustable parameters. This makes the reliability of the calculations totally independent on the coordination environment. Second, the availability of information on the nature of bonding and its changes following variations of temperature or introduction of defects. The results presented for copper oxide clusters, silicon doped fullerenes and disordered networks have in common the quantitative characterization of structural properties. The case of copper oxide clusters highlights the importance of structural optimization and the role of temperature in reaching a correct structural assignment. The identification of small CuO_2 molecules is prohibitive since different shapes coexist in the photoelectron spectra. By using molecular dynamics we were able to determine under which conditions of temperature and spin state some isomers can be observed. In the case of CuO_6 the disagreement between the prediction extracted from the experimental analysis of the spectra and the results of calculations for CuO_6 is particularly striking. This clearly demonstrates that educated guesses are insufficient to describe structural properties. With increasing size, the idea of building blocks is quite useful to rationalize the shapes of the clusters and their energetic ordering. Unusual species, as the CuO_3 ozonide, can be found in large proportions. The stability of fullerenes doped with silicon, first observed in mass spectrometry measurements, is found to be due to the introduction of a chemically distinct Si site. Structural and electronic effects are fully correlated, with the cage distortion accompanied by substantial electronic charge transfer from Si to C atoms. It is worth stressing that while the mere deformation induced in the fullerene by the Si atoms might have been found via interatomic potentials as well, bonding effects can only be observed within a scheme providing explicit information on the electronic structure. In disordered systems the nature and the variation with time of the bonding configurations have been described in the case of a network-forming system, liquid SiSe$_2$. When a specific structural motif is predominant, as in the case of the tetrahedron SiSe$_4$, intermediate range order manifests through specific features in the diffraction data taken in reciprocal space. Accurate atomic-scale simulations are needed to retrieve corresponding structural patterns in real space. However, network-forming systems do not show significant differences in the connectivity and in the composition when studied with potentials. Going beyond the limitations of this approach, one is able to classify network-forming systems according to the kind of connectivity and the amount of chemical disorder. In liquid and amorphous $SiSe_2$ edge-sharing connections are more frequent than the corner-sharing ones. Focusing on the first-principles models for two close systems, liquid $SiSe_2$ and liquid $GeSe_2$, it appears that fluctuations of concentrations on intermediate range distances are more important in the first case. This allows to correlate them to the higher chemical order found in liquid $SiSe_2$. On the experimental side, clusters and disordered systems share a non-trivial determination of structural properties. As shown in this contribution, atomic-scale simulations based on density functional theory is a well-suited tool to find out about atomic positions and their thermal behavior.

References

[7.1]   W.A. de Heer, Rev. Mod. Phys. **65**, 611 (1993).
[7.2]   L.-S. Wang and H. Wu, Zeit. für Physikalische Chemie **203**, 45 (1998).
[7.3]   P.S. Salmon, Proc. R. Soc. London A **437**, 591 (1992).
[7.4]   P.S. Salmon, Proc. R. Soc. London A **445**, 351 (1994).
[7.5]   D. Marx and J. Hütter, 'Ab Initio molecular dynamics: theory and imple-mentation', in *Modern Methods and Algorithms of Quantum Chemistry*, J. Grotendorst (Ed.), John von Neumann Institute for Computing, Jülich, NIC Series, Vol.1, 301-449 (2000)
[7.6]   M. Wilson and P.A. Madden, J. Phys. Condens. Matter **5**, 6833 (1993).
[7.7]   S.M. Foiles, M.I. Baskes and M.S. Daw, Phys. Rev. B **33**, 7983 (1986).
[7.8]   F.H. Stillinger and T.A. Weber, Phys. Rev. B **31**, 5262 (1985).
[7.9]   R. Car and M. Parrinello, Phys. Rev. Lett. **55**, 2471 (1985).
[7.10] J.M. Zuo, M. Kim, M.O'Keefe and J.C.H. Spence, Nature **401**, 49 (1999).
[7.11] R.E. Dickerson, H.B. Gray and G.P. Haight, *Chemical Principles* (W.A. Ben-jamin inc., Menlo Park, California, 1974).
[7.12] P. Vashishta, R.K. Kalia, G.A. Antonio and I. Ebbsjö, Phys. Rev. Lett. **62**, 1651 (1989).
[7.13] R.P. Feynmann, Phys. Rev. **56**, 340 (1939).
[7.14] W. Kohn and L.J. Sham, Phys. Rev. **140**, A1133 (1965).
[7.15] S. Nosé, Mol. Phys. **52**, 255 (1984); W. G. Hoover, Phys. Rev. A **31**, 1695 (1985).
[7.16] P. Blöchl and M. Parrinello, Phys. Rev. B **45**, 9413 (1992).
[7.17] Y. Pouillon and C. Massobrio, Chem. Phys. Lett. **331**, 290 (2000).
[7.18] Y. Pouillon and C. Massobrio, Chem. Phys. Lett. **356**, 469 (2002).
[7.19] M. Celino and C. Massobrio, Comp. Mat. Science **24**, 28 (2002).
[7.20] M. Celino and C. Massobrio, Comp. Physics Comm. **147** 166 (2002).
[7.21] J. P. Perdew, J. A. Chevary, S. H. Vosko, K. A. Jackson, M. R. Pederson, D. J. Singh and C. Fiolhais, Phys. Rev. B **46**, 6671 (1992).

156   C. Massobrio et al.

[7.22] A.D. Becke, Phys. Rev. A **38**, 3098 (1988).
[7.23] C. Lee, W. Yang, and R.G. Parr, Phys. Rev. B **37**, 785 (1988).
[7.24] D. Vanderbilt, Phys. Rev. B **41**, 7892 (1990).
[7.25] K. Laasonen, A. Pasquarello, R. Car, C. Lee and D. Vanderbilt, Phys. Rev. B **47**, 10142 (1993).
[7.26] I.M.L. Billas, C. Massobrio, M. Boero, M. Parrinello, W. Branz, F. Tast, N. Malinowski, M. Heinebrodt and T.P. Martin, J. Chem. Phys. **111**, 6787 (1999).
[7.27] P. Császár and P. Pulay, J. Molec. Struc. **114**, 31 (1984).
[7.28] J. Hutter, H.P. Lüthi and M. Parrinello, Comp. Mat. Science **2**, 244 (1994).
[7.29] F. Tassone, F. Mauri and R. Car, Phys. Rev. B **50**, 10561 (1994).
[7.30] Code CPMD, version 3.0f, developed by J. Hutter, P. Ballone, M. Bernasconi, P. Focher, E. Fois, S. Goedecker, D. Marx, M. Parrinello, M. Tuckerman, at MPI für Festkörperforschung and IBM Zurich Research Laboratory (1990-1997).
[7.31] H. Wu, S.R. Desai and L.S. Wang, J. Chem. Phys. **103**, 4363 (1995).
[7.32] H. Wu, S.R. Desai and L.S. Wang, J. Phys. Chem. **101**, 2103 (1997).
[7.33] K. Hedberg, Science **254**, 410 (1991).
[7.34] W. Andreoni, F. Gigy and M. Parrinello, Chem. Phys. Lett. **190**, 159 (1992).
[7.35] W. Andreoni, A. Curioni, K. Holczer, K. Prassides, M. Keshavarz-K., J.-C. Hummelen and F. Wudl, J. Am. Chem. Soc. **118**, 11335 (1996).
[7.36] R. West, Angew. Chemie **99**, 1231 (1987).
[7.37] R.E. Dickerson, H.B. Gray and G.P. Haight, *Chemical Principles* (W.A. Benjamin inc., Menlo Park, California, 1974).
[7.38] T. Guo, C. Jin and R.E. Smalley, J. Phys. Chem. **95**, 4948 (1991).
[7.39] N. Kurita, K. Kobayashi, H. Kumahora and K. Tago, Phys. Rev. B **48**, 4850 (1993).
[7.40] S.R. Elliott, Phys. Rev. Lett. **67**, 711 (1991).
[7.41] C. Massobrio, A. Pasquarello and R. Car, Phys. Rev. Lett. **80**, 2342 (1998).
[7.42] C. Massobrio, A. Pasquarello and R. Car, J. Am. Chem. Soc. **121**, 2943 (1999).
[7.43] I.T. Penfold and P.S. Salmon, Phys. Rev. Lett. **67**, 97 (1991).
[7.44] I. Petri, P. S. Salmon and H.E. Fischer, Phys. Rev. Lett. **84**, 2413 (2000).
[7.45] S. Susman, R.N. Johnson, D.L. Price and K.J. Volin, Mat. Res. Soc. Symp. Proc. **61**, 91 (1986).
[7.46] A. Pradel, G. Taillades, M. Ribes, H. Eckert, J. Non-Cryst. Solids **188**, 75 (1995).
[7.47] C. Rau, P. Armand, A. Pradel, C.P.E. Varsamis, E.I. Kamitsos, D. Granier, A. Ibanez, E. Philippot, Phys. Rev. B **63**, 184204 (2001).
[7.48] P. Boolchand, W.J. Bresser, Nature **410**, 1070 (2001).
[7.49] P. Walsh, W. Li, R.K. Kalia, A. Nakano, P. Vashishta, S. Saini, Appl. Phys. Lett. **78**, 3328 (2001).
[7.50] M. Tenhover, M.A. Hazle, R.K. Grasselli, Phys. Rev. Lett. **51**, 404 (1983).
[7.51] M. Tenhover, M.A. Hazle, R.K. Grasselli, C.W. Thompson, Phys. Rev. B **28**, 4608 (1983).
[7.52] J.E. Griffiths, M. Malyj, G.P. Espinoso, J.P. Remeika, Phys. Rev. B **30**, 6978 (1984).
[7.53] S. Susman, R.W. Johnson, D.L. Price, K.J. Volin in: Defects in Glasses, eds. F.L. Galeener, D.L. Griscom and M.J. Weber, vol. 61 Symp. Proc. (Mat. Res. Soc. Pittsburgh, PA, 1986) p. 91.

[7.54] S.R. Elliott, *Physics of Amorphous Materials*, Longman Group UK Limited, Essex, 1990.

[7.55] M. Tenhover, R.D. Boyer, R.S. Henderson, T.E. Hammond, G.A. Shreve, Solid St. Comm. **65**, 1517 (1988).

[7.56] L.F. Gladden, S.R. Elliott, Phys. Rev. Lett. **59**, 908 (1987).

[7.57] L.F. Gladden, S.R. Elliott, J. Non-Cryst. Solids **109**, 211 (1989).

[7.58] L.F. Gladden, S.R. Elliott, J. Non-Cryst. Solids **109**, 223 (1989).

[7.59] L.F. Gladden, J. Non-Cryst. Solids **123**, 22 (1990).

[7.60] W. Li, R.K. Kalia, P. Vashishta, Phys. Rev. Lett. **77**, 2241 (1996).

[7.61] W. Li, R.K. Kalia, P. Vashishta, Europhys. Lett. **35**, 103 (1996).

[7.62] K. Jackson, S. Grossman, Phys. Rev. B **65**, 012206 (2001).

[7.63] M. Celino, PhD dissertation, Université Louis Pasteur Strasbourg, 2002.

[7.64] G. Antonio, R.K. Kalia, A. Nakano and P. Vashishta, Phys. Rev. B **45**, 7455 (1992).

[7.65] For the explicit relationship between the three sets of partial structure factors commonly used (Faber-Ziman, Ashcroft-Langreth and Bhatia-Thornton) see Y. Waseda, *The structure of Non-Crystalline Materials*, (McGraw-Hill, New York, 1980)).

[7.66] P. Vashishta, R.K. Kalia, J.P. Rino, and I. Ebbsjö, Phys. Rev. B **41**, 12197 (1990).

[7.67] H. Iyetomi, P. Vashishta and R.K. Kalia, Phys. Rev. B **43**, 1726 (1991).

[7.68] I.T. Penfold, P.S. Salmon, Phys. Rev. Lett. **67**, 97 (1991).

[7.69] C. Massobrio, M. Celino and A. Pasquarello, J. Phys. Condens. Matter to be published.

[7.70] C. Massobrio, A. Pasquarello and R. Car, Phys. Rev. B **64**, 144205 (2001).

[7.71] M. Celino and C. Massobrio, to be published.

[7.72] M. F. Thorpe, in *Insulating and Semiconducting Glasses* edited by P. Boolchand (World Scientific, Singapore, 2000), p.95.

8 Magnetism, Structure and Interactions at the Atomic Scale

V.S. Stepanyuk[1] and W. Hergert[2]

[1] Max Planck Institute of Microstructure Physics, Weinberg 2, 06120 Halle, Germany
[2] Martin-Luther-University Halle-Wittenberg, Department of Physics, Von-Seckendorff-Platz 1, 06120 Halle, Germany

Abstract. An efficient scheme is developed to study magnetism and structure as well as interaction between supported particles on the atomic scale. Starting by ab initio calculations of the electronic structure in the framework of density functional theory, interaction potentials for molecular dynamics simulations of metallic nanostructures supported on metallic surfaces are carefully optimized.

The two methods are shortly explained. Examples for the application of the methods are given. Mainly electronic and structural properties of Co nanostructures on Cu(001) and Cu(111) surfaces are investigated.

8.1 Introduction

The essence of nanoscience and technology is the ability to understand and manipulate matter at the atomic level. Structures behave differently when their dimensions are reduced to dimensions between 1 and 100 nm. Such structures show novel physical and chemical properties, due to their nanoscopic size.

In the frontier field of nanomagnetism, understanding of the relationship between magnetism and structure plays a central role. During the past few years experimental investigations of metallic nanostructures in the initial stage of heteroepitaxial growth revealed a lot of information which asks for a consistent theoretical explanation. Some important effects experimentally observed recently are:

- Surface alloying is found also for metals immiscible in bulk form (i.e. Co on Cu(001)). [8.1, 8.2]
- Burrowing of Co clusters into Au, Cu and Ag surfaces has been observed. [8.3, 8.4]
- It was observed, that the motion of adatoms on top of islands is not the same as on a flat surface. [8.5]
- Fast island decay in homoepitaxial growth was observed by Giesen *et al.* [8.6–8.9]
- By using STM (scanning tunnelling microscope) adsorbate manipulation techniques, it is possible to construct atomic-scale structures on metal surfaces and to study artificially confined quantum systems. [8.10]

V.S. Stepanyuk and W. Hergert, Magnetism, Structure and Interactions at the Atomic Scale,
Lect. Notes Phys. **642**, 159–176 (2004)
http://www.springerlink.com/

To discuss all the effects from theoretical point of view, to get a deep under-standing of the underlying physics, it is absolutely necessary to investigate the real structure of the system as well as the electronic and magnetic struc-ture of the nanosystems, because these aspects are strongly interconnected on the atomic scale.

Our combination of the Korringa-Kohn-Rostoker (KKR) Green's func-tion(GF) method with a molecular dynamics (MD) scheme allows us to study the effects mentioned above in detail.

We will discuss the methods briefly. The magnetic properties of metallic nanostructures are discussed. We start from an ideal lattice structure and take into account step by step imperfections, mixing and relaxations. The effect of quantum interference and the implications for long-range interactions and self-organization are discussed next. Finally, we introduce the new concept of mesoscopic misfit and discuss the consequences for strain fields, adatom motion and island decay.

8.2 Theoretical Methods

8.2.1 Calculation of Electronic Structure

Our calculations are based on the density functional theory and multiple-scattering approach using the Korringa-Kohn-Rostoker Green's function method for low-dimensional systems [8.11]. The basic idea of the method is a hierarchical scheme for the construction of the Green's function of adatoms on a metal surface by means of successive applications of Dyson's equation. We treat the surface as an infinite two-dimensional perturbation of the bulk.

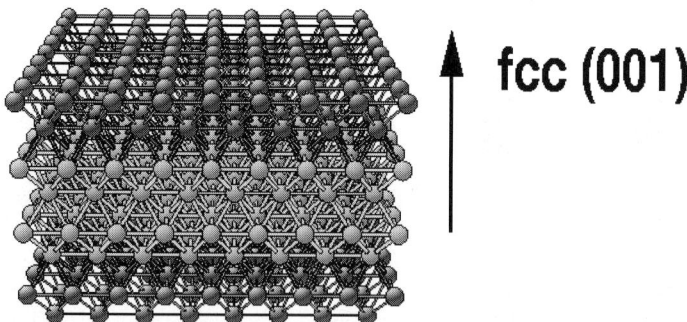

Fig. 8.1. Structure to calculate the surface Green's function for the (001) surface of the fcc-structure (blue -decoupled half-crystals, brown - vacuum layers).

For the construction of the ideal surface the nuclear charges of several monolayers are removed, thus creating two half crystals being practically uncoupled. Taking into account the 2D periodicity of the ideal surface, we calculate the structural Green's function by solving a Dyson equation self-consistently:

$$G_{LL'}^{jj'}(\mathbf{k}_\parallel, E) = \mathring{G}_{LL'}^{jj'}(\mathbf{k}_\parallel, E) + \sum_{j''L''} \mathring{G}_{LL''}^{jj''}(\mathbf{k}_\parallel, E)\Delta t_{L''}^{j''}(E)G_{L''L'}^{j''j'}(\mathbf{k}_\parallel, E) \tag{8.1}$$

Here $\mathring{G}$ is the structural Green's function of the bulk in a $\mathbf{k}_\parallel$-layer representation $(j, j'$ - layer indices). The $\mathbf{k}_\parallel$ wave vector belongs to the 2D Brillouin zone. $\Delta t_L^j(E)$ is the perturbation of the t matrix to angular momentum $L = (l, m)$ in the j-th layer.

The consideration of adsorbate atoms on the surface destroys the translation symmetry. Therefore the Green's function of the adsorbate adatom on the surface has to be calculated in a real space formulation. The structural Green's function of the ideal surface in real space representation is then used as the reference Green's function for the calculation of the adatom-surface system from an algebraic Dyson equation:

$$G_{LL'}^{nn'}(E) = \mathring{G}_{LL'}^{nn'}(E) + \sum_{n''L''} \mathring{G}_{LL''}^{nn''}(E)\Delta t_{L''}^{n''}(E)G_{L''L'}^{n''n'}(E), \tag{8.2}$$

where $G_{LL'}^{nn'}(E)$ is the energy-dependent structural Green's function matrix and $\mathring{G}_{LL''}^{nn''}(E)$ the corresponding matrix for the ideal surface, serving as a reference system. $\Delta t_L^n(E)$ describes the difference in the scattering properties at site n induced by the existence of the adsorbate atom.

Exchange and correlation effects are included in the local density approximation. The full charge density and the full potential approximation can be used in the calculations. Details of the method and several of its applications can be found elsewhere [8.11].

8.2.2 Molecular Dynamics Simulations

In the last years we developed a method which connects the ab initio electronic structure calculations with large scale molecular dynamics simulations. Our approach is based on fitting of the interaction parameters of potentials for molecular dynamic simulations to accurate first-principle calculations of selected cluster-substrate properties, bulk properties and forces acting on adatoms of the system under investigation. [8.12]

To describe metallic clusters on noble metal substrates, many body potentials in the second moment tight-binding approximation are used. [8.13,8.14]

The cohesive energy E_{coh} is the sum of the band energy E_{B} and the repulsive part E_{R}

$$E_{\text{coh}} = \sum_i \left(E_B^i + E_R^i \right) \tag{8.3}$$

$$E_B^i = -\left(\sum_j \xi_{\alpha\beta}^2 \exp(-2q_{\alpha\beta}(r_{ij}/r_0^{\alpha\beta} - 1)) \right)^{1/2} \tag{8.4}$$

$$E_R^i = \sum_j \left(A_{\alpha\beta}^1(r_{ij}/r_0^{\alpha\beta} - 1)) + A_{\alpha\beta}^0 \right) \exp(-p_{\alpha\beta}(r_{ij}/r_0^{\alpha\beta} - 1)) \tag{8.5}$$

where r_{ij} represents the distance between the atoms i and j, and $r_0^{\alpha\beta}$ is the first-neighbour distance in the α, β lattice structure, while it is just an adjustable parameter in the case of the cross interaction. ξ is an effective hopping integral and depends on the material and $q_{\alpha\beta}$ and $p_{\alpha\beta}$ describe the dependence of the interaction strength on the relative interatomic distance.

Table 8.1. Data used for the fitting of the potential together with the values calculated with the optimized potential. (cohesive energy E_c, bulk modulus B, elastic constants C_{ij} from Cleri *et al.* [8.13], first and second neighbour interaction energies $E_{1,b}^{\text{Co-Co}}$, $E_{2,b}^{\text{Co-Co}}$ from Hoshino *et al.* [8.15] solution energy $E_S^{\text{Co in Cu}}$ from Drittler *et al.* [8.16] and binding energies of small Co clusters $E_{1,\text{on Cu(001)}}^{\text{Co-Co}}$, $E_{1,\text{in Cu}}^{\text{Co-Co}}$, $E_{\text{on Cu(100)}}^{\text{trimer}}$, $E_{\text{on Cu(100)}}^{2\times2\text{island}}$ are calculated using the KKR Green's function method.

	quantity	data	fitted value
Cu	a_{Cu}	3.615 Å	3.614 Å
(fcc)	E_c	3.544 eV	3.545 eV
	B	1.42 Mbar	1.42 Mbar
	C_{11}	1.76 Mbar	1.76 Mbar
	C_{12}	1.25 Mbar	1.25 Mbar
	C_{44}	0.82 Mbar	0.82 Mbar
Co	a_{Co}	2.507 Å	2.515 Å
	E_c	4.386 eV	4.395 eV
	B	1.948 Mbar	1.989 Mbar
	C_{11}	3.195 Mbar	3.337 Mbar
	C_{12}	1.661 Mbar	1.426 Mbar
	C_{13}	1.021 Mbar	1.178 Mbar
	C_{33}	3.736 Mbar	3.665 Mbar
	C_{44}	0.824 Mbar	0.646 Mbar
Co-Cu	$E_S^{\text{Co in Cu}}$	0.4 eV	0.38 eV
	$E_{1,b}^{\text{Co-Co}}$	-0.12 eV	-0.18 eV
	$E_{2,b}^{\text{Co-Co}}$	0.03 eV	-0.05 eV
	$E_{1,\text{on Cu(001)}}^{\text{Co-Co}}$	-1.04 eV	-1.04 eV
	$E_{1,\text{in Cu}}^{\text{Co-Co}}$	-0.26 eV	-0.35 eV
	$E_{\text{on Cu(100)}}^{\text{trimer}}$	-2.06 eV	-1.96 eV
	$E_{\text{on Cu(100)}}^{2\times2\text{ island}}$	-3.84 eV	-3.86 eV

We will explain the method for the system Co/Cu(001). Co and Cu are not miscible in bulk form. Therefore the determination of the cross interaction is a problem. A careful fitting to accurate first-principles calculations of selected cluster substrate properties solves the problem. The result is a manageable and inexpensive scheme able to account for structural relaxation and including implicitly magnetic effects, crucial for a realistic determination of interatomic interactions in systems having a magnetic nature. After determination of the Cu-Cu parameters, which are fitted to experimental data only [8.14], the Co-Co and Co-Cu parameters are optimized simultaneously by including in the fit the results of first-principles KKR calculations. To this purpose, we have taken the solution energy of a single Co impurity in bulk Cu $E_S^{\text{Co in Cu}}$ [8.16], energies of interaction of two Co impurities in Cu bulk [8.15] $E_{1,b}^{\text{Co-Co}}$, $E_{2,b}^{\text{Co-Co}}$ and binding energies of small supported Co clusters on Cu(001) - $E_{1,\text{on Cu(001)}}^{\text{Co-Co}}$, $E_{1,\text{in Cu}}^{\text{Co-Co}}$ (terrace position), $E_{\text{on Cu(100)}}^{\text{trimer}}$, $E_{\text{on Cu(100)}}^{2\times2 \text{ island}}$. The set of data used to define the potential and the corresponding values calculated by means of the optimized potential are given in Table 8.1. The bulk and surface properties are well reproduced.

The method, discussed so far has been further improved. We are able to calculate forces on atoms above the surface on the ab initio level. The forces are also included in the fitting procedure. This gives a further improvement of the potentials used in the MD simulations. It should be mentioned that our method allows also to use only ab initio bulk properties from KKR calculations. Therefore, we can construct ab initio based many-body potentials.

8.3 Magnetic Properties of Nanostructures on Metallic Surfaces

Using the KKR Green's function method we have studied the properties of 3d, 4d and 5d adatoms on Ag(001), Pd(001) and Pt(001) systematically. [8.17, 8.18] One central point of investigation was the study of imperfect nanostructures. We have investigated the influence of Ag impurities on the magnetism on small Rh and Ru clusters on the Ag(001) surface. [8.19] The change of the magnetic moments could be explained in the framework of a tight-binding model. Nevertheless it was observed that the magnetism of Rh nanostructures shows some unusual effects. [8.20] An anomalous increase in the magnetic moments of Rh adatoms on the Ag(001) surface with decreasing interatomic distance between atoms was observed, whereas for dimers of other transition metals the opposite behaviour is found.

In this chapter we will discuss some selected results for the real, electronic and magnetic structure of metal nanostructures on noble metal surfaces. We will concentrate our discussion on one special system: Co nanostructures on Cu surfaces. Although a special system is investigated general conclusions can be drawn.

8.3.1 Metamagnetic States of 3d Nanostructures on the Cu(001)Surface

The existence of different magnetic states like   high spin ferromagnetic (HSF), low spin ferromagnetic (LSF) and antiferromagnetic (AF) states is well known for bulk systems.

A theoretical investigation of Zhou *et al.* [8.21] shows that up to five different magnetic states are found for γ Fe. (LSF, HSF, AF, and two ferrimagnetic states). Different theoretical investigations have shown, that energy differences between the magnetic states can be of the order of 1 meV. In such a case magnetic fluctuations can be excited by temperature changes or external fields. Magneto-volume effects play also an important role in the theory of the Invar effect. [8.22]

Lee and Callaway [8.23] have studied the electronic and magnetic properties of free V and Cr clusters. They found that for some atomic spacings as many as four or five magnetic states exists for a V_9 or Cr_9 cluster. The typical low and high spin moments are 0.33 μ_B and 2.78 μ_B for the V_9 cluster.

We have calculated the electronic and magnetic properties of small $3d$ transition metal clusters on the Cu(001) surfaces. Dimers, trimers and tetramers, as given in Fig. 8.2, are investigated. All atoms occupy ideal lattice sites. No relaxation at the surface is taken into account. [8.24, 8.25]

While larger clusters might show a non-collinear structure of the magnetic moments, such a situation is not likely for the clusters studied here. Dimers and tetramers have only one non-equivalent site in the paramagnetic state.

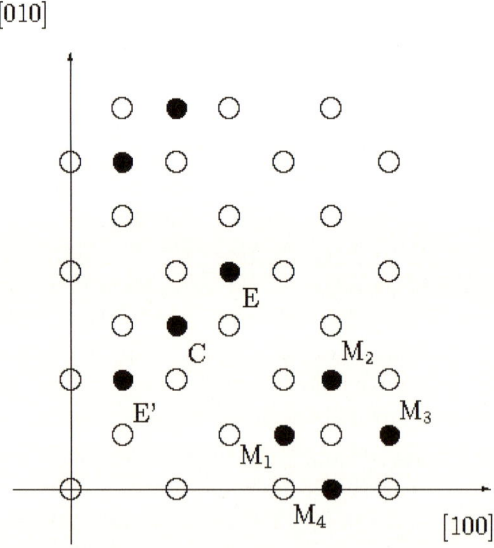

Fig. 8.2. Metallic nanostructures (Dimer, trimer and tetramer) on the fcc(001) surface.

The trimers have two non-equivalent sites (C - center, E, E' - edge positions). Ferromagnetic states of the trimers, either low spin (LSF) or high spin (HSF) states, have parallel moments at the sites C and E, E', but the moments have different sign at C and E, E' in the antiferromagnetic state (AF). The atoms at the edge positions (E, E') have the same moment ($M_E = M_{E'}$) for LSF, HSF and AF states. Another possible magnetic state, which is compatible with the chemical symmetry of the system is an antisymmetric (AS) one. The magnetic moment at the central atom of the trimer is zero and the moments at the edge positions have different sign ($M_E = -M_{E'}$).

We concentrate our discussion on the multiplicity of magnetic states to V and Mn. For the single V adatom only a high spin state with a moment of 3.0 μ_B is obtained. For the V_2 dimer we find both a ferromagnetic and an antiferromagnetic state with moments of 2.85 and 2.58 μ_B respectively. The antiferromagnetic state has the lowest energy being about 0.2 eV/atom lower than the ferromagnetic one.

The magnetic moments for all the different magnetic states of the V and Mn trimers are summarized in Table 8.2. All the magnetic states have a lower total energy than the paramagnetic state. The AF state is the ground state of the V trimer. The energy difference between the AF and LSF state in V is about 8 meV/atom. The LSF state is more stable than the HSF state. The ground state of the Mn trimer is also the antiferromagnetic state. The energy difference between the ground state and the HSF state is only 2 meV/atom. This energy difference corresponds to a temperature difference of 25 K. A transition between the two states caused by temperature changes or an external field leads to a change of the total moment of the Mn trimer of 7.8 μ_B. Such a strong change of the total moment, controlled by an external parameter opens a new field for an experimental proof of the theoretical results.

Table 8.2. Magnetic moments (in μ_B) for the atoms of the trimers V_3 and Mn_3 on Cu(001).

state	V_3			Mn_3		
	M_E	M_C	$M_{E'}$	M_E	M_C	$M_{E'}$
HSF	2.85	2.58	2.85	4.03	3.83	4.03
LSF	2.63	1.41	2.63	4.04	0.01	4.04
AF	2.63	-2.02	2.63	3.99	-3.88	3.99
AS	2.62	0.00	-2.62	3.98	0.00	-3.98

We have shown, that metamagnetic behaviour exists in supported clusters. It is shown, that the energy differences between different magnetic states can be small, which can lead to a change of the magnetic state of the cluster by an external parameter. The energy differences between different magnetic states will strongly depend on the cluster size. Therefore such *ab inito* calculations can help to select interesting systems for experimental investigations.

8.3.2 Mixed Co-Cu Clusters on Cu(001)

The magnetic properties of Co nanostructures on Cu substrate can be strongly influenced by Cu atoms. For example, Cu coverages as small as three hundredths of a monolayer drastically affect the magnetization of Co films. [8.26] Experiments and theoretical studies demonstrated that magnetization of mixed clusters of Co and Cu depends on the relative concentration of Co and Cu in a nonobvious way. Quenching of ferromagnetism in Co clusters embedded in copper was reported. [8.27] Calculations by means of our MD method showed that surface alloying is energetically favourable in the case of Co/Cu(001) and mixed Co-Cu clusters are formed in the early stages of heteroepitaxy. Recent experiments [8.2] suggest that mixed Co-Cu clusters indeed exist.

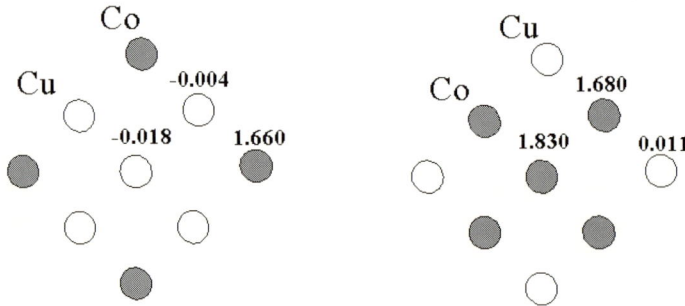

Fig. 8.3. Spin polarization of Co-Cu mixed clusters on Cu(001). Magnetic moments in Bohr magnetons are given for all inequivalent site.

We have studied all possible mixed configurations in 3×3-atoms islands on Cu(001) surfaces. [8.28] We observe a small induced moment at the Cu atoms in the island and a decrease of the moments at the Co atoms in comparison with the 3×3 Co-island. A stronger reduction of the Co moments is achieved, if the Co_9 cluster is surrounded by a Cu brim and capped by a Cu cluster. A reduction of 14 % is obtained for the average moment of the Co_9 cluster. This effect should have a strong influence on the properties of the Co-Cu interface in the early stages of growth. Coating of Co clusters with Cu atoms has been found recently in experiments. [8.4]

8.3.3 Effect of Atomic Relaxations on Magnetic Properties of Adatoms and Small Clusters

Possible technological applications of supported magnetic clusters are connected with the magnetic anisotropy energy (MAE), which determines the orientation of the magnetization of the cluster with respect to the surface. Large MAE barriers can stabilize the magnetization direction in the cluster and a stable magnetic bit can be made. Ab initio calculations have predicted very large MAE and orbital moments for $3d$, $5d$ adatoms and $3d$ clusters on Ag(001). [8.29–8.31]

The interplay between magnetism and atomic structure is one of the central issues in physics of new magnetic nanostructures. Performing ab initio and tight-binding calculations we demonstrate the effect of atomic relaxations on the magnetic properties of Co adatoms and Co clusters on the Cu(001) surface. [8.32] We address this problem by calculating magnetic properties of the Co adatom and the Co_9 cluster on the Cu(100) surface. First, we calculate the effect of relaxations on the spin moment of the Co atom using the KKR Green's function method. (cf. Fig. 8.4)

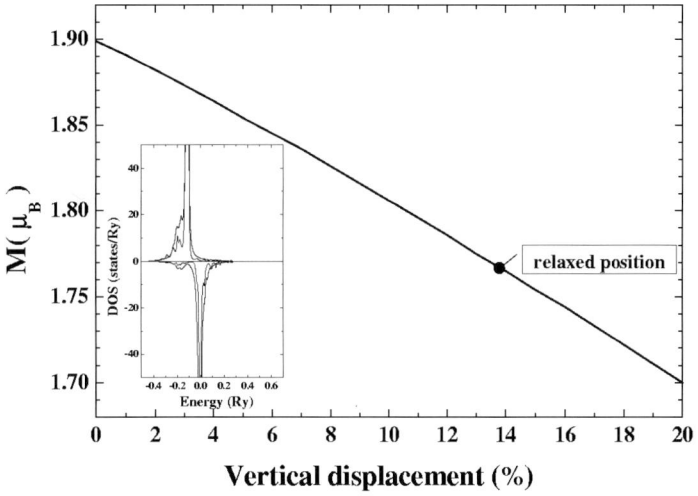

Fig. 8.4. The dependence of the spin magnetic moment of the Co adatom on the distance from the Cu substrate. The relaxed position of the adatom is indicated: 0% correspond to Co at a Cu interlayer separation above the Cu surface. The magnetic moments per atom are given in Bohr magnetons. Inset: ab initio (thick line) and TB (thin line) results for the d component of the local density of states (LDOS) of the Co adatom for unrelaxed position.

We employ the tight binding electronic Hamiltonian with parameters chosen to fit the KKR local densities of electronic states and the local magnetic moments of Co overlayers and small clusters on Cu(100) to calculate the MAE. To evaluate the MAE the intraatomic spin-orbit coupling is presented by the operator $\xi \mathbf{L} \cdot \mathbf{s}$ where ξ is the spin-orbit coupling parameter.

The results for the orbital moments and the MAE of the Co adatoms in the unrelaxed and relaxed positions are presented in Table 8.3. Calculation of

Table 8.3. Magnetic orbital moments and magnetic anisotropy energy of a single Co adatom on the Cu(100) surface: L_Z^m and L_X^m are the orbital moments for magnetization along the normal Z and in-plane X direction; the electronic part of the mangetic anisotropy energy ΔE (meV) is presented.

	Unrelaxed geometry	Relaxed geometry
L_Z^m	1.06	0.77
L_X^m	1.04	0.80
ΔE	1.70	-0.37

the MAE reveal that for the unrelaxed position above the the surface, out-of plane magnetization for the Co adatom is more stable. The relaxation of the vertical position of the adatom by 14% shortens the first nearest neighbour Co-Cu separation from 2.56 to 2.39 Å and has a drastic effect on MAE. We find that the relaxation of the Co adatom leads to in-plane magnetization.

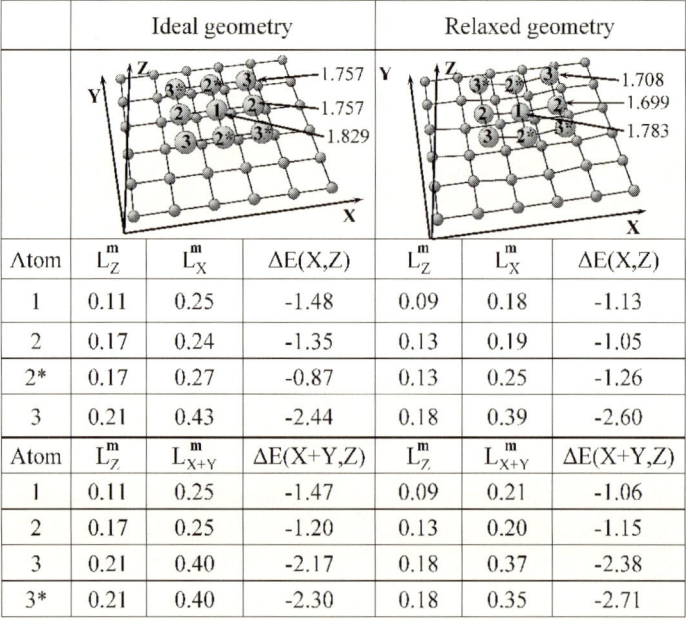

Atom	L_Z^m	L_X^m	$\Delta E(X,Z)$	L_Z^m	L_X^m	$\Delta E(X,Z)$
1	0.11	0.25	-1.48	0.09	0.18	-1.13
2	0.17	0.24	-1.35	0.13	0.19	-1.05
2*	0.17	0.27	-0.87	0.13	0.25	-1.26
3	0.21	0.43	-2.44	0.18	0.39	-2.60
Atom	L_Z^m	L_{X+Y}^m	$\Delta E(X+Y,Z)$	L_Z^m	L_{X+Y}^m	$\Delta E(X+Y,Z)$
1	0.11	0.25	-1.47	0.09	0.21	-1.06
2	0.17	0.25	-1.20	0.13	0.20	-1.15
3	0.21	0.40	-2.17	0.18	0.37	-2.38
3*	0.21	0.40	-2.30	0.18	0.35	-2.71

Fig. 8.5. Magnetic properties of Co$_9$ cluster on the Cu(100) surface in unrelaxed and relaxed geometries; spin magnetic moments in Bohr magnetons are shown for each atom in the cluster; orbital magnetic moments and electronic part of the MAE are presented in the table for the normal Z, in-plane X and $X+Y$ directions of the magnetization. For the unrelaxed cluster the the average MAE is -1.74 meV/Co atom for X-direction; for $X+Y$ direction these values are -1.69 meV/Co atom. For the relaxed cluster, the above energies are -1.79 meV/Co atom for X direction and -1.76 meV/Co atom for $X+Y$ direction.

The results for the Co_9 cluster are presented in Fig. 8.5. We find that the spin moments of atoms of the cluster are close to moments of the Co adatom (see Fig. 8.4) and the Co monolayer (1.7 μ_B). We find that spin and orbital magnetic moments are strongly affected by the relaxation. For example, the orbital moment L_X^m of the central atom (cf. Fig. 8.5) is reduced by 30 %. This effect is caused by the strong reduction of all first NN Co-Co distances to about 2.41 Å (2.56 Å for the unrelaxed structure). The stability of the in-plane magnetization is reduced by relaxations for the central atom and for atom 2, while it is enhanced for the corner atom and for atom 2*. For all the atoms in the cluster the MAE is found to be considerably larger than the MAE of the single Co atom in relaxed geometry.

For both, the relaxed and unrelaxed Co_9 cluster, the magnetization along X is slightly more stable than along $X + Y$. However, MAE from particular atoms have an inhomogeneous distribution and possibility of noncollinear magnetization cannot be ruled out.

8.4 Quantum Interference and Interatomic Interactions

Surface-state electrons on the (111) surfaces of noble metals form a two-dimensional (2D) nearly free electron gas. Such states are confined in a narrow layer at the surface. An electron in such a state *runs* along the surface, much like a 2D plane wave. The quantum interference between the electron wave travelling towards the scattering defect and the backscattered one leads to standing waves in the electronic local density of states (LDOS) around

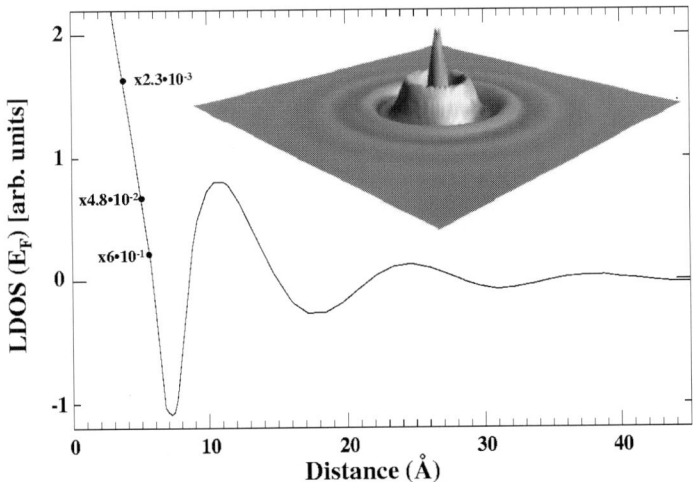

Fig. 8.6. Calculated standing waves in the LDOS around single Co atom on Cu(111).

the defect. [8.33] These standing waves are the energy-resolved Friedel oscillations. The scanning tunnelling microscope (STM) images taken at low bias directly reflect the oscillations in the LDOS close to E_F.

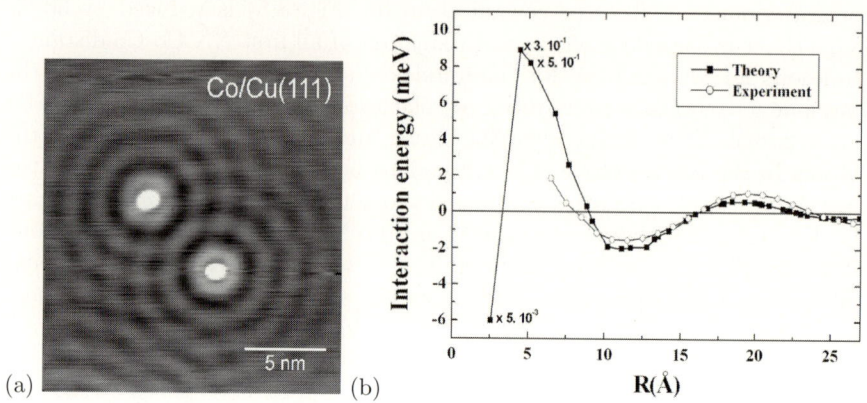

Fig. 8.7. (a) Constant current STM image of two Co adatoms on Cu(111) which interact via the standing waves ($I = 2nA$, $V = -50mV$, $T = 6K$); b) Experimental and calculated interaction energies between two Co adatoms on Cu(111).

We will discuss our recent ab initio studies of long-range adsorbate interactions caused by the quantum interference of surface-state electrons. [8.35]

While we concentrate on a particular system, Co adatoms on Cu(111), our results are of general significance because they show that the quantum interference on metal surfaces can strongly affect the growth process of the transition-metal nanostructures.

Our calculation for the Cu(111) surface gives a surface-state Fermi wavelength $\lambda_F = 29$ Å. The scattering of surface state electrons by Co atoms leads to quantum interference patterns around the adsorbate. Figure 8.6 shows the calculated LDOS at E_F displaying the $\lambda_F/2 \approx 15$ Å period oscillations. The concentric rings surrounding the the Co adatom (cf. Fig. 8.6, inset) are standing waves due to the quantum interference.

Now we turn to the discussion of the long-range interactions between Co adatoms on Cu(111) An example of a STM image of two Co adatoms at a distance of about 60 Å from each other is shown in Fig. 8.7(a). [8.36] One can see that the atoms share LDOS oscillations with each other. Thus, the adsorbates should interact via Friedel oscillations. The experimental results and calculations (cf. [8.35]) for the interaction energies are presented in Fig. 8.7(b) and they show that the interaction energy is oscillatory with a period of about 15 Å. The ab initio results are in good quantitative agreement with experiment.

The long-range interactions caused by the quantum interference provide a mechanism which leads to self-assembly of one-dimensional structures on Cu(111). (cf. [8.35])

8.5 Strain and Stress on the Mesoscale

8.5.1 The Concept of Mesoscopic Misfit

If some material is grown on a substrate with a different bond length the lattice mismatch at the interface leads to strain fields. Strain can be relieved through the introduction of defects in the atomic structure, such as dislocations, or by an atomic rearrangement. Usually strain relaxations are predicted on the basis of the macroscopic lattice mismatch between the two materials. However, if the deposited system is of mesoscopic size of several 100 atoms, its intrinsic bond lengths are different from the bond length in the bulk materials. For the Co/Cu(001) interface the *macroscopic* mismatch m_0 between Co and Cu defined as $m_0 = (a_{Cu} - a_{Co})/a_{Cu}$ (a_{Cu} and a_{Co} are the lattice constants) is only $\approx 2\%$. Several recent experiments have suggested that strain relaxations for submonolayer coverage [8.37] or even for a few monolayers [8.38] cannot be explained by the macroscopic misfit between bulk materials.

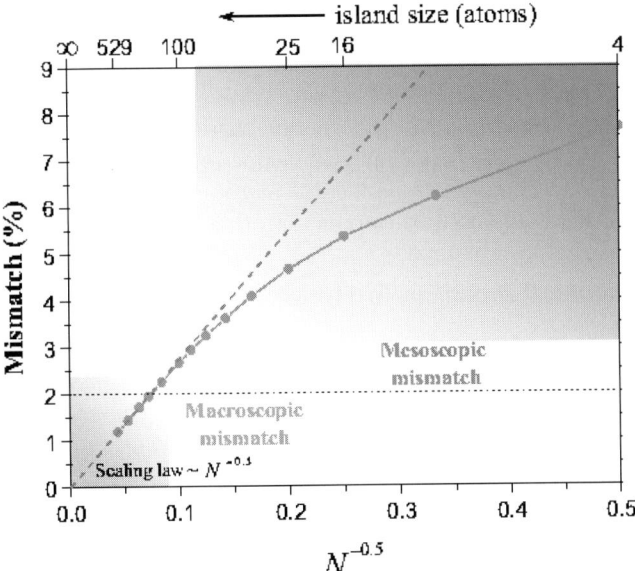

Fig. 8.8. Size-dependent mismatch $m = (r_b^{Cu} - r^{Co})/r_b^{Cu}$ for the Co square islands on Cu(001) (r_b^{Cu} - first bond length for Cu bulk; r^{Co} - average bond length in Co islands.

In order to get a deeper insight into the local strain relaxations on an atomic scale, the equilibrium geometries of plane square Co islands of different sizes (up to 600 atoms) on Cu(001) are calculated by computing the forces at each atomic site and relaxing the geometry of islands and the substrate atoms using our many-body potentials fitted to ab initio results. [8.39]

In Fig. 8.8 we show the change in the mismatch with the size of Co islands. It is seen, that the *mesoscopic* mismatch between small Co islands and the substrate is considerably larger than the mismatch calculated from the lattice constants of the two materials. Only for Co islands incorporating more than 200 atoms the local strain can be described by the macroscopic mismatch. We found that both the mesoscopic and macroscopic mismatch depend on the size of the islands and for islands larger than 60 atoms mismatch scales like $N^{-0.5}$ (N - number of atoms in islands). Such scaling behaviour is determined by the relaxations of the edge atoms of the islands whose number changes as $\sqrt{N}$. One very fundamental issue predicted by these results is the possible strong impact of the size dependent mismatch on the local strain field. The substrate can dynamically respond to the growth of islands and can exhibit a strong inhomogeneous strain distribution during the growth process.

8.5.2 Strain and Adatom Motion on Mesoscopic Islands

The mesospic mismatch and mesoscopic strain depend strongly on the size of the clusters. Therefore we expect that also the barriers for hopping diffusion depend on the size of the clusters. The calculations show, that the barriers for hopping on the small Co islands (16-50 atoms) are found to be about $\approx 20\%$ lower than those on the large islands (100-500 atoms) (cf. Fig. 8.9). The diffusivity D is related to the hopping rate of single adatoms by $D = D_0 \cdot exp(-E_d/kT)$, where E_d is the energy barrier for hopping, D_0 is the prefactor. We found that D_0 is nearly the same for all islands, therefore the diffusion coefficient D on small Co islands at room temperature is found to be about two orders of magnitude larger than that on large Co islands. [8.40]

8.5.3 Mesoscopic Relaxation in Homoepitaxial Growth

Up to now we have discussed strain effects in the heterogeneous system Co on Cu(100). The investigations suggest, that such strain effects should play also an important role in homoepitaxy. But only recently strain effects in ho-moepitaxy have been discussed in the framework of the concept of mesoscopic mismatch. [8.41, 8.42]

Motivated by the experiments of Giesen *et al.* [8.6–8.9] we concentrate on double layer Cu islands on Cu(111). We reveal that islands and substrate atoms exhibit unexpected strong relaxations.

Now we turn to the discussion of the effect of mesoscopic mismatch in double layers of Cu islands on Cu(111). In this case, the scenario of meso-scopic relaxations is more complicated compared to the flat substrate. Both

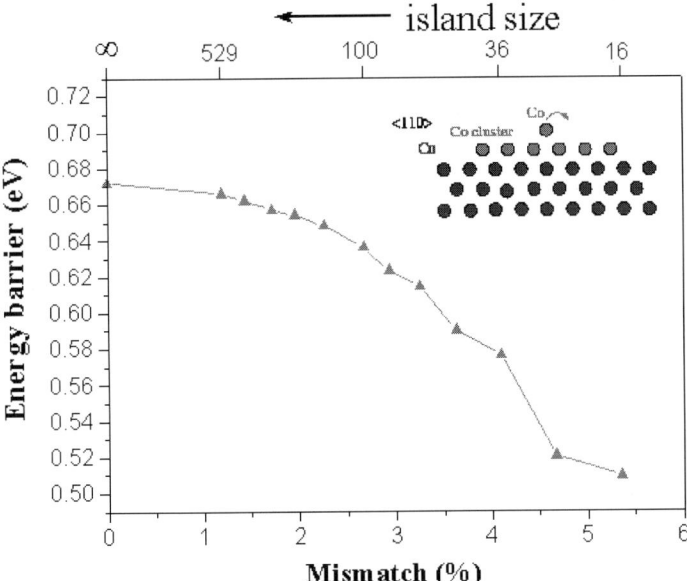

Fig. 8.9. Strain dependence of energy barrier for hopping diffusion on top of Co square islands. Since the strain depends on island size, the activation barrier for diffusion depends also on island size (see upper horizontal scale).

the upper and the lower islands exhibit strain relaxations. There are two kinds of step edges: (100) microfaceted step A and (111) microfaceted step B. Due to the relaxation of the edge atoms, the average bond lengths near the island edges at both A and B steps are reduced compared to the center. Therefore, we expect that the mesoscopic mismatch between the upper island and the lower island depends on the distance between the edges and may be different for the step A and the step B. For example, our calculations for a Cu dimer for different positions on the Cu_{271} island reveal that when the dimer approaches the edge of the island, mismatch between the dimer and the island changes abruptly and differently for A and B steps. These results suggest that the shape of double layer islands and atomic relaxations in islands and the substrate underneath may depend on the distance between the edges of islands. To prove this, we perform calculations for the double layer Cu island when a close contact between the edges occurs. Results shown in Fig. 8.10 reveal that the atoms at the edge of the lower island and the substrate underneath are pushed up, while atoms of the upper island and the substrate under the large island are pushed down. The strain relief at the edge of islands and in the substrate leads to the shape variation in islands as they approach the edge. We believe that a strongly inhomogeneous displacement pattern in the islands and in the substrate can affect the interlayer mass transport at the edge.

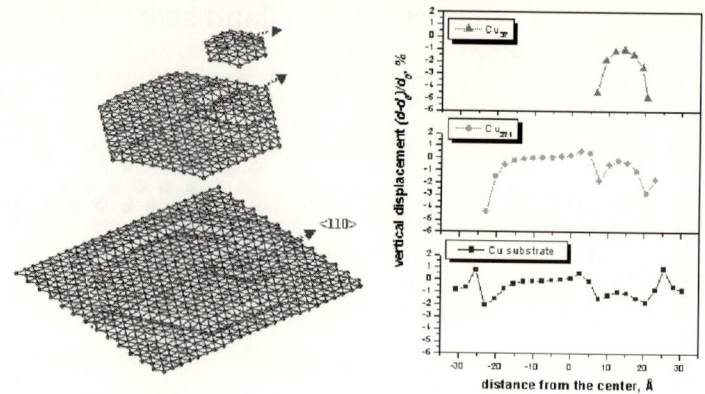

Fig. 8.10. The shape of the double layer Cu island and the substrate for a close contact between the island edges. The vertical displacement of Cu atoms in the upper and the lower islands, and in the substrate are shown for < 110 > direction.

Acknowledgements

We are grateful to J. Kirschner, Š. Pick, A.N. Baranov, P. Bruno, and K. Kern for their contributions to our investigations, helpful suggestions and stimulating discussions.

References

[8.1] J. Fassbender, R. Allenspach, and U. Düring, Surf. Sci **383**, L742 (1997).
[8.2] F. Nouvertne, G. Günterrodt, R. Pentcheva, M. Scheffler, Phys. Rev. B **60**, 14382 (1999).
[8.3] S. Padovani, F. Scheurer, and J.P. Bucher, Europhys. Lett. **45**, 327 (1999).
[8.4] C.G. Zimmermann, M. Yeadon, K. Nordlund, J.M. Gibson and R.S. Averback, Phys. Rev. Lett. **83**, 1163 (1999).
[8.5] S.C. Wang and G. Ehrlich, Phys. Rev. Lett. **67**, 2509 (1991).
[8.6] M. Giesen, G. Schulze Icking-Konert, and H. Ibach, Phys. Rev. Lett. **80**, 552 (1998).
[8.7] M. Giesen, G. Schulze Icking-Konert, and H. Ibach, Phys. Rev. Lett. **82**, 3101 (1999).
[8.8] M. Giesen, and H. Ibach, Surf. Sci. Lett. **464**, L697 (2000).
[8.9] M. Giesen, Prog. Surf. Sci.**68**, 1 (2001).
[8.10] M.F. Crommie, J. Electron Spectrosc. Relat. Phenom. **109**, 1 (2000); S. Crampin and O.R. Bryant, Phys. Rev. B **54**, R17367 (1996).
[8.11] V.S. Stepanyuk, W. Hergert, R. Zeller, and P.H. Dederichs, Phys. Rev. B **53**, 2121 (1996); V.S. Stepanyuk, W. Hergert, P. Rennert, K. Wildberger, R. Zeller and P.H. Dederichs, Phys. Rev. B **59**, 1681 (1999); K. Wildberger, V.S. Stepanyuk, P. Lang, R. Zeller, and P.H. Dederichs, Phys. Rev. Lett. **75**, 509 (1995); K. Wildberger, P.H. Dederichs, P. Lang, V.S. Stepanyuk, and R. Zeller, Research Report,Berichte des Forschungszentrum Jülich, No **3022**, ISSN 0944-2952 (1995).

[8.12] N.A. Levanov, V.S. Stepanyuk, W. Hergert, D.I. Bazhanov, P.H. Dederichs, A. Katsnelson, and C. Massobrio Phys. Rev. B **61**, 2230 (2000).

[8.13] V. Rosato, B. Guillope, and B. Legrand, Philos. Mag. A **59**, 321 (1989).

[8.14] F. Cleri and V. Rosato, Phys. Rev. B **48**, 22 (1993).

[8.15] T. Hoshino, W. Schweika, R. Zeller, and P.H. Dederichs, Phys. Rev. B **47**, 5106 (1993).

[8.16] B. Drittler, M. Weinert, R. Zeller, and P.H. Dederichs, Phys. Rev. B **39**, 930(1989).

[8.17] V.S. Stepanyuk, W. Hergert, P. Rennert, K. Wildberger, R. Zeller, and P.H. Dederichs, Phys. Rev. B **54**, 14121 (1996).

[8.18] V.S. Stepanyuk, W. Hergert, K. Wildberger, R. Zeller, and P.H. Dederichs, Phys. Rev. B **53**, 2121 (1996).

[8.19] V.S. Stepanyuk, W. Hergert, P. Rennert, K. Wildberger, R. Zeller, and P.H. Dederichs, Phys. Rev. B **59**, 1681 (1996).

[8.20] V.S. Stepanyuk, W. Hergert, P. Rennert, J. Izquierdo, A. Vega, and L.C. Balbás, Phys. Rev. B **57**, R14020 (1998).

[8.21] Yu-mei Zhou, Nuen-qing Zhang, Lie-ping Zhong, and Ding-sheng Wang, J. Magn. Magn. Mater. **145**, L237 (1995).

[8.22] E.F. Wassermann, Journal of Mag. Mag. Mat. **100**, 346 (1991).

[8.23] Keeyung Lee and J. Callaway, Phys. Rev. B **48**, 15358 (1993); Keeyung Lee and J. Callaway, Phys. Rev. B **49**, 13906 (1993).

[8.24] V.S. Stepanyuk, W. Hergert, P. Rennert, K. Wildberger, R. Zeller, P.H. Dederichs, Solid State Commun. **101**, 559 (1997).

[8.25] V.S. Stepanyuk, W. Hergert, K. Wildberger, S. Najak, P. Jena, Surf. Sci. Lett. **384**, L892 (1997).

[8.26] W. Weber, C.H. Back, A. Bischof, D. Pescia, and R. Allenspach, Nature (London) **374**,788 (1995).

[8.27] D.A. Eastham, Y. Qiang, T.H. Maddock, J. Kraft, J.-P. Schille, G.S. Thompson, H. Haberland, J. Phys.: Condensed Matter **9**, L497 (1997).

[8.28] V.S. Stepanyuk, A.N. Baranov, D.I. Bazhanov, W. Hergert, A.A. Katsnelson, Surface Sci. **482-485**, 1045 (2001).

[8.29] P. Gambardella,, S. Rusponi, M. Veronese, S.S. Dhesi, C. Grazioli, A. Dallmeyer, I. Cabria, R. Zeller, P.H. Dederichs, K. Kern,C. Carbone, and H. Brune, Science **300**, 1130 (2003).

[8.30] B. Lazarovits, L. Szunyough, and P. wWeinberger, Phys. Rev. B **65**, 10444 (2002).

[8.31] B. Nonas, I. Cabria, R. Zeller, P.H. Dederichs, T. Huhne, and H. Ebert, Phys. Rev. Lett. **86**, 2146 (2001).

[8.32] Š. Pick, V.S. Stepanyuk, A.N. Baranov, W. Hergert, and P. Bruno, Phys. Rev. B **68**, 104410 (2003).

[8.33] S. Crampin, J. Elect. Spectr. Rel. Phenom. **109**, 51 (2000).

[8.34] P. Hyldgaard, and M. Persson, J. Phys: Cond. Matter. **12**, L13 (2000).

[8.35] V.S. Stepanyuk, A.N.Baranov, D.V. Tsivlin, W. Hergert, P. Bruno, N. Knorr, M.A. Schneider, and K. Kern, Phys. Rev. B **68**, xxxx (2003).

[8.36] N. Knorr, H. Brune, M. Epple, A. Hirstein, M.A. Schneider, and K. Kern, Phys. Rev. B **65**, 115420 (2002).

[8.37] D. Sander, R. Skomski, C. Schmidthals, A. Enders, and J. Kirschner, Phys. Rev. Lett. **77**, 2566 1996.

[8.38] A. Grossmann, W. Erley, J.B. Hannon, and H. Ibach, Phys. Rev. Lett. **77**, 127 1996.

[8.39] V.S. Stepanyuk, D.I. Bazhanov, A.N.N. Baranov, W. Hergert, P.H. Dederichs, J. Kirschner, Phys. Rev. B62, 15398 (2000).

[8.40] V.S. Stepanyuk, D.I. Bazhanov, W. Hergert, J. Kirschner, Phys. Rev. B 63, 153406 (2001).

[8.41] O.V. Lysenko, V.S. Stepanyuk, W. Hergert, and J. Kirschner Phys. Rev. Lett. 89, 126102 (2002).

[8.42] O.V. Lysenko, V.S. Stepanyuk, W. Hergert, and J. Kirschner Phys. Rev. B 68, 033409 (2003).

9 Molecular Dynamics Simulations in Biology, Chemistry and Physics

P. Entel[1], W.A. Adeagbo[1], M. Sugihara[1], G. Rollmann[1], A.T. Zayak[1], M. Kreth[1], and K. Kadau[2]

[1] Institute of Physics, University of Duisburg–Essen, 47048 Duisburg, Germany
[2] Los Alamos National Laboratory, T–11, MS B262, Los Alamos, NM 87545, USA

Abstract. We review recent progress in understanding fundamental processes in biology, chemistry and physics on the basis of ab initio and molecular dynamics simulations. The first step of the visual process involving the excitation of bovine rhodopsin after absorption of light is taken as an example from biochemistry to demonstrate what is nowadays possible to simulate numerically. The act of freezing of water has recently been simulated, for the first time successfully, by scientists from chemistry. Martensitic transformation in bulk and nanophase materials, a typical and hitherto not completely solved problem from solid state physics, is used to illustrate the achievements of multimillion atoms simulations.

9.1 Molecular Dynamics as a Multidisciplinary Numerical Tool

Molecular dynamics (MD) has proved to be an optimum numerical recipe applicable to problems with many degrees of freedom from quite different fields of science. The knowledge of the energy or potential landscape of interacting particles, like electrons and atoms, enables one to calculate the forces acting on the particles and to study the evolution of the system with time. As long as classical mechanics is appropriate to describe the dynamics of the individual constituents (i.e. atoms or molecules), the Newtonian equations of motion can be related to the statistical mechanics of the (classical) particles by using the equipartition theorem, i.e. by combining the equations

$$m_i \ddot{\mathbf{r}}_i = \mathbf{F}_i, \tag{9.1}$$

$$0 = \tfrac{3}{2} N k_B T - \left\langle \sum_i \tfrac{1}{2} m_i \dot{\mathbf{r}}_i^2 \right\rangle \tag{9.2}$$

for $i = 1 \ldots N$ particles. Although the equipartition theorem holds only for classical particles, to the authors' knowledge the combination of classical and statistical physics has also been used to simulate small molecules at low temperatures without critically discussing so far the limitations of 9.1 and 9.2 when applying to very small quantum mechanical systems. The forces in 9.1 are then simply related to the gradients of the potential energy surface (PES) of either the classical or the quantum mechanical N-particle system. For pedagogical reasons let us recall the classical constant-temperature case. Omitting the statistical average in 9.2 means that the temperature is a measure

P. Entel, W.A. Adeagbo, M. Sugihara, G. Rollmann, A.T. Zayak, M. Kreth, and K. Kadau,
Molecular Dynamics Simulations, Lect. Notes Phys. **642**, 177–206 (2004)
http://www.springerlink.com/

of the instantaneous kinetic energy in the system (isokinetic simulation) and the equations of motions to be solved are [9.1]

$$m_i\ddot{\mathbf{r}}_i = \mathbf{F}_i - \left(\frac{\sum_j \mathbf{F}_j\dot{\mathbf{r}}_j}{\sum_j m_j\dot{\mathbf{r}}_j^2}\right)m_i\dot{\mathbf{r}}_i, \tag{9.3}$$

which are derived from 9.1 and 9.2 by applying the Gaussian principle of least mean square action, hence the name "Gaussian thermostat". It is to note that the friction-like term in 9.3 is deterministic and gives rise to time-reversal invariant trajectories.

Using a different approach Nosé reproduced the canonical phase-space distribution, so that the kinetic energy can fluctuate with a distribution proportional to

$$\exp\left(-\frac{1}{\beta}\sum_i \frac{\mathbf{p}_i^2}{2m_i}\right). \tag{9.4}$$

The canonical distribution is generated with deterministic and time-reversal invariant trajectories of the Hamiltonian

$$H = \sum_i \frac{\mathbf{p}_i^2}{2m_is^2} + (\tilde{N}+1)k_BT\ln s + \frac{p_s^2}{2Q} + V(q) \tag{9.5}$$

with the time-scale variable s, its conjugate momentum p_s and the "heat bath mass" Q; $\tilde{N}$ is the number of degrees of freedom. Nosé proved that the microcanonical distribution in the augmented set of variables is equivalent to a canonical distribution of the variables q, p', where $p' = p/s$ are the scaled momenta [9.2]. Although a single harmonic oscillator is not sufficiently chaotic to reproduce the canonical distribution, it is interesting to see the result if the method is applied to it. In this case the result depends on Q: For larger values of Q the trajectories fill in a region between two limiting curves [9.3]. For an N-particle system the results (expectation values, i.e. time averages, of observables) are independent of Q and hence of H, the corresponding method of controlling the temperature during the simulations is called "Nosé-Hoover thermostat". The extension of the method with respect to isobaric (constant pressure) simulations by making use of the virial theorem [9.2] allows to deal with many experimental set-ups, where the experiments can be simulated by using the so-called (NPT) ensemble. Finally for the description of structural changes (for example, of condensed matter materials) fluctuations of the size and geometry of the simulation box can be taken into account by using the Parinello-Rahman scheme [9.4]. For further details like, for example, algorithms to solve the equations of motion or a discussion of the statistical nature of the trajectories, we refer to the literature [9.5].

We close the introductory remarks by mentioning some new trends in connection with molecular dynamics. One advantage of molecular dynamics is that the simulations involve the physical time in contrast to other techniques like Monte Carlo. This, however, limits the time that can be spanned to the pico- or nanosecond range if an integration step of the order of one femtosecond is used. This means that many interesting events like conformational changes of large biomolecules, protein folding, solid state reactions and transformations cannot be followed directly. One possibility to circumvent this difficulty is to rely on hybrid methods like a combination of molecular dynamics and kinetic Monte Carlo, for example, in the simulation of growth processes in solid state physics [9.6]. Another way is to freeze in the rapid oscillations, for example, of light atoms like hydrogen etc. and to retain only the dynamics involved with the heavier constituents. Finally there are new developments under way in many places to find optimum tools for the simulations of rare events which likewise can be done by a subtle modification of the potential function [9.7]. For a separation of different time scales in view of biological systems see also [9.8, 9.9].

Apart from the time problem, the extension to larger length scales is another important challenge which has recently been addressed by coupling the region of atomistic simulations to some effective elastic medium [9.10]. Of particular interest is the problem to extend the size of biological systems playing a role in signal transduction. The present state-of-the-art with respect to such simulations is to describe important reactive parts of the molecules quantum mechanically while the rest is treated by molecular mechanics with empirical potentials. This so-called QM/MM technique, which is based on partitioning the Hamiltonian into

$$H = H_{\mathrm{QM}} + H_{\mathrm{MM}} + H_{\mathrm{QM-MM}}, \tag{9.6}$$

allows, for example, to deal with the chromophore of rhodopsin on a quantum mechanical basis, whereas the whole sequence of amino acids of the protein as well as part of the lipid membrane into which the latter is embedded, can be described by model potentials (here H_{QM} is the Hamiltonian to be treated by ab initio methods, H_{MM} is the purely classical Hamiltonian to be treated by force field methods and $H_{\mathrm{QM-MM}}$ is the Hamiltonian describing the interaction between the quantum mechanical and classical region). Conceptually this procedure is straightforward, difficulties may arise when trying a rigorous treatment of the coupling of the two regions [9.11]. This mixed QM/MM approach can be extended to describe the dynamics of the first excited singlet state. This was done recently on the basis of using pseudopotentials in the quantum mechanical region and the simulation package AMBER in the MM region [9.12]. We have done alike simulations, in particular for rhodopsin, which will be discussed below. Instead of using pseudopotentials, our method relies on a density-functional based tight-binding approach [9.13], which allows to treat a much larger region quantum mechanically with less computational efforts. In the latter approach the MM region is described by using CHARMM [9.14]. Compared to ab initio quantum chemical approaches, the latter two methods have the advantage that a relatively large QM region can be treated with sufficient accuracy.

The last remark concerns ground state and excited states properties of a quantum mechanical system in view of the normally used Born-Oppenheimer approximation to separate the dynamics of electrons and atoms. The dynamics of the electrons is then determined by the zero-temperature solution of the corresponding Schrödinger equation for the momentary distribution of atoms, while the atoms can be handled at finite temperatures in the molecular dynamics simulations. While this may be a reasonable approximation when simulating ground-state properties of close-packed metallic systems, its accuracy might be questioned when dealing with excited-state properties of biological systems. In additon the crossover from excited electronic states back to the ground sate requires a complicated mixing of basis sets. Both problems are so far not solved unambigiously. For a discussion of the extension of conventional molecular dynamics for simulations not confined to the Born-Oppenheimer surface we refer to [9.15].

9.2 Simulation of Biochemical Systems

In the following we discuss two molecular systems, cyclodextrin interacting with bulk water and the protein rhodopsin with internal water, for which exemplary the molecular simulations help to deepen our understanding of the experimental data being available. We start the discussion with water which is the most abundant substance on earth and the only naturally occuring inorganic liquid. Due to the distinctive properties of H_2O, its interaction with larger molecules, proteins or even the DNA is in many cases decisive for their properties: In the absence of water (and its hydrogen bonding network) proteins lack activity and the DNA loses its double-helical structure. For recent reviews concerning fundamental aspects of water (from bulk water to biological, internal and surface waters, the phase diagram of ice, etc.) we refer to [9.16–9.19]. The different dynamics of biological water compared to bulk water is, for example, determined by the protein-solvent interaction: Typical consequences are multiple-time-scale processes, time-dependent diffusion coefficients and excess low-frequency vibrational modes (glassy behavior of water, see [9.19]). In the context of methods concentrating on ab initio MD of liquids and applied to water in particular see, for example, [9.20]. Much recent work is connected to the simulation of proton transfer and solvation of ions in water or high pressure ice [9.21–9.24]. This is also of interest with respect to the pH value of different waters affecting organisms living in water (pH $= -\log_{10}[H^+]$, where $[H^+]$ is the molar concentration (mol/L); pH ranges from 0 to 14, with 7 being neutral; pHs less than 7 are acidic while pHs greater than 7 are basic; normal rainwater has a pH value of 5.7).

9.2.1 Molecular Dynamics Simulation of Liquid Water

Useful tools to sample the structure and dynamics of water are the radial distribution function (RDF), the time correlation functions and diffusion coefficients. Here we briefly discuss results for the RDFs defined by

$$g_{\alpha\beta}(\mathbf{r}) = \frac{1}{\rho^2}\left\langle \sum_{i \neq j} \delta(\mathbf{r}_{i\alpha})\delta(\mathbf{r}_{j\beta} - \mathbf{r})\right\rangle, \tag{9.7}$$

which is related by Fourier transformation to the (intermolecular) structure factor,

$$S_{\alpha\beta}(\mathbf{k}) = 1 + \rho \int d^3 r\, e^{i\mathbf{k}\mathbf{r}}\left[g_{\alpha\beta}(\mathbf{r}) - 1\right], \tag{9.8}$$

where α, β = O, H; i, j refer to the molecules (note that in the case of g_{OH} and g_{HH} we have, in addition, intramolecular contributions from $i = j$); the X-ray data give the O–O pair correlations [9.25] while the neutron data provide all pair correlations including the intramolecular correlations [9.26]); $\rho = N/V$, N is the number of water molecules. $S_{\alpha\beta}(\mathbf{k})$ is directly related to X-ray or neutron scattering intensities.

Figure 9.1 shows calculated RDFs of bulk water obtained from Car-Parrinello (CP) MD simulations [9.27] in comparison to the most recent results of X-ray [9.25] and neutron scattering experiments [9.26]. In the simulations the standard CP method [9.28] has been employed within a plane-wave-basis density functional theory, using the BLYP [9.29] gradient-corrected exchange-correlation energy functional and the norm-conserving Troullier-Martins pseudopotential [9.30] (with a plane-wave cutoff of 80 Ry for the "good" results and Γ point only in the Brillouin sampling; for further details and discussion of previous simulations see [9.27]). There is remarkable agreement of the calculated RDFs with most recent both X-ray [9.25] and neutron [9.26] scattering experiments.

We have checked the RDFs by simulations using two other methods, VASP [9.31] and DFTB [9.13]. For the bulk density of water and room temperature we obtain the same good agreement with experiment as in the CP simulations when using VASP while the DFTB calculations yield a smaller and higher first peak in $g_{OO}(\mathbf{r})$. However, the simulations also show that for slightly different densities than bulk density or insufficient system sizes and simulation times position and height of the first peak in $g_{OO}(\mathbf{r})$ are not so well reproduced. One should also note that with respect to simulations using empirical water models in many cases a relatively high and sharp first peak, contrary to experiments, is obtained [9.25, 9.26] (for a survey of molecular models of water see [9.32]). Thus ab initio MD simulations are very useful to obtain correct information about the structure (and its temperature dependence) of water, they are also useful for the interpretation of the experimental data.

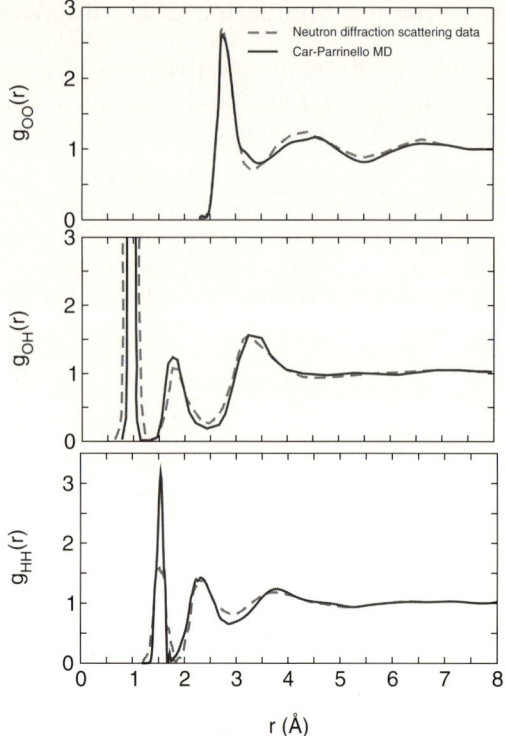

Fig. 9.1. The oxygen-oxygen (top), oxygen-hydrogen (middle) and hydrogen-hydrogen (bottom) radial distribution functions obtained from a Car-Parrinello molecular dynamics simulation of a system of 64 H_2O molecules (solid lines) [9.27] compared to results of neutron diffraction scattering [9.26] (dashed lines). The total simulation time was 11 ps and the average ionic temperature was 307 K. The figure was adapted from [9.27].

Of all the RDFs, $g_{OO}(\mathbf{r})$ is perhaps the most informative. Calculation of the coordination number defined by

$$N_c = 4\pi\rho \int\limits_0^{r_{min}} dr\, r^2\, g_{OO}(r), \tag{9.9}$$

(where r_{min} is the location of the first minimum in $g_{OO}(r)$) gives the number of nearest neighbor water molecules. The recent X-ray data give $N_c = 4.7$ [9.25] while the CP simulations yield an average number of H_2O molecules in the first coordination shell of about 4.0. A coordination number $N_c < 5$ indicates that liquid water preserves much of its tetrahedral, ice-like structuring, while differences from the ice-like structuring can be found from the hydrogen-bonding patterns, see, for example, [9.33]. The ratio of the distances of the first two peak positions in $g_{OO}(r)$ is about 1.64 close to the ideal value of $2(2/3)^{1/2} = 1.633$ in an ice Ih-type lattice. The increasing width with increasing distance is an indication of increasing disorder in the liquid. The first two peaks in $g_{\mathrm{OH}}(r)$ and $g_{\mathrm{HH}}(r)$ correspond to the average intramolecular distances of O–H and H–H, respectively. For further discussions we refer to [9.27].

The RDFs of water and ice from 220 to 673 K and at pressures up to 400 MPa have recently been discussed on the basis of neutron scattering data [9.26]. It is interesting to note that in the ice formation there is still substantial disorder in the hydrogen bonding pattern as can be checked from the width of the RDFs. MD simulations of the phase transition, i.e., freezing of water to ice, are more difficult to achieve than melting of ice. There have only been a few successful MD runs of free (i.e., not confined) water which show ice nucleation and subsequent percolation of the nucleus throughout the simulation box containing 512 water molecules [9.34]. Due to the complex global potential-energy surface, a large number of possible network configurations are possible. This causes large structural fluctuations showing up in the simulations hindering the system to find an easy pathway from the liquid to the frozen state (in spite of the fact that water molecules forming tiny ice-like clusters with four-coordinated hydrogen bonds have by 2 kJ/mol lower potential energy than that of other water molecules [9.34]). Results of MD simulations of ice nucleation are shown in Fig. 9.2.

The constant-temperature MD simulations have been done for 512 molecules in the simulation box with a time step of 1 fs. The TIP4P model for water has been employed, which is a flat 4-center model with a potential energy consisting of Coulomb and Lennard-Jones terms, whereby the oxygen charge is shifted to a fourth site located closer but equidistant from the two hydrogen atoms [9.35].

The essential four stages in the freezing process are discussed in detail in [9.34]. The simulations show that the number of six-member rings N_{6R} fluctuates during the simulations but only increases after an initial nucleus has formed and the ice-growth process has set in, see Fig. 9.2b and following. The authors investigated also the so-called Q_6 parameter [9.36], which may serve to characterize in how far the hydrogen bonds are coherently (icosahedrally) ordered. From the simulations it seems to follow that neither N_{6R} nor Q_6 are suitable order parameters to describe the entire freezing process. Obviously coherent icosahedral orientational correlations are an imperfect solution to characterize tetrahedral packing of water molecules as in ice [9.36].

The reverse process, melting of ice, is easier to simulate. In the following we briefly present results of melting of water clusters simulated by our group. Numerous studies have been devoted to understand the dynamics of small clusters of water since the beginning of simulation studies in the 1970s and to elucidate the nature of the pseudo-first order melting transition.

Our aim is to simulate the melting of water clusters $(\mathrm{H_2O})_n$ of selected sizes (shown in Fig. 9.3) and how their properties evolve with size from ab initio type of calculations using the DFTB method [9.37]. Some of the bigger

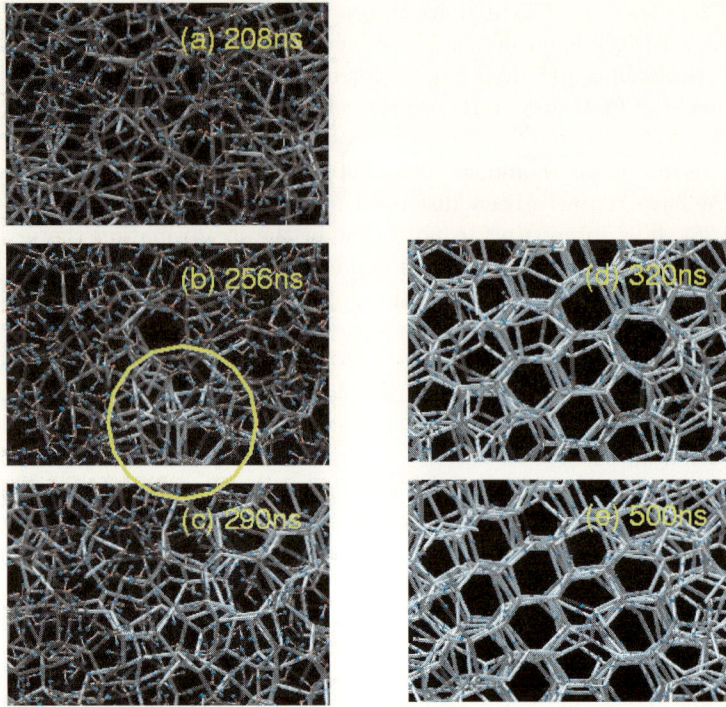

Fig. 9.2. Change of the hydrogen bond network structure of water with the simulation time. During the freezing (for times $t > 320$ ns) the gradual formation of a perfect honeycomb structure becomes visible (lower panel) accompanied by a considerable decrease of the potential energy or loss of kinetic energy of the water molecules. The MD simulation of water freezing is performed for 512 molecules (with density 0.96 g/cm^3) in the simulation box with periodic boundary conditions and involves thermalization at a high temperature followed by quenching (at time $t = 0$) to a low temperature of 230 K (supercooled water). After 256 ns a polyhedral structure consisting of long-lasting hydrogen bonds forms spontaneously acting as an initial nucleus, see the circled region in (b). The lines indicate hydrogen bonds among water molecules. The brightest blue lines are those with hydrogen-bond lifetimes of more than 2 ns. Reprinted by permission from Nature copyright 2002 Macmillan Publishers Ltd. (http://www.nature.com/nature) [9.34].

clusters were modeled from the existing smaller units of $n = 3$, 4, 5 and 6. The preference is given to those with lowest energy. A fair search for the minima of these structures was carrried out before the structures were used as the starting geometry in the MD simulation runs. The results of our simulation are compared with some of the results available for the existing clusters from simulation with a model classical pairwise additive potential (CPAP) [9.38]. We have, however, to point out that results to be found in the literature for the apparent global minimum structures of water clusters differ for some of

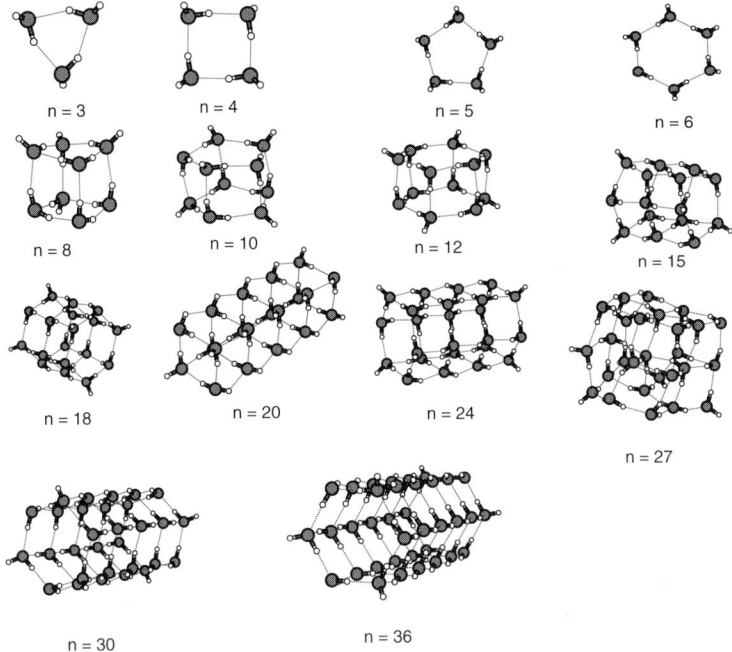

Fig. 9.3. Initial configuration for some selected $(H_2O)_n$ clusters. The number of molecules in each of the clusters is indicated by n.

the clusters (compare, for example, the global minmum structures of water clusters calculated using the TIP3P and TIP4P potentials [9.39]). Our final relaxed structures for some of the smaller clusters such as $n = 3, 4, 5, 6, 8$ and 20 are in agreement with some of the relaxed geometries of Wales and Lee [9.38, 9.40, 9.41].

The melting temperature can then be obtained from the inflexion point of the calorific curve (energy versus temperature). However, it is sometimes difficult to see the abrupt change in the energy. Hence, instead Lindemann's criteria of melting (along with the former condition) is used, which is obtained from the root-mean-square bond-length fluctuations δ_{OO} of oxygen in each of the clusters as given by

$$\delta_{OO} = \frac{2}{N(N-1)} \sum_{i<j} \frac{\sqrt{\langle r_{ij}^2 \rangle - \langle r_{ij} \rangle^2}}{\langle r_{ij} \rangle}, \qquad (9.10)$$

where the brackets denote time averages, and r_{ij} is the distance between the oxygen atoms i and j. The summation is over all N molecules. According to this criterion, melting is caused by a vibrational instability when δ_{OO} reaches a critical value. We paid attention of how to define the melting temperature T_m for a particular cluster. A cluster of water may have different isomeriza-

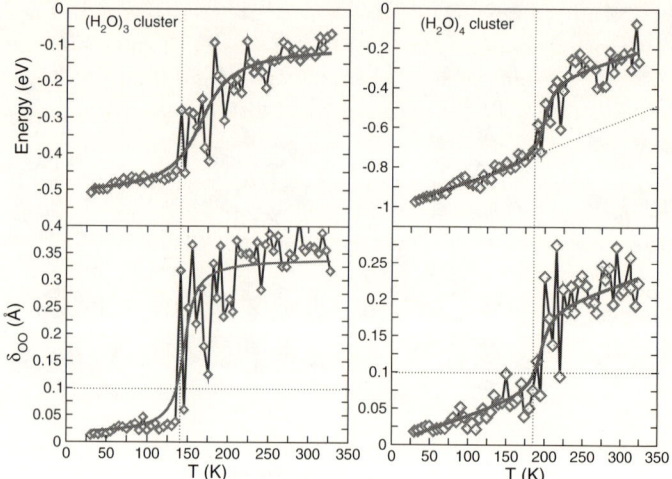

Fig. 9.4. Calculated energy per $(H_2O)_n$ cluster together with the root mean square fluctuations in the O–O bond lengths versus temperature ($n = 3, 4$). The vertical dotted lines mark the approximate melting tmperatures.

tion in which there is interconversion of a water cluster from one form of isomer to another when the hydrogen-bond breaking and reforming is substantial, leading to new structures before melting. This behavior shows up as fluctuations in $\delta_{OO}(T)$. Figure 9.4 shows the behavior of the energy and of δ_{OO} versus temperature for the case of two small clusters.

The size dependence of the melting temperature obtained with the DFTB method compared to calculations using model potentials ([9.38,9.42], a pair-

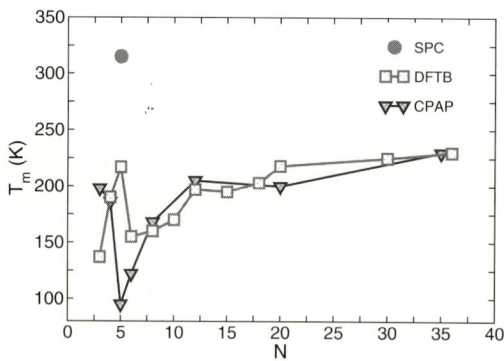

Fig. 9.5. The melting temperature against the number of water molecules. Both results of our DFTB calculations and simulations using the CPAP [9.38] classical potential are plotted. Also shown is the T_m value for the pentamer cluster obtained from a point charge model [9.42]. Remarkable are the rather high melting temperatures for the small clusters.

wise additive and point charge model, respectively) is shown in Fig. 9.5. Both calorific curve and Lindemann's criterion from $\delta_{OO}(T)$ show some agreements in locating T_m. For some of the clusters it becomes difficult to see clearly the transition point in $\delta_{OO}(T)$ because of the fluctuations.

9.2.2 Simulation of β-Cyclodextrin-Binaphtyl and Water

Cyclodextrins are truncated doughnut-shaped cyclic oligosaccharide molecules consisting of α-1,4 linked D-glucose units with a hydrophobic interior surface and a hydrophilic external surface. The most prominent and abundant of the cyclodextrins are α ($[C_{36}H_{60}O_{30}]$), β ($[C_{42}H_{70}O_{35}]$) and γ ($[C_{48}H_{80}O_{40}]$) cyclodextrins with six, seven and eight glucose units, respectively. They are produced by degradation of amylose by glucosyltransferases, in which one or several turns of the amylose helix are hydrolysed off and their ends are joined together to form cyclic oligosaccharides called cyclodextrins or cycloamylases. They have the ability to form inclusion compounds, acting as hosts, by allowing other molecules (guests) into their hydrophobic cavity [9.43–9.45]. In various sizes and chemical characteristics they are being used in pharmaceutical chemistry as drug delivery systems, in chromatography and as enzyme catalysis models or assistants in protein folding [9.46].

Of particular interest of these mentioned applications are inclusion of β-cyclodextrin (BCD) complexes with organic binaphthyl derivatives (BNP) (2,2'-dihydroxy-1,1'-binaphthyl, a chiral molecule, which exists in an enantiomeric form). We have done structure optimization and MD simulation for one of the enantiomeric pairs (R) of 2,2'-dihydroxyl-1,1'-binaphthyl. The optimization of the complex in the crystalized form without solvent water was first carried out in order to have the initial information about how the complexation pathway should proceed. After this, we solvated the whole complex in water which provides the driving force for the complexation and further stabilization of the complex. The most obvious candidate properties from simulation that might indicate the stability of these complexes are a number of properties that describe the structure over time. One of these is the evolution of the geometry of the guest molecule inside the cavity of the BCD and the formation of the hydrogen-bonded network of the complex. The energy of various trial configurations of the guest molecule inside the cavity of the host as well as geometrical evolution of the guest molecules has been calculated. The role of water as the solvent in the stabilization of the complex is also reported. Details of the simulations are discussed with respect to the UV/Vis and CD spectra in [9.47]. Here we simply present two results.

Figure 9.6 shows the results of structure optimization and MD simulation of BCD in liquid water and the analogous simulation with the binaphtyl guest molecule (the starting geometry for BCD was taken from the Crystallographic Database [9.48]). We have employed the DFTB method which allows to handle such a large number of atoms in reasonable computational time. As the Figure shows, without the guest molecule, the hydrophobic interior of

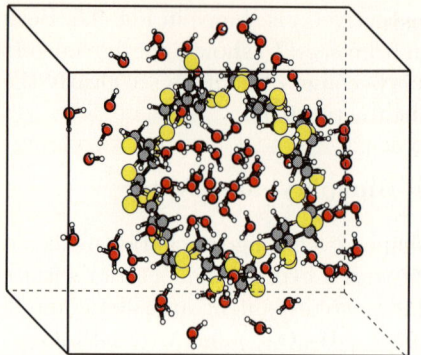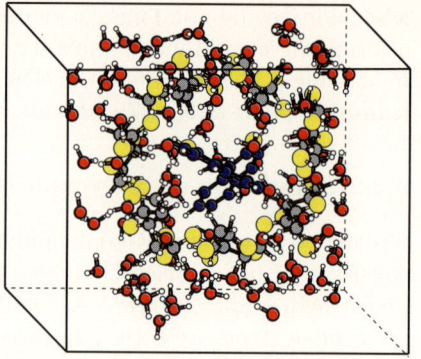

Fig. 9.6. Left: The relaxed structure of β-cyclodextrin in water (79 water molecules) inside the simulation box of dimension $25 \times 25 \times 25$ Å^3 from DFTB MD simulations. The inside of β-cyclodextrin is hydrophobic. It can take a maximum of 7 water molecules in equilibrium. Right: The relaxed structure of β-cyclodextrin plus binaphtyl guest molecule in water (119 water molecules) inside the simulation box of dimension $25 \times 25 \times 25$ Å^3 from DFTB MD simulations. The water inside the hydrophobic cavity of cyclodextrin is easily displaced by an active guest molecule like binaphtyl. The starting geometry for β-cyclodextrin, 'SIBJAO', was taken from the Crystallographic Database [9.48].

BCD can accomodate a maximum of 7 water molecules. This water is easily displaced if an active guest molecule like binaphtyl is used. The most stable conformation of this complex is due to the hydrogen bonding formation of the active agent guest molecule with the torus-like macro ring of the host BCD leading to the formation of the stable adduct in the lipophilic cavity of the biopolimeric matrix.

The role of the solvent, water, in stabilizing the complex becomes obvious from the changes of the structure of water around the complex.

The calculated RDF between the BCD hydroxyl oxygen atoms (OH) and the water atoms reveals the existence of two shells of water molecules around BCD at $r = 3.0$ and 5.3 Å, respectively, in agreement with the calculation reported in [9.49] using the DLPOLY(2) program. From the calculated RDF between BCD hydroxyl (OH) and glycosidic oxygen atoms we find that in the equilibrium conformation the hydroxyl groups of the BCD molecule link together with the glycosidic oxygens via the formation of hydrogen bonds, as evidenced by the peaks of the RDF.

Furthermore we observe the formation adduct of the host (BCD)-BNP complex with the formation of hydrogen bonding between the hydroxyl group of the BNP and the glycosidic oxygen atoms of BCD as indicated by the first peak in the corresponding RDF at around 2.85 Å. This indicates the presence of biphenyl aromatic rings of binaphtyl inside the hydrophobic cavity of BCD. Water molecules form a network of hydrogen bonds with both the primary and secondary hydroxyl groups.

9.2.3 Simulation of Bovine Rhodopsin

Rhodopsin is the visual pigment of the vertebrate rod photoreceptor cells in the retina, which is responsible for light/dark vision. It is composed of the 40-kDa apoprotein opsin (348 amino acids) and its chromophore, 11-*cis*-retinal. The retinal is covalently bound to the protein via a protonated Schiff base (pSb) linkage to the side chain of Lys296 [9.50]. Upon illumination of light $\sim$ 500 nm, the chromophore photoisomerizes from 11-cis to all-trans in less than 200 fs [9.51]. This is followed by a conformational change in the protein and after a series of distinct photointermediates has been passed through [9.52–9.54], an active receptor conformation is formed by deprotonation of the Schiff base (Sb), which corresponds spectroscopically to the blue-shifted metarhodopsin II (meta-II) with 380 nm absorption maximum [9.55]. This meta-II state allows binding to a G-protein transducin (Gt) to the cytoplasmic surface of the receptor [9.56]. The currently proposed model is shown in Fig. 9.7, although it may not reflect intermediates which have similar absorbance spectra but different protein structures [9.57].

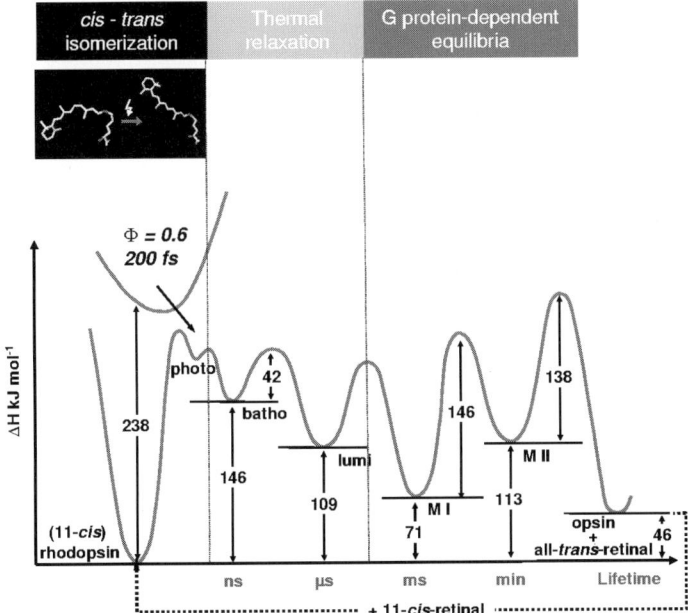

Fig. 9.7. The energy landscape of the visual cycle of rhodopsin. A 2.8 Å high-resolution structure is available only for 11-*cis* rhodopsin in the ground state. No high-resolution structures are availabe for the intermediates. Reprinted by permission from CMLS copyright 1998 Birkhäuser Publishers Ltd. (http://www.birkhauser.ch) [9.57].

Rhodopsin can be isolated from the retina of a number of species [9.58] and it is one of the most extensively studied G-protein-coupled receptors (GPCRs). Bovine eyes are readily available from meat packing plants and are a major source of rhodopsin. Due to the obstacles to obtain three-dimensional crystals, until recently structural data were very limited. A break-through was achieved by Tetsuji Okada who developed an efficient protocol of purifying the protein and using high concentration of zinc led to dissolutions yielding three-dimensional crystals useful for X-ray crystallographic analysis [9.59, 9.60], which allowed a 2.8 Å resolution of the structure [9.61]. This is the first high-resolution structure of a member of the GPCRs family and will trigger a major (re)modeling wave through the entire GPCRs superfamily. Till now three data sets of coordinates of rhodopsin are available from the Brookhaven protein data bank [9.61–9.63].

We report here of ab initio molecular dynamics simulations which are based on one of the new crystal structures (1F88). The new crystal structure has shown that the counterion Glu113 is located somehow near the Sb protonated nitrogen; however, the mechanism of the stability of the proton is not clear from the crystal structure. Using VASP we have investigated the stability of the proton. Our results show that one of the amino acids, Thr94, which stays near the counterion, has, in addition to the stabilizing role of a water molecule, influence on the stability of the protonated state [9.64]. This as well as results for the chromophore in the binding pocket consisting of 27 amino acids [9.65] and results for the whole protein using QM/MM will be briefly discussed in the following.

We note that Lys296 (lysine), to which the chromophore in rhodopsin is attached but otherwise free, and Glu113 (glutamine), the counterion, are two of the key amino acid residues responsible for the structure and function of the retinal chromophore in rhodopsin. Spectroscopy data [9.66] predict that Glu113 is located near C12 of the retinal polyene (upon absortion of light and electronic excitation the chromophore photoisomerizes with a twist around the C12–C13 bond) and the counterion is not directly associated with the protonated Schiff base [9.67]. It is also known that Thr94 (threonine) is an important residue (for example, the mutation of Thr94 can cause night blindness [9.68]). The new crystal structure has revealed that Thr94 is located near the counterion. The results of our MD simulations have shown that the absence of threonine and/or one water molecule makes the protonation state of the chromophore unstable. The calculation was done with VASP for a minimal 1F88 model consisting of 11-*cis*-retinal pSb, Lys296, Glu112, Thr94 and one water molecule, which results in a protonated state stabilized by a hydrogen-bonding network [9.64].

Calculations for a larger part of the binding pocket (also for the 1F88 structure) were done using the DFTB method. The initial planar geometry of the chromophore is marked by yellow in Fig. 9.9 (the positions of the amino residues of the crystal structure are marked by thin lines) while the resulting optimum geometry of the 6-s-*cis*, 11-*cis*-retinal pSb is marked by

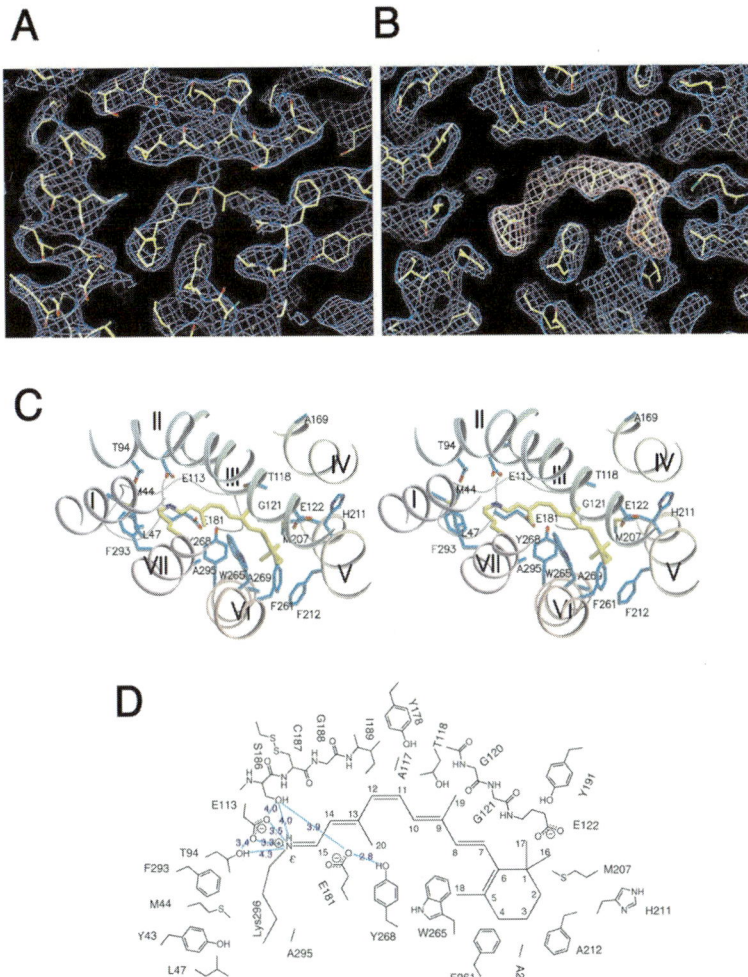

Fig. 9.8. Environment of the 11-*cis*-retinal chromophore. (A) Experimental electron density with 3.3 and (B) with 2.8 Å resolution; (C) shows schematically the side chains around the chromophore while (D) shows the binding pocket of the protein with the amino acids within 4.5 Å distance of the chromophore. Reprinted by permission from Science copyright 2000 AAAS (http://www.sciencemag.org) [9.61].

green (the fully relaxed positions of the amino residues are shown by thick lines). We observe that Thr94 is hydrogen bonded to the oxygen atom of Glu113 in agreement with the previous investigations of the minimal model with VASP. The figure also shows that the β-ionone region is very tightly packed with many aromatic residues.

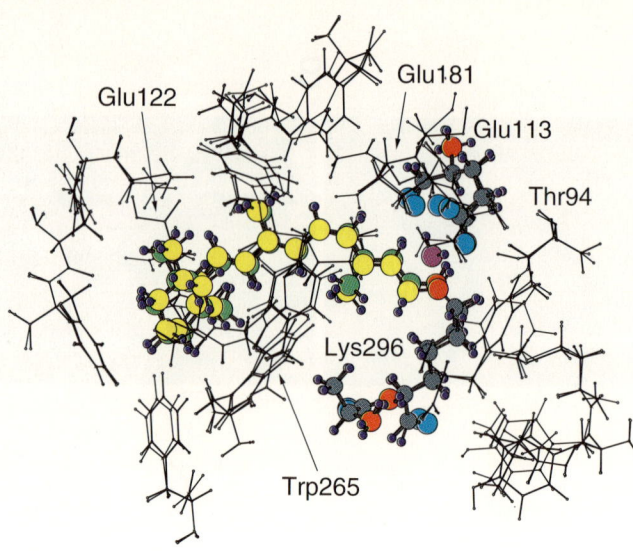

Fig. 9.9. Overlay of the minimized (green) and initial planar (yellow) structures of the 11-*cis*-retinal chromophore. Thin and thick lines correspond to the pocket and the minimized conformations and the initial crystal structure [9.61], respectively.

The QM/MM calculations were done with the CHARMM package [9.14] (in which the quantum mechanical region containing the chromophore and part of Lys296 (altogether 60 atoms) was calculated using DFTB, while the outer region, containing the remaining part of the protein, was calculated using the CHARMM force-field method [9.64]). For the connection between the region treated quantum mechanically and the outer region we have used the link atom method: We have cut a single bond of Lys296 between C_β and C_γ and inserted a dummy hydrogen atom, see Fig. 9.10. The link atom is not seen by the atoms in the classical regions. The entire system with additional water molecules consists of 5578 atoms (348 amino acids with missing residues inserted and 19 water molecules).

Although a quantum mechanical treatment of the entire system would be desirable, it cannot be done in reasonable computer time. In the simulations we used the refined 2.6 Å resolution structure 1L9H of Okada [9.63]. Röhrig *et al.* have also performed MM calculations for the entire protein [9.69], however their conclusion of a deprotonated Glu181 is inconsistent with our results and also contradicts the experimental result [9.70] showing that the protonated state of Glu181 is stable due to the hydrogen bonded network. This hydrogen bonded network extends to the counterion Glu113 and has an important function in the rhodopsin photocycle [9.70]. The stability of the 1L9H structure of rhodopsin was examined by MD simulations using the canonical ensemble and the Nosé thermostat [9.64].

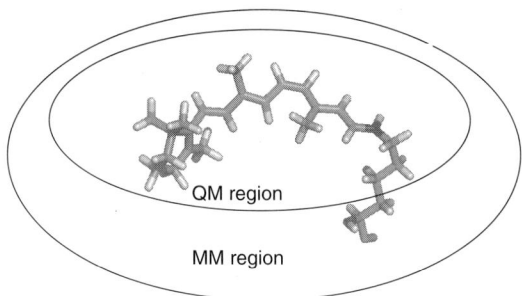

Fig. 9.10. The link of the QM to the MM region which has been used in the calculations of the protein rhodopsin. The link is between the C_β and C_γ atoms of Lys296, where a dummy hydrogen atom is inserted.

9.3 Simulation of Chemical Reactions in the Gas Phase

The study of chemical reactions by using computational methods is of great technological interest because it can provide useful information about details of a reaction which are – at least not yet – feasible in experiment. Today it is often not possible to directly observe a chemical reaction in real time, so experimentalists have often only got information about reactants and products. It is easy to imagine that, if one knew exactly how a reaction proceeded (the 'reaction pathway'), one could optimize the conditions under which reactions occur and increase productivity.

Whereas in conventional transition state theory one can obtain information about activation energies and thereby rate constants of a reaction to high accuracy, ab initio MD allows one to simulate the reaction itself at finite temperature. This immediately provides valuable information about possible reaction pathways. As an example we present results of MD simulations of the abstraction reaction $SiCl_4 + H \rightleftharpoons SiCl_3 + HCl$. Due to the high reactivity of the radicals involved this reaction plays an important role in the formation of nano- and microcrystalline silicon films and nanoparticles from gas phase precursor molecules $SiCl_4$ and H_2, which are commonly used, for example, in chemical vapor deposition (CVD) processes.

Figure 9.11 shows snapshots from an MD simulation of this reaction. The calculations were performed on the basis of DFT in combination with plane waves and ultrasoft pseudopotentials in periodic boundary conditions using VASP. The exchange correlation energy was calculated by using the generalized gradient approximation in a functional form proposed by Perdew and Wang in 1991 [9.71]. The size of the supercell was $12 \times 12 \times 20$ Å^3 to ensure that no interaction between atoms and their periodic images occured. To simulate possible reaction conditions, the $SiCl_4$ molecule was first placed alone into the cell and equilibrated at a temperature of 2000 K. After that the H atom was put into the simulation box at a distance of 6 Å from the nearest Cl atom and given an initial velocity of 0.1 Å/fs in the direction of the

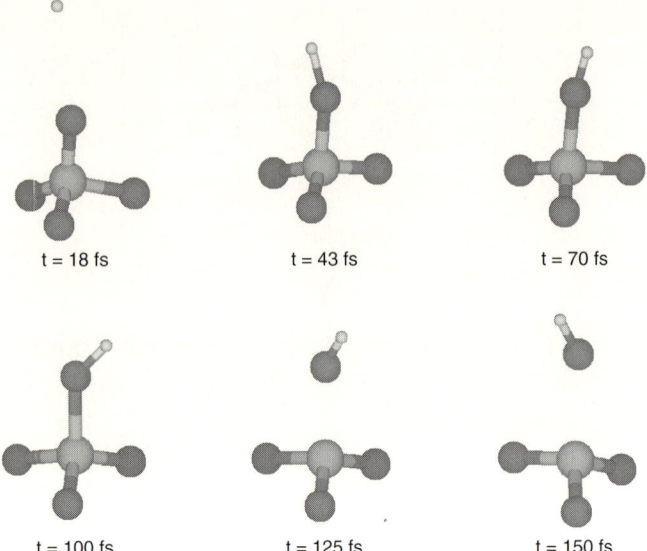

Fig. 9.11. Snapshots of an MD simulation of the reaction SiCl$_4$ + H $\rightleftharpoons$ SiCl$_3$ + HCl. The SiCl$_4$ molecule has been equilibrated at 2000 K. The relative velocity of the reactants is 0.1 Å/fs. The fluctuations of the Si–Cl bonds are clearly visible.

molecule. From the figure one can see that the Si-Cl bonds are first fluctuating around their equilibrium values. When the H atom has reached the molecule, the corresponding Si-Cl bond is elongated (at t = 43 fs). This stretching is increased by the H-Cl interaction and eventually the bond breaks (t = 125 fs), resulting in a SiCl$_3$ radical and the HCl molecule, which are moving apart (t = 150 fs).

From this simulation we learn that the two relevant parameters of the reaction are the distances between Si–Cl and Cl–H (Cl being the abstracted Cl atom). In Fig. 9.12 a contour plot of the PES of the reaction in terms of these two parameters is shown. Contour lines are drawn at a distance of 100 meV. The energy barrier was calculated to 420 meV, the corresponding transition state (TS) is clearly visible as the saddlepoint in the diagram. Together with the PES the trajectories of MD simulations of the reaction are drawn. The solid line corresponds to the (successful) simulation of which the snapshots are depicted in Fig. 9.11. The other lines belong to simulations, where the starting conditions were only slightly different. The only change was the instantaneous initial velocity of the involved Cl atom. For the solid curve it was 0.004 Å/fs towards the arriving H atom, in case of the curves with the long dashes it was zero (the velocity of the H atom was increased to 0.11 Å/fs for the curve with the dots) and -0.004 Å/fs (away from the H atom) for the curve with the short dashes. The result of this minor change is tremendous: When the Cl atom does not move towards the H atom, the

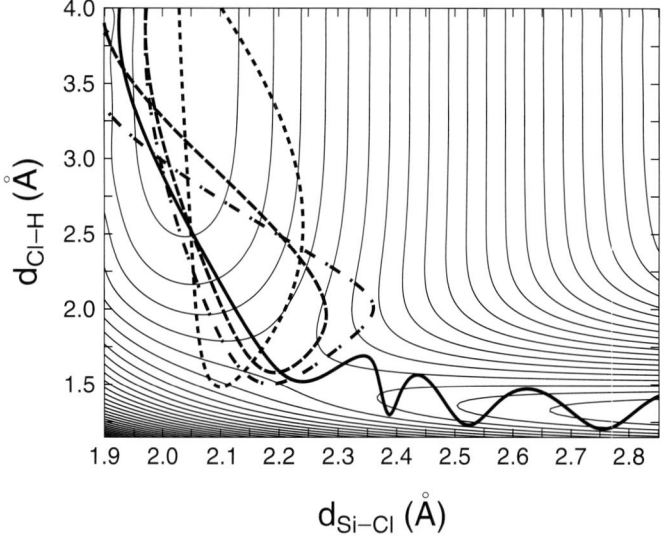

Fig. 9.12. Potential-energy surface of the reaction $SiCl_4 + H \rightleftharpoons SiCl_3 + HCl$ in terms of the two relevant parameters, the Si–Cl and Cl–H bond lengths. Contour lines are drawn at a distance of 100 meV. The trajectories of different MD runs are also included. The solid line corresponds to the simulation of which the snapshots are shown in Fig. 9.11. Initial velocities of the scattered Cl atom: 0.004 Å/fs in the direction of the incoming H atom (solid line), 0.0 Å/fs (long-dashed line), -0.004 Å/fs (short-dashed line).

Si–Cl bond is not stretched, when the hydrogen atom has arrived, and the force between the Cl and the H atom is not strong enough to result in the breaking of the Si–Cl bond. The hydrogen atom is then just scattered back, which becomes very clear from the trajectories on the PES. This can also not be changed by increasing the impact velocity, the reaction then just proceeds deeper into the repulsive part of the PES.

We see now that the momentary vibrational state of the $SiCl_4$ molecule is crucial for the outcome of the reaction, the translational kinetic energy of the reactants is not that decisive. Even if this energy is large enough to overcome the barrier, in many cases the system is not able to cross the TS. This is related to the fact that the TS is located towards the side of the products. Molecular dynamics has once again proved to be a valuable tool in the examination of the bahavior of molecular systems.

9.4 Simulation of Structural Transformations in Solids and Particles

As examples from solid state physics we consider successful ab initio calculations and MD simulations of the phase transformations in Fe-Ni and Ni-Mn-Ga alloys. Martensitic instabilities, shape memory and magnetovolume effects in iron-based alloys are of scientific and technological interest. In spite of the large amount of experimental and theoretical work there have remained unsolved problems [9.72]. The Heusler compound Ni_2MnGa and related alloys also exhibit displacive, diffusionless structural transformations. The Heusler alloys have attracted much interest because in these systems the martensitic transformation occurs below the Curie temperature which leads to magnetic shape-memory effects controllable by an external magnetic field [9.73].

In addition we have considered the crossover from the bulk materials to nanoparticles and have investigated the change of the structural transformation with the size of the particles. As last example we show that the melting of Al clusters depends on the size of the clusters; the simulation yields a smooth melting curve different from the case of melting of water clusters. Scaling of the austenitic/martensitic transformation and melting temperatures with the inverse of the cluster diameter is observed.

9.4.1 Simulation of the Phase Diagram of Fe-Ni and Ni-Mn-Ga Alloys

Figures 9.13 and 9.14 show the simulated phase diagrams in comparison to the experimental ones. Early ab initio calculations using the KKR CPA method gave already most of the structure of the theoretical phase diagram for Fe-Ni by plotting energy differences between the phases as a function of the composition [9.74]. Refined KKR CPA calculations give improved results [9.75]. It is remarkable how, for example, the energy differences between the nonmagnetic and magnetic ab initio solutions mimic the temperature variation of the experimental Curie temperature with the composition. Tendencies of the experimental transformation temperatures are also reproduced by the MD simulations [9.76, 9.77] employing the so-called EAM potentials [9.78]. The transformation temperatures are not so well reproduced for compositions near elemental iron, which can be attributed to the fact that the EAM potentials have been built to reproduce the elastic properties of the ferromagnetic ground states of the two elements without construction of an explicit magnetic term, which might be important when varying the temperature from the magnetic ground state to the high-temperature parmamagnetic state in the simulations.

The experimental Ni-Mn-Ga phase diagram has been taken from [9.79] (premartensitic phase transformation) and [9.80] (Curie temperature and

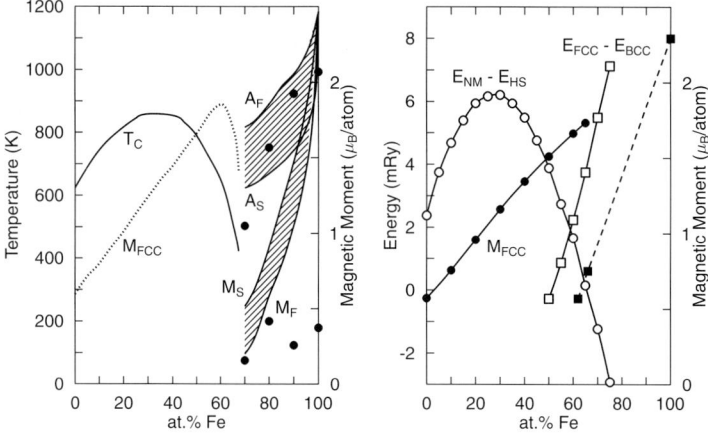

Fig. 9.13. Left: The experimental phase diagram showing the Curie temperature T_c and the magnetization M_{FCC} of the iron-nickel alloys in the fcc structure. $A_{S,F}$ and $M_{S,F}$ denote the experimental austenite and martensite start and final temperatures, respectively, on the iron-rich side. Filled circles: Results of molecular-dynamics simulations employing model potentials (upper dots: $\alpha \rightarrow \gamma$ transition for increasing temperature; lower dots: $\gamma \rightarrow \alpha$ transition for decreasing temperature [9.76, 9.77]. Right: The theoretical phase diagram showing the energy differences between the nonmagnetic (NM) and ferromagnetic (HS: high spin) ground states, $E_{NM} - E_{HS}$, of the alloys in the face centered cubic (fcc) structure and the energy differences between the fcc and body centered cubic (bcc) structures on the iron-rich side: Open squares mark single-site KKR CPA calculations by [9.74] and filled squares are results of multi-site KKR CPA calculations by [9.75]. The filled circles are the ab initio results for the magnetic moments.

martensitic transformation). The theoretical phase diagram has been simulated using a phenomenological model based on Ginzburg-Landau theory [9.81–9.83]. Also in this case there is reasonable agreement between experimental and theoretical data.

Detailed ab initio investigations of the vibrational properties using VASP show that, in contrast to the case of Fe-Ni, the Ni-Mn-Ga alloys show pronounced phonon softening [9.84, 9.85] in agreement with experimental data [9.86]. The calculated phonon dispersion relations seem to show that besides Kohn anomalies originating from nesting properties of the Fermi surfaces of minority and majority spin electrons [9.85], hybridization effects involving the optical phonons of Ni and acoustical phonons could give a different explanation for the occurrence of the 3M, 5M and 7M martensitic structures [9.84]. Further *ab inito* simulations are currently performed in order to investigate the important influence of shuffling of atoms on premartensitic and martensitic transformations.

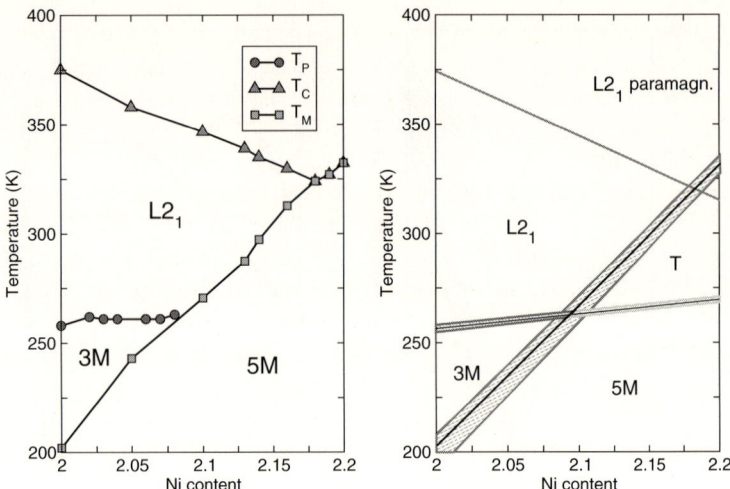

Fig. 9.14. Experimental (left) and theoretical (right) phase diagrams for $Ni_{2+x}Mn_{1-x}Ga$. Data for the premartensitic transformation temperature T_P have been taken from [9.79], data for T_c and T_M from [9.80]. The theoretical diagram has been obtained by phenomenological modeling [9.81–9.83]. Thick lines denote phase transformations, thin lines show the boundaries of the stability areas; broad areas (simultaneous stability of two different phases) are around first order phase transitions. $L2_1$ is the high-temperature austenite structure, 3M the premartensitic structure and T is the tetragonal non-modulated martensitic structure.

9.4.2 Simulation of the Structural Transformation in Fe-Ni Particles

We have employed MD simulations using EAM potentials in order to characterize the crossover from the bulk properties of Fe-Ni alloys to the properties of Fe-Ni nanoparticles. Metallic nanoparticles show interesting effects such as systematic changes of elastic and vibrational properties with the size of the particles. Increasing hardness with decreasing grain size due to dislocation immobilization at the grain boundaries (known as Hall-Petch effect) and increasing softness for still smaller sizes due to grain boundary sliding (reverse Hall-Petch effect) have been discussed in the frame of MD simulations for Cu [9.87]. Interestingly Young's modulus or the hardness of the particles scale with the inverse square root of the diameter of the particles [9.88].

Of particular interest are structural transformations in the nanoparticles. Multimillion atoms simulations have shown that perfect spherical Fe-Ni particles without any defects do not undergo a martensitic transformation when cooling the particles from the high temperature fcc phase to low temperatures. This means that the austenitic phase is stable at least during the time of the MD runs and that we do not observe the homogeneous nucleation of martensite. This is in agreement with the experimental observation that the austenite is stabilized in nanometer-sized particles, i.e., there is no doubt that the reduction in size results in the stabilization of austenite [9.89]. Although the same authors speculate that the increase of surface areas with decreasing size and lattice softening (in combination with defects) may enhance the martensitic transformation in nanometer-sized particles.

However, the high-temperature austenitic transformation is observed in the MD simulations. If we prepare the spherical Fe-Ni particles (without defects) at low temperatures in the martensitic structure then we find upon heating the martensitic–austenitic transformation (bcc $\rightarrow$ fcc). This transformation (in contrast to the reverse fcc $\rightarrow$ bcc transformation at lower temperature) is entropy driven and seemingly barrierless as simulations of the transformation along the Bain path show [9.77, 9.90]. Of interest is here the change (scaling) of the austenitic transformation temperature with the inverse of the particle diameter shown in Fig. 9.15 for different compositions.

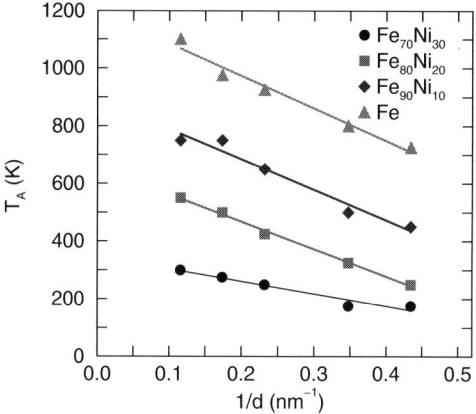

Fig. 9.15. Change of the austenitic transformation temperature with the inverse diameter of the Fe-Ni clusters. The extrapolation of $d \rightarrow \infty$ yields transformation temperatures which lie within the range of experimentally observed austenitic start and final temperatures, see the phase diagram in Fig. 9.13.

In order to observe the heterogeneous nucleation of martensite at defects of the particles we have prepared a simple cubic $Fe_{80}Ni_{20}$ particle, see Fig. 9.16. The corners of the particle act as defects from which the nucleation starts. The cubic particle was prepared in the austenitic fcc structure; the subsequent MD run consisted of 50.000 integration steps at 50 K covering a time range of 50 ps.

A quantitative picture of the particle dynamics is shown in Fig. 9.17 which gives an impression of the variation of the radial distribution function and of the structure with time. The large-scale MD results were obtained by using the SPaSM code [9.91–9.93] on the parallel SUN machine at Los Alamos.

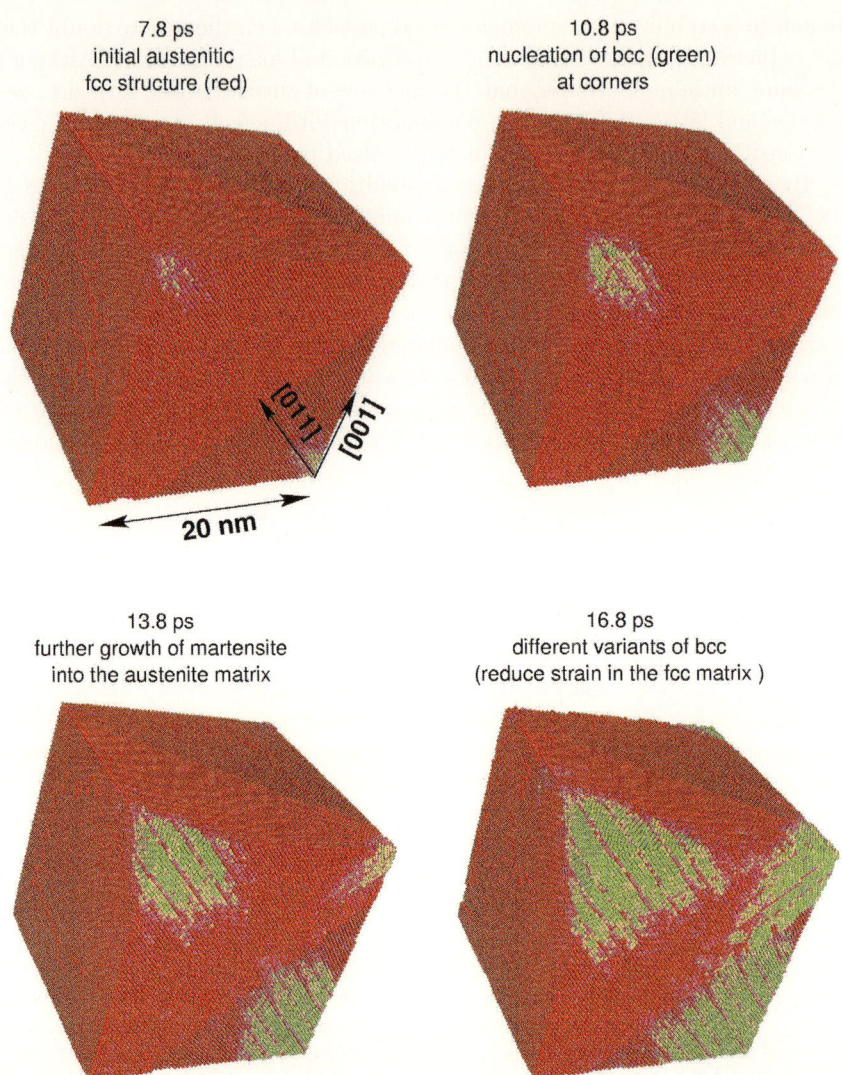

7.8 ps
initial austenitic
fcc structure (red)

10.8 ps
nucleation of bcc (green)
at corners

[011]   [100]

20 nm

13.8 ps
further growth of martensite
into the austenite matrix

16.8 ps
different variants of bcc
(reduce strain in the fcc matrix)

Fig. 9.16. Nucleation of martensite (bcc: green) at the corners of the cubic $Fe_{80}Ni_{20}$ particle at a temperature of 50 K, which grows with a fraction of the sound velocity into the austenitic matrix (fcc: red). The growing martensitic regions consist of twins with twin boundaries marked by thin red lines, the formation of which helps to lower the elastic tension.

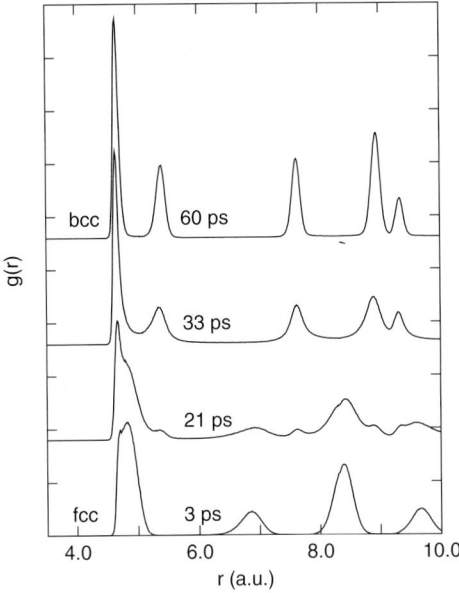

Fig. 9.17. Variation of the radial distribution function of the cubic $Fe_{80}Ni_{20}$ particle in Fig. 9.16 at 50 K with time. The transformation from the fcc to the bcc structure is visible from the splitting of the first maximum as time proceeds.

9.4.3 Simulation of the Melting of Al Clusters

A description of the process of melting is still a challenging problem of statistical physics. Experimental studies show that undercooling of metallic melts is easily achieved while overheating of the solid phase is usually hindered. This asymmetry is associated with the sudden onset of melting (seemingly without an energetic barrier), whereby the solid "sinks" in its own melt which starts from the surfaces of the solid, while for the liquid–solid transition an energy barrier, associated with the formation of nucleation centers, has to be overcome, see the discussion in [9.94]. Only the suppression of the influence of surface effects leads to overheating phenomena, which, however, is not easily achieved in the experiments, but which can be observed in the simulations when employing periodic boundary conditions in order to minimize the effect of surfaces. Furthermore, the problem to designate an order parameter to the solid–liquid transition is an unsettled problem, since the simulations show that there is a continuous change of more than one single local structural symmetry when, for example, one undercools the melt in the simulations to lower temperatures than the melting temperature. However, it has has recently been emphasized that a five-fold symmetry in the liquid phase of lead might be important for the discussion of general aspects of melting [9.95].

Melting of clusters is of particular interest, since here surface effects are dominant for the small clusters. Moreover, the phenomenological melting theory predicts a scaling of the melting temperature with the inverse of the particle diameter like

$$1 - T_m/T_B \propto 1/d, \tag{9.11}$$

where T_m and T_B are the melting temperatures for the cluster and the corresponding bulk material, see, for example, [9.96]. In the MD simulations, when approaching T_m, we observe increasing structural fluctuations with a sudden first-order like transition to the liquid state in the case of Al clusters. The change of the energy of free Al clusters, simulated by using an optimized EAM potential with the help of ab initio results, with different numbers of atoms is shown in Fig. 9.18. The crossover from the solid to the liquid state is associated with a jump in each curve. From the plot of $T_m(1/d)$ we observe indeed a behavior as predicted by the above formula. In fact, this curve is rather smooth. The smoothness is surprising in view of the many structural changes the clusters undergo in picoseconds (in the simulations) just before melting. These structural fluctuations due to the enhanced diffusion of the surface atoms before melting also shows up as additional low-frequency distributions to the vibrational density of states of the particles. A common neighbor analysis shows that around the melting temperature different kinds

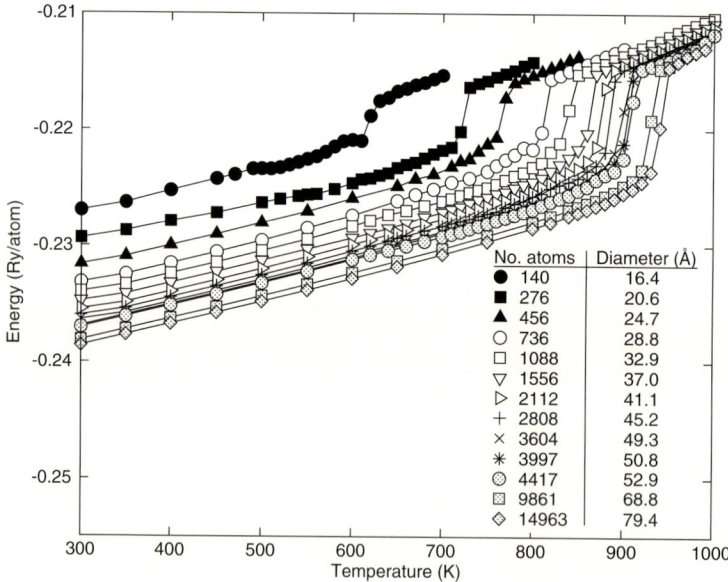

No. atoms	Diameter (Å)
● 140	16.4
■ 276	20.6
▲ 456	24.7
○ 736	28.8
□ 1088	32.9
▽ 1556	37.0
▷ 2112	41.1
+ 2808	45.2
× 3604	49.3
∗ 3997	50.8
⊙ 4417	52.9
⊠ 9861	68.8
◇ 14963	79.4

Fig. 9.18. Change of the energy of free aluminum clusters with different numbers of atoms as a function of the temperature. The sudden change in the energy marks the change from the solid to the liquid phase.

of crystallographic symmetries exist with a dominant fcc contibution in both the solid and liquid states.

On the basis of our simulation results concerning the melting of Al clusters we cannot explain why there are so many oscillations in the experimental melting curve of Na clusters [9.97] (in [9.97] the influence of energy and entropy on the melting of Na clusters is discussed in detail but this cannot explain the oscillation of the melting temperature with the cluster size). We currently employ ab initio MD simulations in order to investigate the melting of Na clusters.

Acknowledgments

We thank Prof. W. Hergert for the invitation to deliver this talk at the Heraeus-2002 Ferienkurs in Halle. Our research on this topic was supported by the Graduiertenkolleg "Struktur und Dynamik heterogener Systeme", the SFB 445 "Nanopartikel aus der Gasphase: Entstehung, Struktur, Eigenschaften" and the Forschergruppe "Retinal Protein Action: A Combination of Theoretical Approaches" through the Deutsche Forschungsgemeinschaft.

References

[9.1] W.G. Hoover, A. Ladd, B. Moran: Phys. Rev. Lett. **48**, 1818 (1982)

[9.2] S. Nosé: J. Chem. Phys. **81**, 511 (1984); S. Nosé: Mol. Phys. **52**, 255 (1984)

[9.3] W.G. Hoover: Phys. Rev. A **31**, 1695 (1985).

[9.4] M. Parrinello: A. Rahman, Phys. Rev. Lett. **45**, 1196 (1980)

[9.5] D. Frenkel, B. Smit: *Understanding Molecular Simulation – From Algorithms to Applications*, Computational Science Series 1, 2nd edn. (Academic Press, San Diego 2002)

[9.6] K.A. Fichthorn, M.L. Merrick, M. Scheffler: Appl. Phys. A **75**, 79 (2002)

[9.7] A. Voter: J. Chem. Phys. **106**, 4665 (1997)

[9.8] J. VandeVondele, U. Rothlisberger: J. Chem. Phys. **113**, 4863 (2000); A. Passerone, M. Parrinello: Phys. Rev. Lett. **87**, 83002 (2001)

[9.9] *Bridging Time Scales: Molecular Simulations for the Next Decade*, Lecture Notes in Physics, LNP 605, ed. by P. Nielaba, M. Mareschal, G. Ciccotti (Springer, Berlin 2002)

[9.10] S.R. Shenoy, T. Lookman, A. Saxena, A.R. Bishop: Phys. Rev. B **60**, R12537 (1999)

[9.11] P. Cartoni, U. Rothlisberger, M. Parrinello: Acc. Chem. Res. **35**, 455 (2002)

[9.12] M. E. Colombo, L. Guidoni, A. Laio, A. Magiskato, P. Maurer, S. Piana, U. Roehrig, K. Spiegel, J. VandeVondele, U. Roethlisberger: Chimica **56**, 11 (2002)

[9.13] M. Elstner, D. Porezag, G. Jungnickel, J. Elsner, M. Haugk, T. Frauenheim, S. Suhai, G. Seifert: Phys. Rev. B **58**, 7260 (1998)

[9.14] Q. Cui, M. Elstner, T. Frauenheim, E. Kaxiras, M. Karpuls: J. Phys. Chem. B **105**, 569 (2001).

[9.15] U. Saalmann, R. Schmidt: Z. Phys. D **38**, 153 (1996)

[9.16] G.W. Robinson, S.-B. Zhu, S. Singh, M.W. Evans: *Water in Biology, Chemistry and Physics – Experimental Overviews and Computational Methodologies*, World Scientific Series in Contemporary Chemical Physics – Vol. 9 (World Scientific, Singapore 1996)

[9.17] *Proceedings of the First International Symposium on Physical, Chemical and Biological Properties of Stable Water* (I_E^{TM}) *Clusters*, ed. by S.Y.Lo, B. Bonavida (World Scientific, Singapore 1998)

[9.18] M. Chaplin: http://www.sbu.ac.uk/water/

[9.19] A. Bizzarri, S. Cannistraro: J. Phys. Chem B **106**, 6617 (2002)

[9.20] M.E. Tuckerman: J. Phys.: Condens. Matter **14**, R 1297 (2002)

[9.21] D. Marx, M.E. Tuckerman, J. Hutter, M. Parrinello: Nature **397**, 601 (1999)

[9.22] U.W. Schmitt, G.A. Voth: J. Chem. Phys. **111**, 9361 (1999)

[9.23] L.M. Ramaniah, M. Parrinello, M. Bernasconi: J. Chem. Phys. **111**, 1595 (1999)

[9.24] M. Benoit, D. Marx, M. Parrinello: Nature **392**, 258 (1998)

[9.25] J.M. Sorenson, G. Hura, R.M. Glaeser, T. Head-Gordon: J. Chem. Phys. **113**, 9149 (2000)

[9.26] A.K. Soper: Chem. Phys. **258**, 121 (2000)

[9.27] S. Izvekov, G.A. Voth: J. Chem. Phys. **116**, 10372 (2002)

[9.28] R. Car, M. Parrinello: Phys. Rev. Lett. **55**, 2471 (1985)

[9.29] A.D. Becke: Phys. Rev. A **38**, 3098 (1988); C. Lee, W. Yang, R.C. Parr: Phys. Rev. B **37**, 785 (1988)

[9.30] N. Troullier, J.L. Martins: Phys. Rev. B **43**, 1993 (1991)

[9.31] G. Kresse, J. Furthmüller: Phys. Rev. B **54**, 11169 (1996)

[9.32] A. Wallqvist, R.D. Mountain: 'Molecular Models of Water: Derivation and Description'. In: *Reviews in Computational Chemistry*, Vol. 13, ed. by K.B. Lipkowitz, D.B. Boyd (Wiley-VCH, New York 1999), pp. 183–247

[9.33] G.A. Jeffrey: *An Introduction to Hydrogen Bonding*, (Oxford University Press, Oxford 1997)

[9.34] M. Matsumoto, S. Saito, I. Ohmine: Nature **416**, 409 (2002) (http://www.nature. com/nature). Figure 2 used by courtesy of Iwao Ohmine, Chemistry Department, Nagoya University, Japan

[9.35] W.L. Jorgensen, J. Chandrasekhar, J.D. Madura: J. Chem. Phys. **79**, 926 (1983)

[9.36] P.J. Steinhardt, D.R. Nelson, M. Ronchetti: Phys. Rev. B **28**, 784 (1983)

[9.37] W.A. Adeagbo, P. Entel: Phase Transitions, submitted

[9.38] A. Vegiri, S.C. Farantos: J. Chem. Phys. **98**, 4059 (1993)

[9.39] H. Kabrede, R. Hentschke: J. Phys. Chem. **107**, 3914 (2003)

[9.40] D.J. Wales, I. Ohmine: J. Chem. Phys. **98**, 7245 (1993)

[9.41] S. Lee, J. Kim, S.J. Lee, K.S. Kim: Phys. Rev. Lett. **79**, 2038 (1997)

[9.42] W.B. Wayne, M.M. Rhodes: J. Chem. Phys. **98**, 4413 (1993)

[9.43] J.L. Atwood, J.E.D. Davies, D.D. MacNicol, F. Vöegtle, J.M. Lehn: *Comprehensive Supramolecular Chemistry*, (Elsevier, Oxford 1996)

[9.44] B.K. Lipkowitz: Chem. Rev. **98**, 1829 (1998)

[9.45] W. Saegner, J. Jacob, K. Steiner, D. Hoffman, H. Sanbe, K. Koizumi, S.M. Smith, T. Takaha: Chem. Rev. **98**, 1787 (1998)

[9.46] J. Szejtli: J. Mater. Chem. **7**, 575 (1997)

[9.47] W.A. Adeagbo, V. Buss, P. Entel: Phase Transitions, submitted

[9.48] J. Ding, T. Steiner, W. Saenger: Acta Cryst. B **47**, 731 (1991)

[9.49] B. Manunza, S. Deiana, M. Pintore, C. Gessa: J. Molec. Struct. (Theochem) **419**, 133 (1997)

[9.50] G. Wald: Science **162**, 230 (1968)

[9.51] R.W. Schoenlein, L.A. Peteanu, R.A. Mathies, C.V. Shank: Science **254**, 412 (1991)

[9.52] S.J. Hug, J.W. Lewis, C.M. Einterz, T.E. Thorgeirsson, D.S. Kliger: Biochemistry **29**, 1475 (1990)

[9.53] B. Borhan, M.L. Souto, H. Imai, Y. Shichda, K. Nakanishi: Science **288**, 2209 (2000)

[9.54] J.E. Kim, D.W. McCamant, L. Zhu, R.A. Mathies: J. Phys. Chem. B **105**, 1240 (2001)

[9.55] A.G. Doukas, R.H. Callender, T.G. Ebrey: Biochemistry **17**, 2430 (1978)

[9.56] T. Okada, K. Palczewski, K.P. Hofmann: Trends Biochem. Sci. **26**, 318 (2001)

[9.57] Y. Shichida, H. Imai, T. Okada: Cell. Mol. Life Sci. **54**, 1299 (1998). Figure 8 used by courtesy of Tetsuji Okada, AIST Tokyo, Japan

[9.58] L. Tang, T.G. Ebrey, S. Subramaniam: Israel Jour. Chem. **35**, 193 (1995)

[9.59] T. Okada, I. Le Trong, B.A. Fox, C.A. Behnke, R.E. Stenkamp, K. Palczewski: J. Struct. Biol. **130**, 73 (2000)

[9.60] T. Okada, K. Palczewski: Current Opinion Struct. Biol. **11**, 420 (2001)

[9.61] K. Palczewski, T. Kumasaka, T. Hori, C.A. Behnke, H. Motoshima, B.A. Fox, I. LeTrong, D.C. Teller, T. Okada, R.E. Stenkamp, M. Yamamoto, M. Miyano: Science **289**, 739 (2000) (http://www.sciencemag.org). Figure 6 used by courtesy of Krzysztof Placzewski, Department of Ophthalmology, University of Washington, USA

[9.62] D.C. Teller, T. Okada, C.A. Behnke, K. Palczewki, R.E. Stenkamp: Biochemistry **40**, 7761 (2001)

[9.63] T. Okada, Y. Fujiyohi, M. Silow, J. Navarro, E.M. Landau, Y. Schichida: Proc. Natl. Acad. Sci. U.S.A. **99**, 5982 (2002)

[9.64] V. Buss, M. Sugihara, P. Entel, J. Hafner: Z. Angew. Chemie, accepted

[9.65] M. Sugihara, V. Buss, P. Entel, M. Elstner, T. Frauenheim: Biochemistry **41**, 15239 (2002)

[9.66] M. Han, B.S. DeDecker, S.O. Smith: Biophys. J. **65**, 6111 (1993)

[9.67] M. Han, S.O. Smith: Biochemistry **34**, 1425 (1995)

[9.68] P. Garriga, J. Manyosa: Biochemistry **528**, 17 (2002)

[9.69] U.F.Röhrig, L. Guidoni, U. Rhothilinsberger: Biochemistry **41**, 10799 (2002)

[9.70] E.C.Y. Yan, M.A. Kazmi, S. De, B.S.W. Chang, C. Seibert, E.P. Marin, R.A. Mathies, T.P. Sakmar: Biochemistry **41**, 3620 (2002)

[9.71] J.-P. Perdew, Y. Wang: Phys. Rev. **45**, 13244 (1992)

[9.72] E.F. Wassermann: 'Invar: Moment-volume instabilities in transition metals and alloys'. In: *Ferromagnetic Materials*, ed. by K.H.J. Buschow, E.P. Wohlfarth (Elsevier, Amsterdam 1990) pp. 237–322

[9.73] A.D. Bozhko, A.N. Vasil'ev, V.V. Khovailo, I.E. Dikshtein, V.V. Koledov, S.M. Seletskii, A.A. Tulaikova, A.A. Cherechukin, V.G. Shavrov, V.D. Buchelnikov: JETP **88**, 957 (1999)

[9.74] M. Schröter, H. Ebert, H. Akai, P. Entel, E. Hoffmann, G.G. Reddy: Phys. Rev. B **52**, 188 (1995)

[9.75] V. Crisan, P. Entel, H. Ebert, H. Akai, D.D. Johnson, J.B. Staunton, Phys. Rev. B **66**, 014416 (2002)

206   P. Entel et al.

[9.76] P. Entel, R. Meyer, K. Kadau, H.C. Herper, E. Hoffmann: Eur. Phys. J. B **5**, 379 (1998)
[9.77] K. Kadau: Molekulardynamik-Simulationen von strukturellen Phasenumwandlungen in Festkörpern, Nanopartikeln und ultradünnen Filmen. PhD Thesis, Gerhard-Mercator-Universität Duisburg, Duisburg (2001)
[9.78] M.S. Daw, M.I. Baskes: Phys. Rev. B **26**, 6443 (1983)
[9.79] V.V. Khovailo, T. Takagi, A.D. Bozhko, M. Matsumoto, J. Tani, V.G. Shavrov: J. Phys.: Condens. Matter **13**, 9655 (2001)
[9.80] A.N. Vasil'ev, A.D. Bozhko, V.V. Khovailo, I.E. Dikshtein, V.G. Shavrov, V.D. Buchelnikov, M. Matsumoto, S. Suzuki, T. Takagi, J. Tani: Phys. Rev. B **59**, 1113 (1999)
[9.81] V.D. Buchelnikov, A.T. Zayak, A.N. Vasil'ev, T. Takagi: Int. J. Appl. Electromagn. Mechan. **1, 2**, 19 (2000)
[9.82] V.D. Buchelnikov, A.T. Zayak, A.N. Vasil'ev, D.L. Dalidovich, V.G. Shavrov, T. Takagi, V.V. Khovailo: JETP **92**, 1010 (2001)
[9.83] A.T. Zayak, V.D. Buchelnikov, P. Entel: Phase Transitions **75**, 243 (2002)
[9.84] A.T. Zayak, P. Entel, J. Enkovaara, A. Ayuela, R.M. Nieminen: cond-mat/0304315 (2003)
[9.85] C.B. Bungaro, K.M. Rabe, A. Dal Corso: cond-mat/0304349 (2003)
[9.86] A. Zheludev, S.M. Shapiro, P. Wochner, A. Schwartz, M. Wall, L.E. Tanner: Phase Transitions **51**, 11310 (1995)
[9.87] J. Schiøtz, F.D.D. Tolla, K.W. Jacobsen: Nature **391**, 561 (1998)
[9.88] R.W. Siegel, G.E. Fougere: 'Mechanical properties of nanophase materials'. In: *Nanophase Materials, Synthesis – Properties – Applications*, NATO ASI Series E, vol. 260, ed. by G.C. Hadjipanayis, R.W. Siegel (Kluwer, Dordrecht 1990), pp. 233
[9.89] K. Asaka, Y. Hirotsu, T. Tadaki: 'Martensitic transformation in nanometer-sized particles of Fe-Ni alloys'. In: *ICOMAT 98: International Conference on Martensitic Transformations at San Carlos de Bariloche, Argentina, December 7–11, 1998*, ed. by M. Ahlers, G. Kostorz, M.Sade, Mater. Sci. Engin. A **273–275**, 257 (1999)
[9.90] K. Kadau, P. Entel, T.C. Germann, P.S. Lomdahl, B.L. Holian: 'Large-scale molecular-dynamics study of the nucleation process of martensite in Fe–Ni alloys'. In: *ESOMAT 2000: Fifth European Symposium on Martensitic Transformations and Shape Memory Alloys at Como, Italy, September 4–8, 2000*, ed. by G. Airoldi, S. Besseghini, J. Physique IV (France) **11**, Pr8–17 (2001)
[9.91] P.S. Lomdahl, P. Tamayo, N.G. Jensen, D.M. Beazley: '50 GFlops molecular dynamics on the CM-5'. In: *Proceedings of Supercomputing 93*, ed. by G.S. Ansell (IEEE Computer Society Press, Los Alamitos, CA, 1993) pp. 520–527
[9.92] D.M. Beazley, P.S. Lomdahl: Computers in Physics **11**, 230 (1997)
[9.93] http://bifrost.lanl.gov/MD/MD.html
[9.94] J.W.W. Frenken, P.M. Maree, J.F. van der Veen: Phys. Rev. B **34**, 7506 (1986)
[9.95] H. Reichert, O. Klein, H. Dosch, M. Denk, V. Honkimäki, T. Lippmann, G. Reiter: Nature **408**, 839 (2000)
[9.96] P. Buffat, J.-P. Borel: Phys. Rev. A **13**, 2287 (1976)
[9.97] M. Schmidt, J. Dongers, Th. Hippler, H. Haberland: Phys. Rev. Lett. **90**, 103401 (2003)

10 Computational Materials Science with Materials Studio®⋆: Applications in Catalysis

M.E. Grillo[1], J.W. Andzelm[2], N. Govind[2], G. Fitzgerald[2], and K.B. Stark[2]

[1] Accelrys GmbH, Inselkammerstr. 1, 82001 Unterhaching, Germany
[2] Accelrys Inc, 9685 Scranton Rd, San Diego, CA, 92121, USA

Abstract. In this article, developments in the primary ab initio quantum mechanics codes (CASTEP and DMol^3) to solve critical problems in materials design and catalysis research are reviewed. A novel, general-purpose internal coordinate optimization scheme in DMol^3 for periodic systems will be presented. Applications of this robust optimizer to a full range of solid-state systems will be discussed. Furthermore, the implementation of a transition state confirmation algorithm based on the nudged elastic band method to validate a transition state by connecting it to the proper reactant and product is explained.

Understanding and controlling the chemical and physical processes behind materials design (*e.g.*, chemical vapor deposition (CVD), corrosion, band-gap engineering in semiconductors) and reactions in heterogeneous catalysis is the principal goal of atomistic modeling and simulation technologies. Due to the impressive methodological developments, density-functional theory (DFT) is accepted in corporate, government and academic research labs as a powerful tool to calculate free energies, as well as the electronic and atomistic structure of medium-sized systems with predictive accuracy. Accelrys has focused a great deal of R&D resources over the past few years to provide fast, robust and accurate DFT codes for handling these systems and their processes.

In this article, developments in the primary ab initio quantum mechanics codes (CASTEP and DMol^3) to solve critical problems in materials design and catalysis research are reviewed. Atomic-level knowledge of structures is crucial to perform predictive simulations, be it of a carbon nanotube tip in a field emission based flat-panel display, or of surface defects in a partial oxidation catalyst. Therefore, geometry optimization of the structure of molecules and crystals is a crucial task in simulations using quantum mechanics codes. A novel, general-purpose internal coordinate optimization scheme in DMol^3 for periodic systems will be presented. Applications of this robust optimizer to a full range of solid-state systems will be discussed. The energies of intermediates and transition states are also critical for catalytic reactions engineering. For reactions at surfaces the dimension of phase space is so high that there may exist more than one transition state. The transition state (TS) search technique implemented in DMol^3, which does not require explicit calculation

---

⋆ Materials Studio is a registered trademark of Accelrys Inc.

M.E. Grillo, J.W. Andzelm, N. Govind, G. Fitzgerald, and K.B. Stark, Computational Materials Science with Materials Studio®, Lect. Notes Phys. **642**, 207–221 (2004)
http://www.springerlink.com/ © Springer-Verlag Berlin Heidelberg 2004

of the Hessian matrix, is presented and validation examples are discussed. Furthermore, the implementation of a transition state confirmation algorithm based on the nudged elastic band method to validate a transition state by connecting it to the proper reactant and product is explained. With input of a reactant, a product and TS, it returns a trajectory containing any alternative minima in the pathway containing these three structures. This approach can be extended to investigate potential energy surfaces containing several minima and maxima for both molecules and solids.

10.1 Introduction

A catalyst is defined as a substrate capable to change the kinetics of a chemical reaction increasing its rate by lowering the energy barrier to activate the reactant molecules [10.1]. Catalysts are not consumed during a chemical reaction, and their activity is defined as the reaction rate for conversion of reactants into products. Understanding at a molecular level the relation between the catalyst surface composition and its activity, assists in the catalyst design in industrial chemical processes.

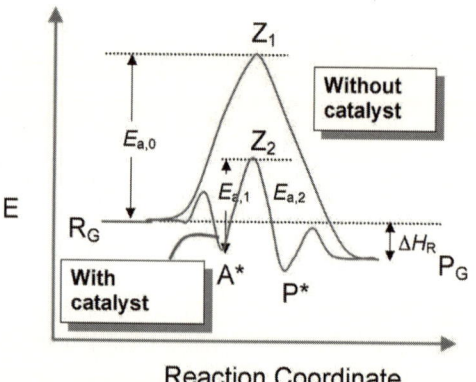

Fig. 10.1. Schematic representation of energy profile for a catalyzed and uncatalyzed reaction.

10.2 Geometry Optimization in Delocalised Internal Coordinates

Structural optimization of materials is a pre-requisite in most studies using atomistic simulation. It is well established by now that internal coordinates are more efficient for optimization of molecular systems than are Cartesian coordinates. In the case of periodic systems the geometry optimization is typically done using Cartesian coordinates. Only recently, the first use of delocalized internal coordinates for periodic systems was reported [10.2, 10.3]. This is because use of internal coordinates for solids is more challenging than for molecules. First, one needs to define a set of necessary primitive internal coordinates such as bonds, angles, torsions that span the entire space of the solid. For an infinite crystal, the number of internal coordinates is, in principle, infinite; however, the number of degrees of freedom required for optimization is just $3N - 3$, where N is the number of the atoms in a single unit cell. Therefore, only a subset of all internal coordinates of the crystal can be chosen for geometry optimization.

According to Baker et al. [10.4], the best combination of internal coordinates that span optimization space for molecules can be constructed by using so-called delocalized coordinate method. We have adopted that method for solid-state calculations. However, the number of primitive internal coordinates is significantly larger for solids, particularly for dense packed systems such as face centered cubic or hexagonal close packed metals. That requires an efficient method of selecting the minimal number of internal coordinates that span optimization space for solids. Such a "pruning algorithm" [10.5] was implemented recently in DMol^3 [10.6]. We will briefly present our method here, referring the interested reader to [10.2, 10.5] for a complete description.

The connection between internal and Cartesian displacements is defined with the usual $\mathbf{B}$ matrix [10.7]. For a molecular systems $\mathbf{B}$ is defined through,

$$B_{\alpha i} = \frac{\partial q_\alpha}{\partial x_i} \qquad (10.1)$$

where q_α are the primitive internal coordinates and x_i are the $3N$ Cartesian coordinates of the atoms. For an infinite periodic crystal there are a finite number of translationally unique primitive internals coordinates, $1 \leq \alpha \leq M$, which form a suitable set of coordinates to describe the system geometry. The generalization of $\mathbf{B}$ for a periodic system is then defined [10.2] as

$$B_{\alpha i} = \sum_R \frac{\partial q_\alpha}{\partial x_{iR}} \qquad (10.2)$$

where the index α runs over a translationally unique set of primitive internal coordinates, i runs over all $3N$ Cartesian coordinates of the atoms in one unit cell and the sum over $\mathbf{R}$ runs over all unit cells in the crystal. $\mathbf{B}$ is $M \times 3N$ dimensional giving the rate of change of primitive internal coordinate q_α on periodic displacement of the i-th atom in the unit cell. The infinite sum over atoms in (10.2) does not pose any problems in practice, because each primitive internal depends at most on the position of four atoms in the crystal. Hence, the terms in sum over $\mathbf{R}$ vanish after only a few cells.

Bakers scheme of construction of the best delocalized internal coordinates requires diagonalization of the $M \times M$ dimensional matrix $\mathbf{G} = \mathbf{B}\mathbf{B}^T$ to obtain the eigenvectors $\mathbf{U}$ with non-zero eigenvalues. However, it is convenient to introduce a closely related $3N \times 3N$ matrix $\mathbf{F}=\mathbf{B}^T\,\mathbf{B}$ [10.2, 10.5]. $\mathbf{U}$ can be obtained directly from the eigenvectors of $\mathbf{F}$. If $\mathbf{FT}=\mathbf{T}\Lambda$, with Λ being the diagonal matrix of eigenvalues and $\mathbf{T}$ the matrix of eigenvectors with non-zero eigenvalues, then the eigenvectors $\mathbf{U}$ of the matrix $\mathbf{G}$ are given by

$$\mathbf{U} = \mathbf{BT}\Lambda^{-\frac{1}{2}} \tag{10.3}$$

We can therefore obtain the matrix $\mathbf{U}$ that is needed to construct the delocalized coordinates by diagonalizing the matrix $\mathbf{F}$, which is much smaller than $\mathbf{G}$, since for solid-state systems M can be much greater than $3N$. This procedure will yield $3N - 3$ vectors in $\mathbf{U}$, which define our non-redundant delocalized coordinates. The non-redundant delocalized internal coordinates c_p^n of the system can be written as a linear combination of primitive internals through

$$c_p^n = \sum_\alpha q_\alpha U_{\alpha p} \tag{10.4}$$

An optimization procedure carried out in the space of non-redundant internal coordinates converges much faster than optimization in Cartesian coordinates.

The natural bonding topology of some crystals yields a set of coordinates that do not span the space of $\mathbf{F}$, and optimization in such an internal coordinate space would potentially fail to relax all degrees of freedom of the system. Examples where this often occurs include graphite, molecular crystals, molecules weakly bound to a surface and crystalline polymers. One solution to this problem is to add additional bonds that connect the disconnected fragments in the unit cell. However, this process is arbitrary and requires user intervention.

We proposed an alternative, completely automatic solution to this problem. It can be proven that the primitive internal coordinates span the optimization space if and only if $\mathbf{F}$ has precisely three zero eigenvalues [6]. If there are more than three zero eigenvalues, supplementary coordinates have to be added. These supplementary coordinates are constructed from the eigenvector space of $\mathbf{F}$ corresponding to the zero eigenvalues after three translational vectors were removed from the space [10.5].

The performance of the new delocalized coordinate optimizer versus the Cartesian based optimizer in DMol^3 was evaluated for many solids [10.2, 10.5]. The computational overhead associated with the optimizer is very small, so the speedup provided by the method can be measured as the number of Cartesian optimization cycles divided by the number of internal coordinate optimization cycles. Typically we find that delocalized internal coordinates are 3-5 times more efficient than Cartesian coordinates. The efficiency of internal coordinates increases with system size, and they are particularly beneficial in the study of systems that can undergo significant rearrangement

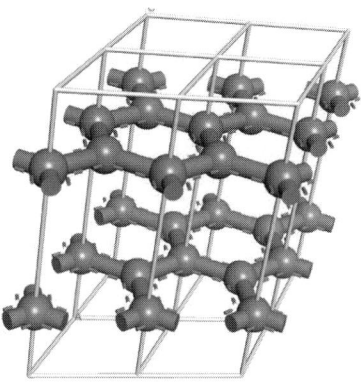

Fig. 10.2. Crystal of graphite. The internal coordinates of this crystal do not span the optimization space

of geometry. Periodic systems that belong to that category include surface reactions, molecular crystals and zeolites. For example, our recent study on Al-substituted chabazite revealed that optimization in delocalized internals converges in 21 cycles, whereas Cartesian coordinates require 96 cycles [10.2].

The need for efficient geometry optimization in studying catalytic reactions is illustrated in the DFT-study of the metal-catalyzed ring opening of maleic anhydride (*e.g.*, selective hydrogenation) [10.8]. This reaction is important for the commercial DuPont process of spandex fibers production. A detailed understanding of the reaction pathway at a molecular level is expected to provide valuable insights into factors controlling catalytic activity and selectivity.

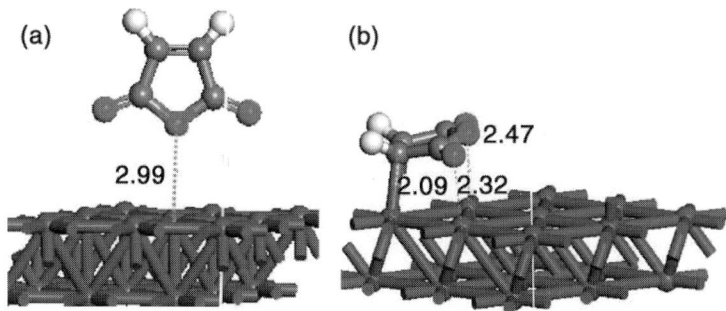

Fig. 10.3. η^1 (a) and di-σ (b) adsorption modes of maleic anhydride on Pd(111).

Two adsorption modes for maleic-anhydride adsorption on Pd(111) were studied: the atop (η^1) configuration, where the molecule is perpendicular to the surface (Fig. 10.3a); and the di-σ configuration (Fig. 10.3b), where the molecule is nearly parallel to the Pd surface.

DFT periodic slab calculations show that the di-σ adsorption configuration (with the ring bound through the olefinic C=C group) is preferred over the atop (η^1) adsorption mode. The calculated adsorption energy of 76 kJ/mol is in good agreement to the measured energy of 90 kJ/mol determined from ultra-high vacuum temperature programmed desorption (UHV-TPD) [10.9]. The di-σ adsorption configuration was found starting from the slightly distorted η^1 mode. Optimization in delocalized internals took 34 cycles to obtain the di-σ structure, whereas optimization using Cartesian coordinates requires 141 cycles to reach the same structure.

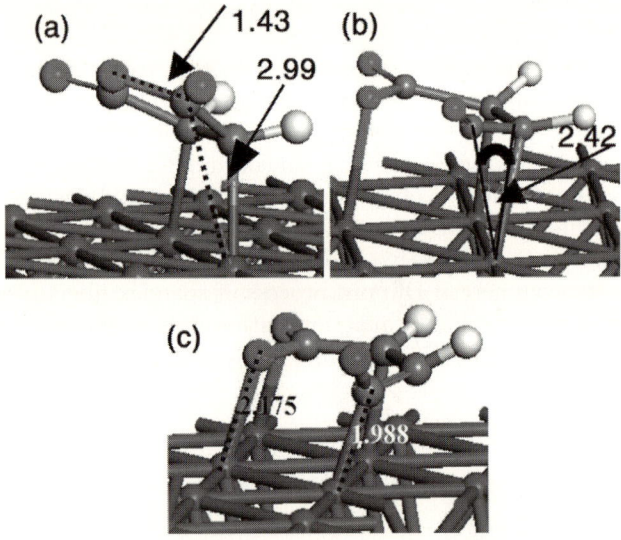

Fig. 10.4. Reactant di-σ (a), transition state (b) and the stable intermediate (c) for the C-O bond breaking in maleic anhydride on Pd(111) surface.

Figure 10.4 shows the reactant di-σ (a), transition state (b), and stable intermediate (c) structures for this reaction. The barrier of 59 kJ/mol is much smaller than for the ring opening of the (η^1) structure. The ring opening of di-σ structure is also an endothermic reaction, and leads to a stable intermediate structure of adsorption energy of about -39 kJ/mol.

10.3 Transition State Searching

The transition state (TS) search algorithm [10.10] implemented in our DFT codes CASTEP and DMol^3 is a generalized scheme based on the traditional linear and quadratic synchronous transit (LST/QST) method [10.10] coupled with previously-proposed conjugate gradient refinement ideas [10.11]. We refer the interested reader to 10 and references therein for a complete description of the method.

The method can be summarized with a flow diagram (Fig. 10.5). The energies of suitable reactant and product structures are first calculated. This is followed by a search for the LST maximum using essentially the method of Halgren and Lipscpomb [10.11]. This maximum energy structure provides an upper bound for the transition state. A conjugate gradient (CG) refinement is initiated for further refinement of the estimated transition state, searching in directions conjugate to the vector connecting reactants to products. CG methods make intelligent use of gradient information, thereby requiring no explicit Hessian to be calculated. As in a conventional optimization, the calculation is considered converged if all the residual forces on the structure fall below a certain threshold.

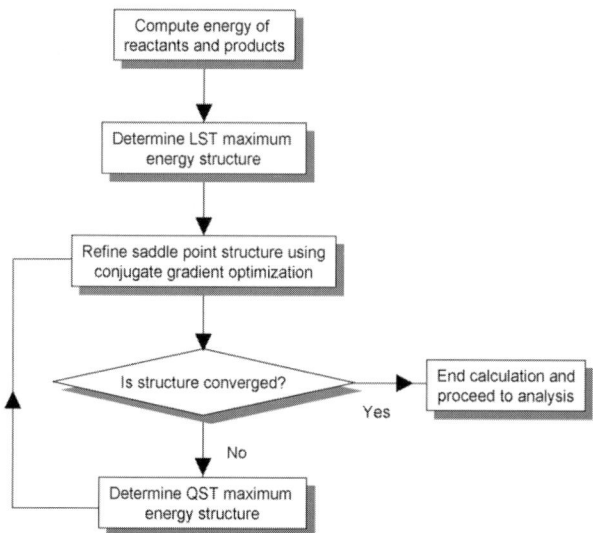

Fig. 10.5. Flow diagram illustrating the transition state search scheme.

If this is not achieved before the number of conjugate directions has been explored, a QST maximum search is performed using the reactant, product, and latest CG geometry. A new CG refinement cycle is started following the QST maximum search. This is continued until the required convergence is attained. A frequency analysis on this converged structure should yield exactly one negative eigenmode, corresponding to the direction by which the system would evolve away from the saddle point.

A number of calculations [10.12–10.15] have been performed using the new transition state search algorithm. For instance, it was used to investigate reaction pathways in the methanol to gasoline conversion process over zeolite catalysts [10.12, 10.13]. Reaction paths and energy barriers for the carbon-oxygen (C-O) bond cleavage and first carbon-carbon (C-C) bond formation were explored using all-electron periodic supercell calculations with DMol^3. Calculations along these lines can help understand mechanism of a reaction and ultimately improve existing catalysts.

The cleavage of the C-O bond of ethanol was studied by calculating the reaction path for the methylation of a zeolitic oxygen at the aluminosilicate Brønsted acid site. A reaction barrier of $\sim$ 54 kcal/mol was calculated for such reaction. This is a concerted SN2-type reaction involving breaking of the C-O bond in methanol and bond formation between the C atom and surface oxygen, with the formation of a water molecule resulting from the transfer of a zeolite proton to the hydroxyl group. A lower activation barrier of 44 kcal/mol was found in the presence of a second methanol molecule. The presence of a second methanol molecule in the zeolite cage causes the spontaneous deprotonation of the zeolite, leading to the formation of the methoxonium ion [10.12, 10.13].

Surface methoxyl species can undergo a reaction with incoming methanol molecule to form the ethanol molecule. The barrier for such reaction is 25 kcal/mol, if the water molecule is present. If the water is not involved in this reaction, the activation energy increases to about 50 kcal/mol. Therefore, water seems to be an important catalytic agent for the initial formation of the first C-C bond [10.12, 10.13].

A recent hypothesis concerning the formation of adsorbed ylide (CH2) species as an initial precursor for the first C-C bond formation was also explored. A very high calculated barrier of 78.5 kcal/mol ruled out the insertion of a ylide group from methoxyl into the Al-O bond. Such reaction would involve a significant rearrangement of the zeolite framework and it is clearly not competitive as compared with reactions involving adsorbed methoxyl species [10.12, 10.13].

10.4 Transition State Confirmation Algorithm

The transition state found by the LST/QST/CG method in Section 10.3 may not be the transition state connecting reactants and products or may be the transition state for a different reaction or reaction step. It is also possible that more than one transition state or other minima exist on a complex reaction path. These additional stationary points may be missed by the LST/QST procedure. Therefore, mapping of a reaction path is an important part of studying reactivity. The simplest way to calculate a reaction path is to start at a saddle point and take successive steps in the direction of the negative gradient. This steepest descent approach leads to a minimum energy path (MEP). If the coordinate system is mass-weighted, this is called an intrinsic reaction coordinate (IRC). The MEP (or IRC) path may be quite complicated and may have several minima. The highest saddle point is of most interest, as the overall reaction rate depends on the height of this reaction barrier. Following the reaction path allows to discover intermediate structures and to connect the reaction barrier to the correct reactant and product. DMol^3 uses the nudged elastic band (NEB) method [10.16] to validate a transition state by connecting it to the proper reactant and product [18]. NEB introduces a fictitious spring force that connects neighboring points on the path to ensure continuity of the path and projection of the force so that the system converges to the MEP.

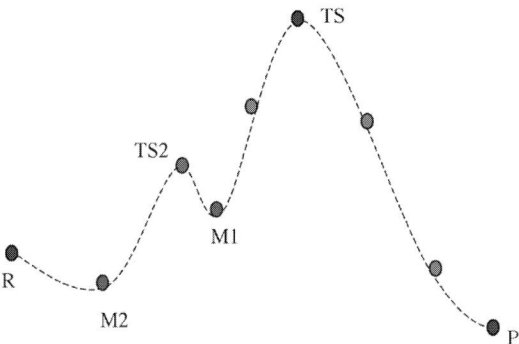

Fig. 10.6. Schematic representation of the minimum energy pathway. R: reactant, P: product, and TS: transition state. M1 and M2 are local minima.

Consider the reaction pathway in Fig. 10.6. The LST/QST algorithm takes the endpoints R and P as input and locates one of the local maxima on the reaction coordinate. Assume, in this example it locates the highest energy barrier, TS. The NEB algorithm will take these three points as input and return a trajectory that contains at least one point in the vicinity of the new minima, in this case M1 or M2. Points are shown in Fig. 10.6 as blue (minima) and red (maxima). In general, the path will also contain a few points that do not correspond to stationary points (green). A conventional NEB is intended to start from the end points and yield the TS and the entire reaction pathway, i.e., many green points. In contrast, the algorithm implemented in DMol^3 is intended to start at the TS, and to locate the alternative stationary points in the direction of reactants and products.

The optimization consists of 2 phases termed macroiteration and microiteration. It is assumed that the points are held loosely in place by springs.

Macroiterations consist of an optimization over all the images; microiterations consist of the constraint optimization of the molecular or crystal in a direction orthogonal to the reaction pathway. The calculation is complete when both the macroiterations and microiterations have converged. The result is an approximation to the MEP based on the number of structures used in the defined trajectory between the reactants and products. The newly found stationary points on the path will not be fully optimized because they have been subjected to constraints. However, they will be sufficiently close to their minimum structures, so that fu rther full optimization only takes very few steps and the energetics will not change substantially.

To illustrate the usefulness of this new NEB algorithm we consider the Diels-Alder reaction [19] of cis-butadiene and ethylene reacting to cyclohexene. The reactants and the product were optimized using DMol^3 with a DNP (double numeric + polarization) basis set and the BLYP functional [20]. A subsequent LST/QST calculation found the transition state structure depicted in Fig. 10.7.

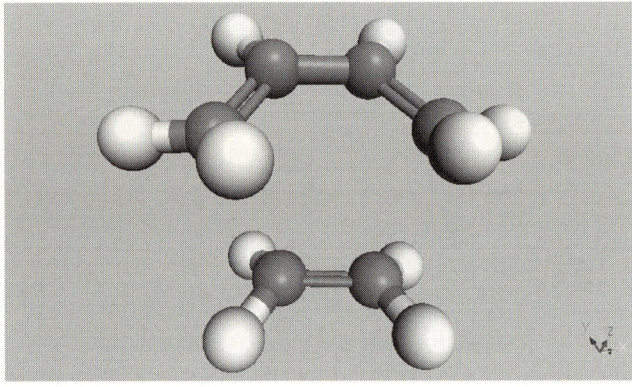

Fig. 10.7. Calculated transition state structure for the Diels-Alder reaction of butadiene and ethylene to cyclohexene.

A normal mode analysis performed on this activated complex resulted in one imaginary frequency of 508i conforming that the LST/QST algorithm successfully found the transition state of this reaction. However, the predicted activation energy was only 17.6 kcal/mol being about 30 % smaller than the experimental value of 25 kcal/mol [10.17]. Such a substantial deviation cannot be attributed to approximations made by the DFT method. Therefore, a NEB calculation was performed in order to check if any minima on the reaction path were missed. The result is displayed in Fig. 10.8.

The graph in Fig. 10.8 shows that the NEB procedure has found a new minimum on the reaction path, at path coordinate 0.23, where the path coordinate is defined between the reactants (0.0) and the products (1.0). The

Energy (Ha)

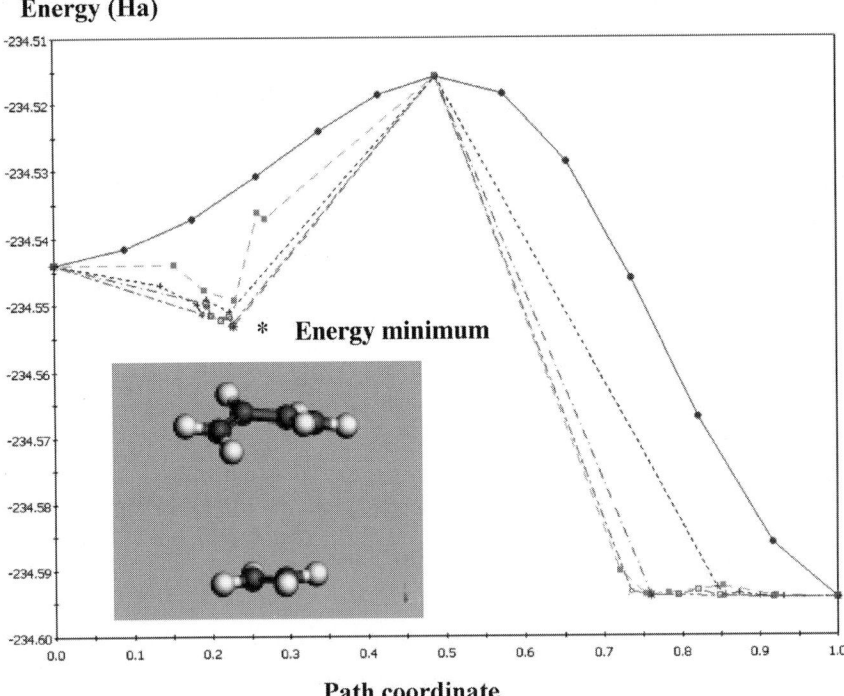

Path coordinate

Fig. 10.8. Resulting graph of the NEB calculation, and geometry of this new minimum structure. A new minimum is found at path coordinate 0.23. Note that the cis-butadiene exhibits a bent structure. The solid curve is the energy *vs* QST path 1. The non solid curves denote the energy *vs* MEP paths.

new minimum structure shown in Fig. 10.8 exhibits a reactant complex in which the cis-butadiene is bent rather than planar as in the reactant input structure. This new minimum corrects the activation energy to 23.9 kcal/mol in good agreement with experiment. A subsequent full optimization of all stationary points on the reaction path refined the activation energy to 23.3 kcal/mol, still in satisfying agreement with experiment.

10.5 Chemical Bonding and Elastic Properties of Corundum-Type Oxides: The Rhodium Oxide Case

Commercial needs for improving durability and performance of supported metal catalysts drive attention in investigating structural changes occurring at high oxidizing environments. In Rh-based three-way emission control catalysts, formation of reducible rhodium sesquioxide (Rh_2O_3) has been associated to the catalyst deactivation [10.18]. Moreover, understanding the nature of the bonding in transition metal oxides posses a challenge to ab initio methods, mainly due to the treatment of the exchange and correlation in partially occupied d bands when the material is reduced.

The band structure and elastic properties of the rhodium sesquioxide (Rh_2O_3) in the corundum structure were calculated using density-functional theory (DFT) and the generalized gradient approximation (GGA) of Perdew, Burke, and Ernzerhof (PBE) [10.19] for the exchange-correlation functional. The calculations were performed using the plane-wave pseudopotential-based code CASTEP [10.20]. The Bulk properties of Rh_2O_3 such as lattice constants (a_0, c_0), bulk modulus (B_0) and its pressure derivative (B') were obtained by calculating the total energy ($E(V)$) as a function of the unit-cell volume (V) and by fitting the calculated energy values to the Birch-Murnaghan equation of state (EOS) [10.21]. The minimum of the Murnaghan curve yields the optimum value of the energy, volume (V_0), lattice parameters (a_0 and c_0), and bulk modulus (B_0) at zero pressure, as reported in Table 10.1.

Table 10.1. Calculated unit-cell edges (a_0, c_0), volume per oxide formula-unit (V_0), internal fractional coordinates ($z(Rh)$, $x(O)$). Percentage errors are reported in the second line for each quantity.

	a_0	(c_0/a_0)	V_0	$z(Rh)$	$x(O)$
Rh_2O_3	5.135 Å	2.710	52.96 Å^3	0.3488	0.5498
[10.22]	+0.2 %	+0.3 %	+1.1 %	+1.4 %	+0.9 %

The bulk modulus as extracted from the Birch-Murnaghan equation of state, $B_0(\partial^2 E/\partial^2 V)_{V_0}$, is calculated to be 2.30 Mbars, and its pressure derivative 5.3. B_0 is related to the form of the function $E(V)$ around V_0. The predicted B_0 value is similar to that measured for other transition metal sesquioxides in the corundum structure as Fe_2O_3 (2.25 Mbars) [10.23] and Cr_2O_3 (2.38 Mbars) [10.23]. Hence, the rhodium sesquioxide is predicted to be as stiff as the corresponding iron and chromium structures under isotropic compression. Sofar, this value is not known experimentally for Rh_2O_3.

CASTEP was also used to calculate the full elastic constant tensor. The calculated elastic constants and bulk modulus obtained as a linear combination of elastic constants are summarized in Table 10.2. CASTEP uses the finite strain technique to predict elastic constants, which applies a given homogeneous deformation (strain) and calculates the resulting stress. The bulk modulus obtained by this technique of 2.29 Mbar agrees well with that obtained by fitting the energy data to the Birch-Murnaghan EOS of 2.11 Mbar.

The band structure and density of states reveal that corundum rhodium is a Mott-Hubbard insulator (see Fig. 10.9) with a metal-to-metal charge transition dominated by the dt_{2g} (valence band edge) and d_{eg} metal (conduction) at an energy of 1.4 eV. This is consistent with the observed semiconduct-

Table 10.2. Calculated elastic constants for corundum Rh_2O_3.

	C_{11}	C_{33}	C_{44}	C_{12}	C_{13}	C_{15}
Rh_2O_3	406	287	181	150	177	-18

ing behaviour of Rh_2O_3 [10.24]. The band structure presents three groups of bands: the low lying bands at about 20 eV correspond to oxygen $2p$ states, the group of bands between 8 and 4 eV originates mainly from oxygen $2p$ states, the bands from about 3 eV up to the Fermi level are of mainly Rh-$4dt_{2g}$ character, and the conduction band consists of mainly oxygen $2p$ states hybridized with Rh-$4deg$ levels.

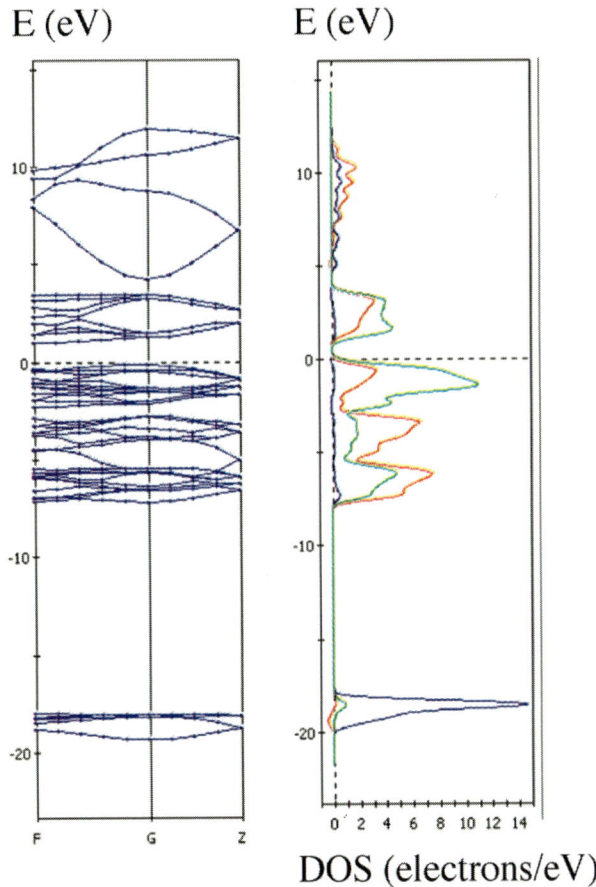

Fig. 10.9. Band structure and density of states (DOS). Red, green and blue curves represent the p-, d- and s-DOS, respectively. The energies are relative to the Fermi level in eV.

10.6 Summary

Industrial research in new materials development or improvement of existing ones for catalysis applications, crystal growth (*e.g.* chemical vapour deposition), and microelectronics are commonly based upon phenomenological theoretical models, and high throughput experimentation. Recent advances in quantum technology provide the materials researcher with tools to thoughtfully develop intelligent descriptors based upon the molecular understanding of the underlying processes. A novel optimization technique for solid-state calculations has been presented, which shows a superior performance for complex problems, *e.g.* surface reactions. A powerful algorithm to optimize the reaction path of a process by DFT within the transition state theory has been described. This method allows not only the evaluation of the energy difference between the particle at a minimum and the saddle point of the potential energy surface (energy barrier), but also, it validates a transition state by connecting it to the actual reactant and product. The constant pressure optimization algorithm in CASTEP allows calculating the equation of state, and the finite strain technique to predict the elastic properties for any kind of existing or novel materials.

Acknowledgment

The authors are grateful to Professor Mathew Neurock for his advice in the project on dissociation of maleic anhydride on Pd(111).

References

[10.1] B.C. Gates. In: *Catalytic Chemistry*, (John Wiley & Sons 1992).

[10.2] J. Andzelm, R.D. King-Smith, G. Fitzgerald: Chem. Phys. Lett. **335**, 321 (2001).

[10.3] K. Kudin, G.E. Scuseria, H.B. Schlegel: J.Chem.Phys. **114**, 2919 (2001).

[10.4] J. Baker, A. Kessi, B. Delley: J. Chem. Phys. **105**, 192 (1996).

[10.5] J. Andzelm, G.Fitzgerald, N.Govind, D. King-Smith: J. Chem. Phys., submitted (2003).

[10.6] B. Delley,: J. Chem. Phys. 92, 508 (1990); J. Phys. Chem. **100**, 6107 (1996).

[10.7] E. B. Wilson, J.C. Decius and P.C. Cross. In: *Molecular Vibrations*, (McGraw-Hill, New York 1955).

[10.8] V. Pallassana, M. Neurock, G. Coulston: J. Phys. Chem. **103**, 8973 (1999).

[10.9] Xu, Goodman, Langmuir **12**, 1807 (1996).

[10.10] N. Govind, M. Petersen, G. Fitzgerald, D. King-Smith, J. Andzelm: Computational Materials Science **28**, 250 (2003)

[10.11] T.A. Halgren, W.N. Lipscomb: Chem. Phys. Lett. **49**, 225 (1977); S. Bell, J.S. Crighton: J. Chem. Phys. **80**, 2464 (1984); S. Fischer, M. Karplus: Chem. Phys. Lett. **194**, 252 (1992); J.E. Sinclair, R. Fletcher: J. Phys. C **7**, 864 (1974).

[10.12] N. Govind, J.W. Andzelm, K. Reindel, G. Fitzgerald: Int. J. Mol. Sci. **3**, 423 (2002).
[10.13] J. Andzelm, N. Govind, G. Fitzgerald, A. Maiti: Int. J. Quant. Chem. **91**, 467 (2003).
[10.14] A. Maiti, N. Govind, P. Kung, D. King-Smith, J. Miller: C. Zhang, G. Whitwell, J. Chem. Phys. **117**, 8080 (2002).
[10.15] M. Petersen: Comp. Mater. Sci (submitted for publication).
[10.16] G. Henkelman, G. Johannesson, H. Jonsson: *Progress in Theoretical Chemistry and Physics*, ed. by S.D. Schwartz, (Kluwer Academic Publishers 2000).
[10.17] K.N. Houk, R.J. Loncharich, J.F. Blake, W.L Jorgensen: J. Am. Chem. Soc. **111**, 9172 (1989).
[10.18] G.L. Kellog: J. Catal. **92**, 167 (1985); G.L. Kellog: Surf. Sci. **171**, 359 (1986).
[10.19] J.P. Perdew, K. Burke, M. Ernzerhof: Phys. Rev. Lett. **77**, 3865 (1996).
[10.20] M.D. Segall, P.L.D. L indan, M.J. Probert, C.J. Pickard, P.J. Hasnip, S.J. Clark, M.C. Payne: J. Phys.: Cond. Matt. **14**, 2717 (2002).
[10.21] F. Birch: J. Geophys. Res. **57**, 227 (1952); F.D. Murnaghan: Proc. Natl. Acad. Sci. U.S.A. **30**, 224 (1944).
[10.22] C.E. Boman: Acta Chem. Scannd. **24**, 116 (1970); K.R. Poeppelmeier, J.M. Newsam, J.M. Brown: J. Solid State Chem. **60**, 68 (1985).
[10.23] L.W. Finger and R.M. Hazen, J. Appl. Phys. **51**, 5362 (1980).
[10.24] G. Bayer and H.G. Wiedemann, Thermochimica Acta **15**, 213 (1976).

11 Integration of Modelling at Various Length and Time Scales

S. McGrother[1], G. Goldbeck-Wood[2], and Y.M. Lam[3]

[1] Accelrys, 9685 Scranton Rd, San Diego, CA, 92121, USA
[2] 334 Cambridge Science Park, Cambridge CB4 0WN, UK
[3] Nanyang Technological University, School of Materials Engineering, Nanyang Avenue, Singapore 639798

Abstract. Materials modelling tools have become increasingly integrated in the R&D portfolio. The unique insights available through simulation of materials at a range of scales, from the quantum and molecular, via the mesoscale to the finite element level, can provide discontinuous scientific advances. These tools are well validated and produce reliable, quantitative information. A key demand of academic and industrial research is that these tools become ever more integrated: integrated at each length and time scale with experimental methods and knowledge as well as integrated across the spectrum of scales in order to capture the multiscale nature of organisation in many materials.

This paper will address recent efforts in this direction. The principal focus will be on the derivation of accurate input parameters for mesoscale simulation, and the subsequent use of finite element modeling to provide quantitative information regarding the properties of the simulated mesoscale morphologies.

In mesoscale modeling the familiar atomistic description of the molecules is coarse-grained, leading to beads of fluid (representing the collective degrees of freedom of many atoms). These beads interact through pair-potentials which, crucially if meaningful data are to be obtained, capture the underlying interactions of the constituent atoms. The use of atomistic modeling to derive such parameters will be discussed. The primary output of mesoscale modeling is phase morphologies with sizes up to the micron level. These morphologies are of interest, but little prediction of the material properties is available with the mesoscale tools. Finite element modeling can be used to predict physical and mechanical properties of arbitrary structures. Details of the link that has been established between Accelrys' Meso-Dyn [11.1] and MatSim's Palmyra-GridMorph [11.2] are given and highlighted with some recent validation work on polymer blends. These results suggest that the combination of simulations at multiple scales can unleash the power of modeling and yield important insights.

11.1 Introduction

There are many levels at which modeling can be useful, ranging from the highly detailed ab initio quantum mechanics, through classical molecular modeling to process engineering modeling. These computations significantly reduce wasted experiment, allow products and processes to be optimized and

S. McGrother, G. Goldbeck-Wood, and Y.M. Lam, Integration of Modelling at Various Length and Time Scales, Lect. Notes Phys. **642**, 223–233 (2004)
http://www.springerlink.com/

permit large numbers of candidate materials to be screened prior to production.

Accelrys offers quantum mechanics, molecular mechanics and mesoscale technologies. These methods cover many decades of both length and time scale (see Table 11.1), and can be applied to arbitrary materials: solids, liquids, interfaces, self-assembling fluids, gas phase molecules and liquid crystals, to name but a few. There are a number of factors, which need to be taken care of to ensure that these methods can be applied routinely and successfully. First and foremost of course are the validity and useability of each method on its own, followed by their interoperability in a common and efficient user environment. These points are taken care of in state-of-the-art packages like the Materials Studio®[1] software [11.3] distributed by Accelrys.

Table 11.1. Comparison of scales of modeling: quantum, classical atomistic simulation and mesoscale modeling

	Quantum	Atomistic	Mesoscale
Length	Angstroms	nm	100s of nm
Fundamental Unit	Electrons/nuclei	atoms	Beads representing many atoms
Time scale	fs	ns	ms
Dynamics	Too expensive	F=ma	Hydrodynamics

Of equal importance of course is the integration of the simulation methods with experiment. In modern materials research and development, one needs to be able to move almost seamlessly from experimental knowledge to simulation and back again, requiring multiple input-output relationships at a range of materials length and time scales. These can take the form of

– Materials QSAR: quantitative structure -activity (property) relationships for materials aim to correlate molecular simulation results with experimental measurements of (macroscale) properties.
– Parameterisation of simulations: accurate materials simulations based on input parameters gained from detailed simulation as well as experimental data.
– Multiscale simulations, based on establishing the appropriate communication between the methods.

In the following, we shall give further detail and examples for each of these cases.

---

[1] Materials Studio is a registered trademark of Accelrys Inc.

11.2 Structure-Activity and Structure-Property Approaches

Quantitative structure activity and property relationships (QSAR/QSPR) have long been used with great success in the life sciences. Based on experimental 'training set' data, correlations can be established between a range of molecular descriptors and biological activity. These correlations may take the form of equations derived by methods such as the Genetic Function Approximation [11.4], or neural networks. QSAR methods have proved to be powerful tools for the design of molecular libraries, investigating similarity and diversity as well as predicting properties.

Not surprisingly, such tools have also been applied successfully in a variety of materials cases as well [11.5, 11.6]. These statistical methods allow experimental information to be mined for important correlations, which can lead to deeper understanding of a material and optimised products. The correlations can be used to help design better materials. These new materials can be screened using the simulation methods and so an effective feedback loop is created which efficiently leads to new materials.

However, the complexity and multiscale nature of many materials and their properties pose particular challenges in the application of QSAR methods, which need to be address in future Materials QSAR tools. Firstly, there are many different materials classes with potentially very different sets of descriptors relevant to them. There is little knowledge so far about which are the most important ones relating for example to the prediction of permeability properties of polymer materials. Secondly, the calculation of the descriptors may involve simulations using methods at various scales, some of which may be computationally expensive.

11.3 Atomistic and Mesoscale Simulations and Their Parameterisation

Quantum, atomistic and mesoscale simulations provide valuable insights into the detailed physico-chemical behaviour of molecules and materials, and there are many properties, which can be determined directly from each, including structure, energies, stability, activity, diversity, solubility, adhesion, adsorption, diffusion, mechanical constants, spectra, and morphology. Ab initio quantum methods have the advantage that they can in principle be used for any element in the periodic table without specific parameterisation. They have been extensively developed so that one is now able to handle systems of a few hundred atoms routinely. For larger systems, however, methods requiring parameterisation are inevitable. In the following, we focus on force field developments for atomistic simulations and parameter determination for mesoscale simulations.

11.3.1 Atomistic Simulation

Fully atomistic simulation (where each atom is uniquely identified) is the core technology of polymer modelling. The methods use molecular mechanics, dynamics and Monte Carlo algorithms to probe the conformational and configurational behaviour of arbitrary materials. Most material properties can be inferred from these techniques, although properties that are fundamentally electronic (polarizability, dielectric constant, rates of chemical reaction, etc) are not the domain of classical simulation. The accuracy of property prediction relies on the force field, that is the mathematical expression used to create the potential function of the interacting components. These force fields comprise terms for: bond stretching, bond bending, torsional twisting, out of plane bending and pair-combinations of these. A typical force-field expression is given in 11.1.

$$
\begin{aligned}
E_{\text{POT}} = &\sum_b D_b \left[1 - e^{-\alpha(b-b_0)} \right] + \sum_\theta H_\theta (\theta - \theta_0)^2 \\
&+ \sum_\phi H_\phi \left[1 + s \cos(n\phi) \right] + \sum_\chi H_\chi \, \chi^2 \\
&+ \sum_b \sum_{b'} (b - b_0)(b' - b_0') + \sum_\theta \sum_{\theta'} F_{\theta\theta'} (\theta - \theta_0)(\theta' - \theta_0') \\
&\sum_b \sum_\theta F_{b\theta}(b - b_0)(\theta - \theta_0) + \sum_\chi \sum_{\chi'} F_{\chi\chi'} \, \chi\chi' \\
&+ \sum_\phi F_{\phi\theta\theta'} \cos\phi(\theta - \theta_0)(\theta' - \theta_0') \\
&+ \sum \varepsilon \left[\left(\frac{r^*}{r} \right)^{12} - 2 \left(\frac{r^*}{r} \right)^6 \right] + \sum q_i q_j / \varepsilon r_{ij}
\end{aligned}
\tag{11.1}
$$

Most force-fields are comparable in their accuracy for the minimum energy structure of simple molecules since they are parameterised to reproduce known behaviour. The true test of a force field is prediction of density and cohesive properties (heat of vaporization, solubility parameter, etc). For these properties the determining factor is the accuracy of non-bonded dispersion and electrostatic interactions (the last two terms in 11.1).

Accelrys has developed its own force field called COMPASS [11.7, 11.8], which stands for 'Condensed-phase Optimized Molecular Potentials for Atomistic Simulation Studies'. It is an ab initio force field because most parameters are initially derived based on data determined by ab initio quantum mechanics calculations. Following this step, parameters are optimized on the basis of experimental data for condensed phase properties. In particular, thermophysical data for molecular liquids and crystals are used to refine the nonbond parameters via molecular dynamics simulations. The result is a highly accurate force field, which gives unsurpassed prediction for density and cohesive

properties of a wide range of organic and some inorganic materials. The COM-
PASS force field is therefore a prime example of how accurate simulation at
one scale (in this case electronic) and experimental data can be combined to
great advantage in parameterisation of models at the next coarser scale (in
this case atomistic).

As an example of the typical $< 1\%$ accuracy in density prediction which
can be achieved with this method, Fig. 11.1 shows the comparison between
experimental and predicted densities for perfluorobutane over a range of temperatures [11.9].

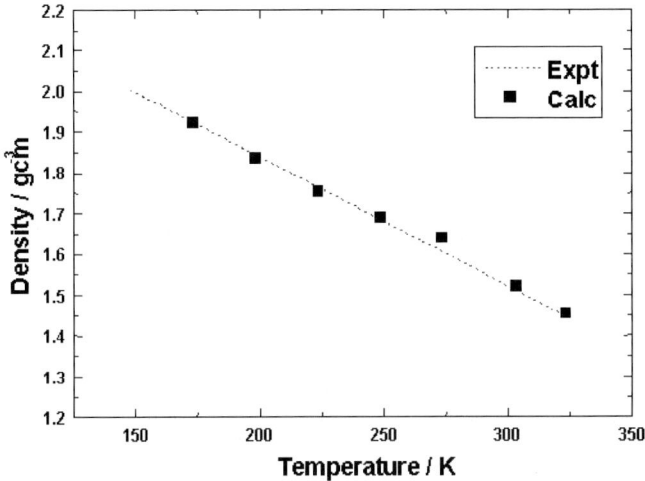

Fig. 11.1. Density versus temperature of perfluorobutane, comparing a fit to experimental data with values calculated from Molecular Dynamics simulations [11.9].

In Fig. 11.2 we show how COMPASS performs for the solubility parameter, which is the square root of the cohesive energy density [11.10]. It is crucial
to be accurate in this parameter, in particular if mixture or diffusivity data
is to be well reproduced. The toluene example shown in Fig. 11.2 [11.9] is
just one of many validations, which show that Molecular Dynamics simulations with the COMPASS force field meet this demand. We can conclude that
COMPASS gives highly accurate data for key properties of bulk materials.

11.3.2 Mesoscale Methods

In classical atomistic modelling, traditional Molecular Dynamics is used to
obtain thermodynamic information about a pure or mixed system. Properties
obtained using these microscopic simulations assume that the system is homogeneous in composition, structure and density, which is a limitation. When
a system is complex, comprising several components, only sparingly miscible,

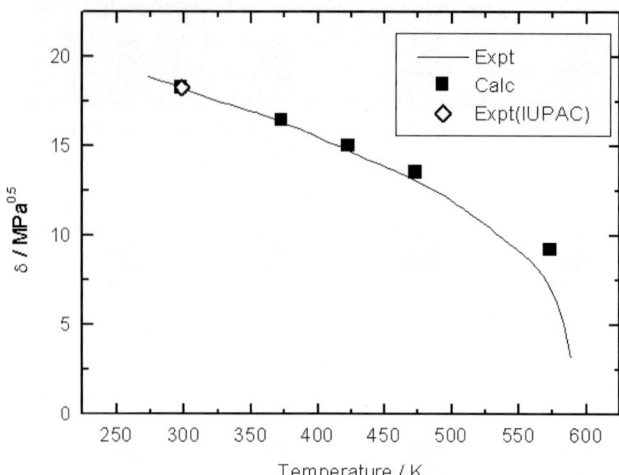

Fig. 11.2. The solubility parameter δ of toluene as a function of temperature. Calculations from molecular dynamics simulations with the COMPASS forcefield [11.9] agree well with a fit through experimental data and the official IUPAC value.

or the chain architecture is such that bulk phase separation is hampered by chemical bonds, exotic phases with remarkable properties can be observed. These so-called 'mesophases' comprise too many atoms for atomistic modeling to realistically describe. Hence coarse-grained methods (see Table 11.1) are better suited to such structures.

The primary techniques for mesoscale modeling are MesoDyn [11.11] and DPD [11.12]. These tools achieve longer length scales by uniting many atoms into a single bead, and longer time scales by integrating out the fast motions of the underlying particles leaving only soft, effective interactions. Complex self-assembling fluids, which have long-range order can be studied with these methods.

– MesoDyn

MesoDyn is a dynamic mean-field density functional theory for complex fluids [11.13]. The free energy comprises an ideal term based on a Gaussian Chain Hamiltonian representation of the polymeric materials, a Gibbs entropy contribution favoring mixing and a non-ideal term accounted for using a mean-field approximation. The key approximation is that in the time regime under consideration the distribution functions are optimized (i.e. the free energy is minimal). Applying appropriate constraints the optimal distribution can be obtained and related back to the free energy. We are left with a simple expression for the non-ideal term (obtained by invoking the random phase approximation-RPA):

$$F_{\text{RPA}}^{\text{nid}}[\rho] = \frac{1}{2} \sum_{IJ} \int_V \varepsilon_{IJ} \, \rho_I(r)\rho_J(r) \, dr \qquad (11.2)$$

Which assumes a local mean-field. However, the mean-field must account for the interchain interactions a non-local mean-field is preferred. A suitable choice leads to:

$$F^{\text{nid}}[\rho] = \frac{1}{2} \int_V \int_V \varepsilon_{IJ} \left(|r - r'| \right) \rho_I(r)\rho_J(r') \, dr \, dr' \qquad (11.3)$$

where $\varepsilon_{IJ} \left(|r - r'| \right)$ is a cohesive interaction defined by the same Gaussian kernel as in the ideal chain Hamiltonian. This parameter is then directly related to a calculable property, namely the Flory-Huggins interaction parameter χ.

– DPD

DPD is a particle based method that uses soft-spheres to represent groups of atoms, and incorporates hydrodynamic behavior via a random noise, which is coupled to a pair-wise dissipation. These terms are coupled so as to obey the fluctuation-dissipation theorem. Groot and Warren [11.12] established the connection between a DPD fluid and a real fluid again relating the bead-bead interaction potential to the Flory-Huggins parameter χ. For a full description of DPD and some of its applications see [11.14] and [11.15].

The two methods overlap, but DPD is preferred where concentrations are low, and MesoDyn is ideal for systems, which comprise polymer melts and blends.

11.3.3 Applications of Mesoscale Modeling

The mesoscale techniques have been used to rationalize complex behaviour of latex emulsions for the paints, coatings and lubricants industries [11.16]. A series of simulations was undertaken to establish the link between latex-particle size distribution and the hydrophilic chain length of the non-ionic surfactants used to stabilize the emulsion. The more uniform the size distribution the more reliable the paint appearance and application rheology. Several MesoDyn calculations were performed with various chain lengths and a system, which led to optimal distribution of the latex particles was found. This was then taken to the laboratory where an improved formulation was established.

In the area of drug delivery DPD and MesoDyn have found many applications including formulation stability, active release profiles, compatibilization, effect of hydrophobic drugs on micelle sizes in a pluronic solution and the role of excipients. These complex problems are difficult to conceptualize, are poorly served by static theories and are critical to the efficacy of a novel drug formulation.

The effect of temperature on self-assembled structures of amphiphilic block copolymers in aqueous solution has been recently studied with Meso-Dyn [11.17]. While, according to Flory-Huggins theory, the interaction parameter has a simple inverse dependence on temperature ($\chi \sim 1/T$), it is well known that corrections need to be applied for most polymer blends and solutions. A more appropriate relationship for polymer solutions is the following

$$\chi = \alpha + \frac{\beta}{T} \tag{11.4}$$

In simple term, α represents a non-combinatorial entropic contribution, and β represents the enthalpic contribution. This expression was fitted to experimental data of interaction parameters between poly (ethylene oxide) and water, as well as poly (propylene oxide) and water, respectively. Typically, these data are determined experimentally by vapour pressure measurements. The resulting $\chi(T)$ could then be used in MesoDyn simulations of Pluronic P85 (triblock copolymer of ethylene oxide, propylene oxide, ethylene oxide, with certain chain lengths) in water for a range of temperatures.

The experimental phase diagram [11.18] shows the striking range of phases exhibited by such a relatively simple system. In particular, it shows that at polymer concentrations above about 20%, a micellar phase is observed at low temperatures, and rods or cylinders are formed above about $60°C$.

MesoDyn simulations were performed at these temperatures, and the resulting morphologies are show in Fig. 11.3(a) and 11.3(b). They are extremely encouraging, with definite evidence of the correct phase evolving at the correct temperature and composition.

11.4 Multiscale Modeling

11.4.1 From the Molecular to the Mesoscale

In order to integrate the molecular level and the mesoscale, the atomistic simulation results can be used to parameterise mesoscale simulation by providing sensible coarse-graining methods and effective interactions between species [11.17, 11.19].

One such example is the work by Vergelati and Spyriouni [11.19]. Their aim was to investigate the compatibility of a polyamide with a poly (vinyl acetate), where the acetate was systematically hydrolyzed towards the poly vinyl alcohol. The authors started on the atomistic level, using Discover Molecular Dynamics with the COMPASS force-field to determine cohesive energy densities of the various mixtures. The Flory-Huggins interaction parameters of the blends could then be calculated and used as input to Meso-Dyn simulations. The bead size parameters for MesoDyn were determined from the molecular weight and characteristic ratios of the polymers. Encouragingly, the length scale and morphology of the phase separation observed

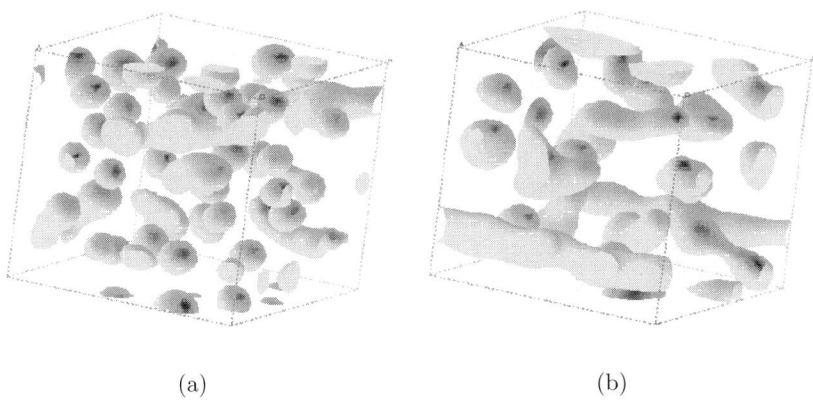

(a)   (b)

Fig. 11.3. (a) MesoDyn simulation of Pluronic PL85 in water at 27% concentration and temperature $15°C$. The isodensity surfaces of the hydrophobic component are shown, clearly revealing the micellar structure. (b) Morphology after increase of temperature to $70°C$, with the appropriate interaction parameter. The spherical micelles coalesce into rods, in line with experimental evidence.

in the simulation was found to match well with TEM results for the real materials.

A clear pathway has therefore been defined and is being more and more established for coarse-graining from the atomistic to the mesoscale. The major hole in the technology remains the reverse mapping from the mesoscale to the atomistic, where no adequate method has been developed.

11.4.2 From Mesoscale to Finite Element Simulation

The structures formed on the nanometer scale give rise to diverse and interesting material properties. As we have seen in the last section, mesoscale methods can be used with confidence to predict such structures. While some properties can be predicted directly from the mesoscale, property prediction given the knowledge of material structure and the property of the pure components that comprise the mixture has been developed widely in Finite Element Methods. An example of such a method, designed to deal with finely textured materials is Palmyra-GridMorph from MatSim [11.2]. Using standard solvers the finite element code can then predict the property for the realistic structured material.

As a test case for this combination of mesoscale and finite element methods we studied the oxygen diffusion through a material designed to act as a gas separation membrane. A binary blend of polystyrene and polybutadiene was simulated with MesoDyn using parameters obtained from atomistic level modeling. These polymers tend to phase segregate and large domains form with little interface. Upon addition of a diblock copolymer of the species (styrene and butadiene), the blend is compatibilized and the interfacial tension is lowered. The resulting morphology is far more complex with much smaller domains, more interfacial zones and frustrated regions. Both of these structures were analyzed for oxygen diffusion using GridMorph [11.20]. The pure component oxygen permaebilities for polystyrene and polybutadiene were obtained using the QSPR method Synthia [11.6, 11.21]. The results are given in Table 11.2.

Table 11.2. Oxygen permeability of two types of blends. Structures were simulated with MesoDyn, and permeabilities calculated for those structures using GridMorph.

System	Oxygen permeability (Dow Units)
Without Compatibilizer	970
With Compatibilizer	1040

The compatibilized blend shows increased permeability of oxygen, which can be attributed to an increase in the number of channels that the oxygen can choose to diffuse through. This study therefore uses atomistically obtained interaction energies and diffusivities to parameterize mesoscale methods and inform finite element tools, in order that mesoscopically calculated structures be analyzed for diffusion rates of the true material. This is an exciting development that we intend to pursue further.

11.5 Conclusion

The power of integrating modeling across different scales and with experimental data has been demonstrated. Combining experimental and simulation data in QSAR/QSPR methods generates valuable correlations and hence knowledge. Combining high quality measurements of some basic quantities (such as densities) with high-level simulations provides a successful parameterisation route for atomistic force field. Classical atomistic simulations with such a force field can then accurately predict material properties over a wide range of temperature, pressure and composition space. Furthermore, these simulations can in turn be used to derive input parameters for mesoscale simulations, while as above, additional experimental data can be used to hone the parameters further. A novel approach is to take the simulated mesoscale morphology as input to finite element methods in order to predict a wide range of material properties based on the morphology obtained. This now gives the modeler a route from the atomistic description of the system to a trust-worthy estimate of the properties of a material, obtained from the underlying molecules in a quantifiable manner.

Acknowledgement

The authors gratefully acknowledge the contribution of Fig. 11.1 and 11.2 by
Dr David Rigby, Accelrys.

References

[11.1] http://www.accelrys.com/mstudio/mesodyn.html
[11.2] http://www.matsim.ch/GridMorphE.html
[11.3] http://www.materials-studio.com/
[11.4] Rogers D. and A.J. Hopfinger, 'Application of Genetic Function Approximation to Quantitative Structure Activity Relationships and Quantitative Structure Property Relationships'. J. Chem. Inf. Comp. Sci., 34, 854-866, 1994.
[11.5] Charles H. Reynolds, J. Comb. Chem. 1999, 1: 297-306
[11.6] Jozef Bicerano, Prediction of Polymer Properties, Third Edition, Marcel Dekker, New York, 2002
[11.7] H. Sun and D. Rigby, Spectrochemica Acta, A53, 1301 (1997).
[11.8] H Sun J. Phys Chem. 102 7338 (1998).
[11.9] D. Rigby, Accelrys, unpublished
[11.10] F. H. Case and J. D. Honeycutt, Trends in Polymer Science, 2, 259 (1994)
[11.11] J.G.E.M. Fraaije, B.A.C. van Vlimmeren, N.M.Maurits, M.Postma, O.A. Evers, C. Hoffmann, P. Altevogt and G. Goldbeck-Wood, J. Chem Phys, 106 4260 (1997)
[11.12] R. D. Groot and P. B. Warren, J. Chem. Phys., 107 4423 (1997)
[11.13] J.G.E.M. Fraaije, J. Chem Phys, 99, 9202 (1993)
[11.14] P. Espanol and P. B. Warren, Europhysics Letters, 30(4), 191 (1995).
[11.15] Y. Kong, C. W. Manke, W. G. Madden and A. G. Schlijper, J. Chem. Phys., 107(2), 592 (1997).
[11.16] http://www.accelrys.com/cases/latex.html
[11.17] Y.M. Lam and G. Goldbeck-Wood, Polymer, in the press (2003)
[11.18] . Mortensen, Europhys. Lett. 19, 599 (1992)
[11.19] T. Spyriouni and C. Vergelati, Macromolecules, 34, 5306 (2001)
[11.20] Work performed by Albert Widdman-Schupak of MatSim GmbH, Switzerland
[11.21] http://www.accelrys.com/cerius2/synthia.html

12 Simulation of the Material Behavior from the Engineering Point of View – Classical Approaches and New Trends

H. Altenbach

Martin-Luther-Universität Halle-Wittenberg, 06099 Halle, Germany

Abstract. The analysis of any engineering structure is based on three steps - the choice of a material model, of a structural model and of an analytical or numerical method. All three items are interlinked, and the improvement, for example, of the structural model demands the improvement of the material behavior model and vice versa. In this contribution is reported on the engineering approaches to the material modelling. The models are mostly phenomenologically that means the real structure of the material is ignored. On the other hand, they are much simpler in comparison with micro-mechanically or physically based equations.

In the first part some general remarks on the principles of material modelling will be given. Three approaches to formulate material equations are presented. The main attention is paid to the inductive approach. In the final part some examples showing the application of the engineering models of material behavior to the analysis of thin-walled structures (beams, plates, shells) are given.

12.1 Introduction

Engineering structural components are made from engineering materials which can be classified as solids. The knowledge of their mechanical behavior and the possibilities to describe the behavior mathematically have a great influence on the accuracy of the structural analysis, on the lifetime predictions and at the same time on the satisfaction of the increasing safety requirements in practice. In addition, the mechanical material behavior model and its mathematical formulation are the basis of any computer aided structural analysis.

The mechanical behavior of engineering materials is not unique. Even in the case of the same material (that means with identical chemical composition, technological treatment, etc.) one obtains different stress-strain diagrams in dependence on the loading level, the temperature and the environmental conditions. The differences in mechanical behavior are more significant if we take into account various materials because for the microstructure of metals, polymers, ceramics, ... one cannot establish similarities. The last fact is well-known from the physics of solids or materials science modelling the behavior of materials on the atomistic or microstructural level and taking into consideration the crystalline structure of metals and alloys or the chains in polymers.

H. Altenbach, Simulation of the Material Behavior from the Engineering Point of View – Classical Approaches and New Trends, Lect. Notes Phys. **642**, 235–257 (2004)
http://www.springerlink.com/ © Springer-Verlag Berlin Heidelberg 2004

It is well-known that in a small range of loading and temperature any material behaves purely elastically and this behavior can be described by the HOOKE's law

$$\sigma = E\varepsilon \qquad (12.1)$$

with σ as the stress, ε as the strain in the stress direction and E as a material property (YOUNG's modulus). Equation (12.1) expresses the individual linear elastic response of each material. In addition, this expression is a phenomenological equation ignoring the microstructure of the material or the physical background. With other words, this equation is based only on macroscopic experimental observations.

Since engineering structures are mostly loaded multi-axially, (12.1) cannot fulfill the demands of a mechanical analysis of structural elements, etc. In this case one has to extend the one-dimensional linear elastic material law to the multi-axial case. This can be made by mathematical approach introducing a tensorial law for the linear elastic behavior as follows (e.g. [12.14, 12.7])

$$\sigma = {}^{(4)}\boldsymbol{E} \cdot\cdot\, \varepsilon^{\text{el}}, \qquad (12.2)$$

where $\boldsymbol{\sigma}$ and $\boldsymbol{\varepsilon}^{\text{el}}$ are the symmetric stress and elastic strain tensors, ${}^{(4)}\boldsymbol{E}$ is the fourth rank HOOKEan tensor expressing the elastic properties of the given material and "$\cdot$" denotes the scalar product. There is only one problem - the non-zero coordinates of ${}^{(4)}\boldsymbol{E}$ must be estimated and verified by tests.

In dependence on the kind of material, on the loading level , on the temperature conditions, etc., various stress-strain (load-elongation) or other (e.g. strain rate vs. time) diagrams can be observed. In this sense one can divide the material behavior into brittle and ductile, into elastic and inelastic, into creep and plasticity, etc. Note that such classification is not unique and so one can find various proposals in the literature (see, for example, [12.40, 12.29]).

For a long time the structural analysis models were grounded on the assumption that the material can be exploit when it shows only elastic behavior. On the other hand, from tests was well-known that the material behavior under special conditions is not only reversible. For example, the increase of the temperature and/or the loading provide inelastic behavior and there are two different types of inelastic behavior: time- or rate dependent (named creep) and spontaneous (named plastic). For the last one from tests it follows that a threshold exists and beyond the threshold the behavior becomes plastic. This threshold is the yield point and with respect due to the difficulties of its experimental establishment this point is defined as the upper limit of the elastic range. In a similar way brittle behavior can be discussed since after approaching a threshold the sudden loss of strength starts.

Due to the fact that the material behavior description is mostly based on experimental observations it must be taken into account microscopic and macroscopic observations. *The microscopic view of material behavior is considered either unnecessary or impractical to engineers who are primarily concerned with the analysis and design of load-carrying structural members* [12.23]. Therefore, for engineering structural analysis problems one needs macroscopic observations. On the other hand, the information on deformation or damage mechanisms follows from material science or material physics, and for this purposes the view inside the material (e.g. on the microstructure) is a powerful tool for establishing mechanism based equations.

The foundations of the phenomenological description are given by simple tests: the tension or the torsion tests. In both cases the dependence between the stress and the strain is registered, and this relation can be either linear or non-linear. Structural elements are mostly loaded in such way that a multi-axial stress and strain state can be expected. That means instead of a scalar constitutive equation a tensorial relation like (12.2) must be introduced. Another possibility to describe the three-dimensional material behavior is based on the use of the so-called equivalent properties concept. The comparison of one-dimensional and multi-dimensional states can be performed, for example, by introduction of scalar-valued equivalent stress and strain quantities instead of the stress and the strain tensor. Considering that a general approach in formulating equivalent quantities based on some physical principles does not exist the equivalence can be established only by introducing some engineering hypotheses.

In many textbooks the material behavior is classified as brittle or ductile. *This classification depends on the test or exploitation conditions because in the reality every material in dependence on the loading, temperature, etc., shows the tendency to be ductile and brittle together* [12.54]. In addition, focussing the attention on the microstructure and the deformation mechanisms one observe great differences for both cases. But in many practical situations this classification is helpful. For example, brittle behavior is a satisfying approximation to analyze the structural elements as elastic and to estimate the failure stress which manifests the loss of stiffness. Otherwise, the ductile behavior assumption allows to take into account the yielding and to calculate the plastic deformation. In both situations one has to estimate the limit stress (yield stress or ultimate strength) which should be observed experimentally and compared with the multi-axial stress state. The differences between the brittle and the ductile behavior are given in the consequences of the limit state estimations. Assuming ductile behavior in some cases the structural element can be exploited even if the stress level arises the yield stress. If we have a material behavior which is approximately brittle the structural element must be taken out from exploitation approaching the limit stress.

Below different approaches in material behavior modelling are presented. The main part is devoted to engineering models using macroscopic observations and which are named in the literature mostly phenomenological models.

12.2 Principles of Modelling

During the last decades the material behavior modelling is a common topic in mechanics, physics and material science. The reason for these activities is the wide range of application fields from modern technologies like electronics, aircraft, spacecraft and automotive industries to traditional areas like mechanical or civil engineering. It must be underlined that in mechanics, material science and physics different sizes (scales) of the research objects are usual. Discussions on the sizes (macroscopic, mesoscopic and microscopic) and the possibilities/limitations of the material modelling on the given size level are presented in various papers and books (e.g. [12.36, 12.12]).

As a typical example, in Fig. 12.1 are shown two principles of modelling creep problems - the microstructural on the left-hand-side and the phenomenological on the right-hand-side (after [12.20, 12.25, 12.45]). Both

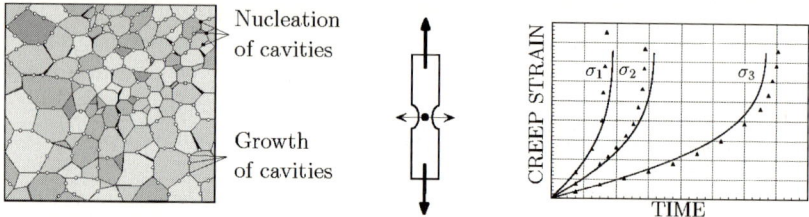

Fig. 12.1. Principles of modelling the creep behavior - microstructural (left) and phenomenological (right)

the microstructural and the phenomenological description of the creep behavior can be applied. The different scales yield various explanations of creep effects. Starting from the microstructure the nucleation and the growth of voids, cavities, etc. are the main observations. Finally, the coalescence of voids and cavities plays an important role for the lifetime predictions. All these effects are ruled by the local deformation and stress states. Similar discussions can be performed for other examples of material behavior. But the extension of the results based on the micromechanical effects to the macroscopic size is connected with some difficulties: the choice of suitable representative volume elements or the homogenization technique (see, for instance, [12.65, 12.52, 12.67, 12.51, 12.66, 12.42]). The phenomenological (macroscopic) modelling ignores the local changes, and the material behavior equations are based on macroscopic observations like creep curves (Fig. 12.1, right). The predictions of the averaged variables (stresses, strains) are in many cases accurate enough for engineering applications, but the predictions of failure states or of the lifetime are sometimes connected with difficulties.

Let us discuss microstructural creep and relevant to the creep process equations. If the starting point is the microstructure, so-called micromechanical models can be worked out [12.55]. These models allow a suitable description of the mechanisms of damage, ageing, hardening, etc. As an example, the following model will be introduced [12.20]

$$A_f = C_f \varepsilon_{eq}^{cr} \left(\frac{\sigma_I}{\sigma_{eq}}\right)^\mu, \qquad \mu = 0.5 - 3,$$

which can be used for the cavitated area fraction A_f description ($\varepsilon_{eq}^{cr}, \sigma_{eq}$ are the equivalent creep strain and the equivalent stress, σ_I is the first principal stress, C_f - a constant of proportionality and μ - the stress state coefficient). Note that this equation can be applied to uniaxial and multi-axial cases. The equivalent properties are proposed as the classical in the VON MISES sense. The problem is that this assumption is not correct in all cases [12.7], and so one have to define suitable equivalent strains and stresses. An equation for the ageing of particulate microstructure Φ was proposed [12.25]

$$\Phi = (1 - l_i/l), \qquad \dot{\Phi} = \left(\frac{K_c}{3}\right)(1 - \Phi)^4$$

with l_i as the initial precipitate spacing, K_c - a constant influenced by l_i and the temperature. The hardening can be represented by [12.25, 12.35]

$$\dot{H} = \frac{h_c \dot{\varepsilon}_{eq}^{cr}}{\sigma_{eq}}\left(1 - \frac{H}{H_*}\right)$$

with h_c, H_* as material constants. Assuming the sinh-law for the secondary creep (A, B are material constants)

$$\dot{\varepsilon}_{eq}^{cr} = A \sinh B\sigma_{eq}$$

one establishes a stress dependence of the creep, and a better modelling for materials with significant tertiary creep is possible [12.55]. Summarizing all equations describing the individual mechanisms HAYHURST and co-workers [12.31, 12.35] have been formulated a so-called mechanism-based multi-axial creep model reflecting all three creep stages

$$\dot{\varepsilon}_{eq}^{cr} = \frac{3}{2}A\frac{s}{\sigma_{eq}}\sinh\left[\frac{B\sigma_{eq(1-H)}}{(1-\Phi)(1-\omega)}\right],$$

$$\dot{H} = \frac{h_c}{\sigma_{eq}}A\sinh\left[\frac{B\sigma_{eq}(1-H)}{(1-\Phi)(1-\omega)}\right]\left(1 - \frac{H}{H^*}\right),$$

$$\dot{\Phi} = \frac{K_c}{3}(1-\Phi)^4,$$

$$\dot{\omega} = DA\left(\frac{\sigma_I}{\sigma_{eq}}\right)^\mu N\sinh\left[\frac{B\sigma_{eq}(1-H)}{(1-\Phi)(1-\omega)}\right]$$

with

$$\varepsilon_{eq}^{cr} = \sqrt{\frac{2}{3}\boldsymbol{\varepsilon}^{cr} \cdot\cdot \boldsymbol{\varepsilon}^{cr}}, \qquad \sigma_{eq} = \sigma_{vM} = \sqrt{\frac{3}{2}\boldsymbol{s} \cdot\cdot \boldsymbol{s}}, \qquad \boldsymbol{s} = \boldsymbol{\sigma} - \frac{1}{3}(\boldsymbol{\sigma} \cdot\cdot \boldsymbol{I})\boldsymbol{I}$$

and $N = 1$ for $\sigma_I > 0$ otherwise $N = 0$. D is a material constant and σ_{vM} the VON MISES equivalent stress. The model is proposed for the description of the uniaxial and the multi-axial creep behavior. In contrast to the classical, purely phenomenological model of the KACHANOV-RABOTNOV type [12.33, 12.56] this model allows to take into account more mechanisms (hardening, damage, ageing). The mechanism-based models must be applied carefully since the results of creep estimations for plates and shells based on the classical and the mechanism-based models are only similar during primary creep [12.15]. The lifetime predictions based on both models are not in a satisfying agreement. This was at first an unexpected result because the uniaxial creep curves approximations are similar for both models. Detailed investigations have shown that the multi-axial stress state was the reason for the disagreements. For the material model we have to use the full three-dimensional stress state representation, whereas for the plate or shell model the stress state is represented in a reduced (two-dimensional) form. These effects are known also in other cases [12.8].

12.3 Phenomenological Models

The phenomenological models are mostly used in continuum mechanics based structural analysis. There are two types of continuum mechanics equations (e.g. [12.23, 12.14, 12.53]):

- The first one is assumed as approximately independent from the specific material response. The equations of this type have the character of fundamental physical principles. They are formulated as balances or generalized balances for mass, momentum, momentum of momentum, energy and entropy.
- The second one reflects the individuality in the behavior of materials caused by their different internal constitution. The relevant equations are called constitutive and evolution equations, and it must be established a precise number of them to supplement the balance laws in such a way that the mechanical problem becomes determinate.

It must be noted that there are no possibilities to introduce a general constitutive equation [12.17] since there does not exist any physical principle on which the constitutive equation can be based. So one has to consider three main ways to establish material-dependent equations exist:

- The deductive derivation of constitutive equations grounded on mathematical techniques and axioms of the *theory of materials*.
- The inductive derivation of constitutive equations considering mathematical techniques and experimental performances (tests).

- The rheological modelling combining elements of the deductive and the inductive approaches.

Let us briefly discuss these three possibilities. The theory of materials (e.g. [12.37, 12.29]) has to answer on the following questions [12.28]:

- How constitutive equations can be derived on a deductive way?
- How material symmetries can be included?
- How internal kinematical constraints can be included?

The derivation of constitutive equations by the deductive approach must be follow at first the physical restrictions (e.g. the 2nd law of thermodynamics must be fulfilled) and at second the functional analysis (e.g. the STONE-WEIERSTRASS' approximation theorem or the RIESZ' representation theorem). For example, two- and three-dimensional deductive theories of viscoelasticity are presented by CHRISTENSEN [12.22] and ROTHERT [12.62] using this approach. In addition, if internal variables [12.24] are included one needs evolution equations describing the internal changes in the material (plasticity, damage, etc.). Following the physics of crystals [12.48] the material symmetries can be included. In general, this results in a reduced number of non-zero coordinates of the tensors of material properties (e.g. the HOOKEan tensor). The third question is related to such problems like the incompressibility condition. The discussion of the constitutive equations must be added by including (or not) of so-called constitutive axioms such as *Causality, Determinism, Equipresence, Material objectivity, Local action, Memory, Physical admissibility*. Using these axioms the theory of simple materials [12.47, 12.26] or the deductive theory of viscoelasticity (see [12.27]) were established. Note that the deductive approach is not so simple in use - this approach can be recommended only for complex material behavior (e.g. particle reinforced polymer suspensions). But the approach has a great advantage: any special variant of a more general theory satisfying all mathematical and physical demands which is embedded fulfills the same mathematical and physical demands.

The inductive deduction is an approach by analogy to the general engineering methods: from the simplest to the complex problems. In contrast to the deductive approach with embedded special cases now for any extension step one has to show the mathematical and physical correctness. Let us demonstrate the inductive approach for the creep behavior. In general, one can consider the additive split of the uniaxial strains ε as follows

$$\varepsilon = \varepsilon^{el} + \varepsilon^{th} + \varepsilon^{inel}$$

with $\varepsilon^{el}, \varepsilon^{th}, \varepsilon^{inel}$ as the elastic, the thermal and the inelastic strains, respectively. The third term can be split into a creep and a plastic (ε^{pl}) part

$$\varepsilon^{inel} = \varepsilon^{cr} + \varepsilon^{pl}$$

Neglecting the thermal and the plastic strains the following simplification yields

$$\varepsilon = \varepsilon^{\text{el}} + \varepsilon^{\text{cr}}$$

Introducing the constitutive relations as

$$f(\varepsilon, \sigma, t, T) = 0,$$

where T denotes the fixed temperature, at prescribed stress history $\sigma(t)$ this equation determines the strain variation $\varepsilon(t)$ and vice versa [12.64]. The identification of this equation is connected with the performance of possible tests, for example, the creep or the relaxation test. For constant uniaxial stresses the creep law can be approximated by separating the stress, time and temperature influences

$$\varepsilon^{\text{cr}} = f_1(\sigma) f_2(t) f_3(T)$$

Representations for the stress function $f_1(\sigma)$, the time function $f_2(t)$ and the temperature function $f_3(T)$ are well-known from literature. A generalization for multi-axial states is possible, for instance, by analogy to the flow theory in plasticity. Below we focus our attention on the stress function. Various time function or temperature function representations are presented in the literature (among others [12.41, 12.63, 12.64]).

The creep behavior can be divided into three stages as shown in Fig. 12.2. The first stage is, characterized by a decreasing creep strain rate. The second stage is the stationary creep with a constant creep strain rate (the creep strains are proportional with respect to the time). The last stage is the tertiary creep with an increasing creep strain rate. The main softening process is realized by damage. Note that some materials show no tertiary creep, others have a very short primary creep period. Bur in all cases we obtain rupture (failure) as the final state.

The modelling of the phenomenological creep behavior based on the division of the creep curve (creep strain versus time at constant stresses) into three parts allows the multi-axial creep description as follows. Let us introduce a set of equations, which contains three different types: an equation for the creep strain rate tensor influenced by hardening and/or damage and two sets of evolution equations for the hardening and the damage variables

$$\dot{\varepsilon}^{\text{cr}} = \frac{\partial F(\sigma_{\text{eq}}, T, H_1, \ldots, H_n, \omega_1, \ldots, \omega_m)}{\partial s},$$

$$\dot{H}_i = H_i(\sigma_{\text{eq}}^H, T, H_1, \ldots, H_q, \omega_1, \ldots, \omega_m), \qquad i = 1, 2, \ldots, q \qquad (12.3)$$

$$\dot{\omega}_k = \omega_k(\sigma_{\text{eq}}^\omega, T, H_1, \ldots, H_q, \omega_1, \ldots, \omega_m), \qquad k = 1, 2, \ldots, m$$

It must be noted that in addition to the problem how to formulate the set of creep equations the identification problem for this set must be solved. The starting point is the identification of the creep equation for the secondary part. Since the creep is stress and temperature dependent this identification

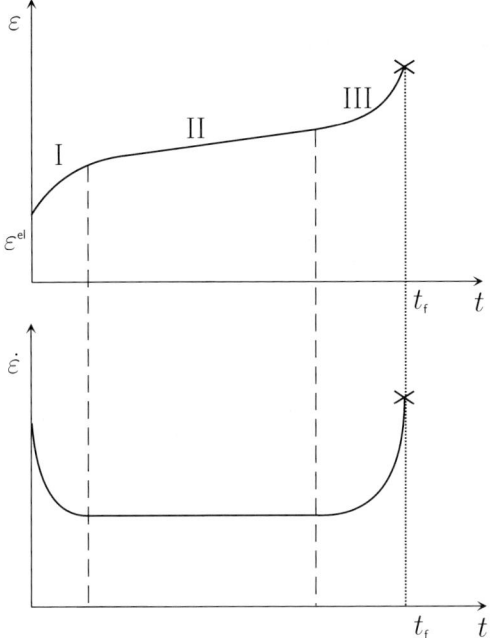

Fig. 12.2. Creep behavior: I - primary creep (hardening), II - secondary creep (hardening and softening equilibrium), III - tertiary creep (softening = degradation), × - rupture

can be realized at fixed temperatures and for constant stress in a very easy way. If such an approach is not satisfying one has to consider a more complex situation and use approaches presented, for example, in [12.27] and valid for a wide range of stresses.

The creep is influenced by hardening and damaging processes and based on the identified secondary creep equation. The equations for primary and tertiary creep can be established by extension of the secondary creep equations. The identification can be performed by analogy using additional experimental results. This approach is presented in several textbooks [12.56, 12.40, 12.64, 12.63, 12.39].

Finally let us briefly discuss the method of rheological modelling which is based on the idea to formulate phenomenological material equations for simple cases which are consistent in the sense of thermomechanics and to connect in a first approximation the basic models of the elastic HOOKE element, the viscous NEWTON element and the plastic ST. VENANT element in series or in parallel to describe a more complex material behavior. The classical method of rheological modelling was restricted to small strains, homogeneous and isotropic materials grounded on two assumptions

– The volume dilatation is pure elastic.
– The significant differences in the material equations for various materials are based on shear effects only.

New developments demonstrate (e.g. [12.53]) that the method of rheological modelling can be extended to large deformations.

12.4 Classical and Nonclassical Material Behavior Models

Classical and nonclassical material behavior can be observed for elasticity, creep, plasticity, damage, etc. Examples of experimental results are summarized in various papers and monographs. A simple definition of classical and nonclassical behavior can be given as [12.7]:

Definition 1. *All material models for isotropic materials of which the specific material parameters of the equivalent stress can be identified with the help of one basic test (mostly uniaxial tension test) are called classical models. If more than one test is necessary for the parameter identification we have a nonclassical or generalized model.*

Definition 2. *All material models for anisotropic materials for which the identification of the specific material parameters is possible with the help of one basic test, but in different directions, are defined as classical models, otherwise the models are called nonclassical or generalized models.*

Below we focus the attention on the classical and the non-classical creep.

Creep mechanics is a branch of solid mechanics (continuum mechanics) with a history of more than one hundred years. Due to the practical problems the first summary of creep problems had been reported in the nineteenth century by ANDRADE (1910). From the very beginning the literature on creep problems was divided into two parts - one connected with engineering (mostly phenomenological) theories oriented to the designer needs, the other containing approaches describing creep in terms of physics.

The next big step after ANDRADE was done by BAILEY (1929) and NORTON (1929) replacing HOOKE's law in the elastic analysis of structural mechanics by a power law relation between creep rate and stress. In contrast to the elastic case, even in the uniaxial case this new constitutive relation can be characterized by two differences: at first, this relation is nonlinear and at second, one needs more material parameters. As was pointed out by, for instance, BETTEN [12.18] the power law can be based on images about the mechanisms in the material (diffusion controlled creep). Later the uniaxial power law was generalized to multi-axial states by ODQVIST (1933) and BAILEY (1935). The generalization was presented in tensorial form using some invariants. In addition, other approximations of experimental creep data were introduced: the sinh-law, the exponential law, etc.

Until World War II the dominant majority of applications were related to metals and alloys. With the increasing use of plastics (polymeric materials) it was necessary to model the creep behavior of these materials. REINER (1945) and RABOTNOV (1948) contributed two new directions in the creep mechanics: the rheological modelling and the modelling based on integral equations. The main ideas of both directions are summarized in [12.58, 12.57]. At the same time the mathematical theory of creep using the tensor function representation was finished.

During the first half of the twentieth century the main attention in the phenomenological description was focused on primary and secondary creep. In 1958 LAZAR KACHANOV published his famous paper [12.33] devoted to the problem of modelling tertiary creep. His idea to modify the equations of secondary creep introducing a new internal variable (in KACHANOV's paper this is the continuity, later this variable was substituted by the dual variable named damage) was the starting point for the development of a new branch in the mechanics of deformable solids: continuum damage mechanics (CDM).

The creep-damage equations are the basis of many investigations of structural mechanics problems. Let us underline that the lifetime prediction and the solution of other problems related to the safer exploitation of machines, etc. are only possible by the use of the CDM-knowledge collected during the last 40 years. Now we can consider that based on a great number of theories many practical problems are solved. In addition, numerous experimental data are published, and the phenomenological approaches are founded with the help of materials science and the physics of solids [12.61, 12.40, 12.44].

Let us assume that the material behavior is isotropic and the temperature is constant. The creep behavior can be presented by the following uniaxial state (constitutive) equation for secondary creep - the NORTON's law

$$\dot{\varepsilon}^{\mathrm{cr}} = a\sigma^n, \tag{12.4}$$

where a and n are material constants depending on the temperature. This state equation can be easily extended to the case of primary and tertiary creep. For example, by KACHANOV [12.33, 12.34] the isotropic damage was introduced as follows. In a certain section of a body with the unit normal $\boldsymbol{n}$ the initial area of the undamaged material is denoted by A_0. As the result of tertiary creep the detoriation of the initially undamaged area starts. If A can be identified such as the actual area, an internal variable (damage) can be introduced

$$\omega = \frac{A_0 - A}{A_0} \quad (0 \le \omega \le 1)$$

and the power law can be modified introducing the actual stress as $\sigma/(1-\omega)$

$$\dot{\varepsilon}^{\mathrm{cr}} = a\left(\frac{\sigma}{1-\omega}\right)^n$$

This equation must be supplemented by an evolution equation for the damage variable

$$\dot{\omega} = b \left(\frac{\sigma}{1-\omega} \right)^k$$

For engineering applications one has to generalize the uniaxial equations to three-dimensional. Following ODQVIST and HULT [12.50] the stationary creep behavior can be described by a tensorial equation as

$$\dot{\boldsymbol{\varepsilon}}^{\mathrm{cr}} = \frac{3}{2} a \sigma_{\mathrm{eq}}^{n-1} \boldsymbol{s} \tag{12.5}$$

This equation is founded on the following five hypotheses [12.49]: incompressible material, creep rate independent of superimposed hydrostatic pressure, existence of a flow potential, isotropic material (co-axiality of tensors of stress and strain rate) and NORTON's law holds in the special case of uniaxial stress. It was assumed that the equivalence between the uniaxial and the three-dimensional stress states can be given in the VON MISES sense that means

$$\sigma_{\mathrm{eq}}^2 = \sigma_{\mathrm{vM}}^2 = \frac{3}{2} \boldsymbol{s} \cdot\cdot \boldsymbol{s}$$

The equivalence of the three-dimensional and the one-dimensional creep equations (12.4) and (12.5) was demonstrated, for example, in [12.2].

By analogy the tensorial equation for the stationary creep behavior can be extended to the case of tertiary creep taking into account damage only. The starting point is the strain equivalence principle introduced, for instance, in [12.39]. Now we have

$$\dot{\boldsymbol{\varepsilon}} = \frac{3}{2} a \left(\frac{\sigma_{\mathrm{vM}}}{1-\omega} \right)^{n-1} \boldsymbol{s} \qquad \text{with} \qquad \dot{\omega} = b \left(\frac{\sigma_{\mathrm{vM}}}{1-\omega} \right)^k$$

This is the simplest form of a possible extension of the tensorial secondary creep equation to the case of creep-damage.

Now we introduce the non-classical creep behavior. The following facts can be established experimentally: the tension and the compression test show results which differ significantly, the equivalent creep strains vs. time curves are not identical for various stress states, an influence of the hydrostatic pressure on the creep behavior can be obtained, the volumetric creep depends on the kind of loading, etc. Summarizing the experimental results one concludes that the kind of loading (kind of stress state) results in changes of the creep behavior which cannot described by any classical theory discussed above.

The starting point of the formulation of a non-classical creep law for isotropic materials is the flow theory with

$$\dot{\boldsymbol{\varepsilon}}^{\mathrm{cr}} = \dot{\eta} \frac{\partial \Phi}{\partial \boldsymbol{\sigma}}$$

Here $\dot{\eta}$ is a scalar factor and Φ denotes the creep potential. The creep potential is a function of the stress tensor, the temperature and internal (hidden) variables [12.24, 12.56]. Instead of the stress tensor dependence we can take into account the equivalent stress

$$\Phi = \Phi(\sigma_{eq}, q_1, \ldots, q_n)|_{T-\text{fixed}}$$

with $q_k (k = 1, \ldots, n)$ as a set of internal variables describing the hardening, the softening or the damage behavior. In addition to the creep law, we have to introduce evolution equations for internal variables.

Assuming isotropic creep behavior and neglecting internal processes we get a reduced form for the creep potential $\Phi = \Phi(\sigma_{eq})$ with $\sigma_{eq} = \sigma_{eq}(I_\sigma, II_\sigma, III_\sigma)$. The I_σ, II_σ and III_σ denote a set of possible stress tensor invariants. The choice of the invariants is not unique. Various sets are proposed by several authors, (cf. [12.7, 12.21, 12.68]). A systematic investigation of this problem is presented, for example, by BETTEN [12.18].

Let us introduce the basic invariants

$$I_1 = \sigma \cdot\cdot I, \quad I_2 = \sigma \cdot\cdot \sigma, \quad I_3 = (\sigma \cdot \sigma) \cdot\cdot \sigma$$

Finally, an equivalent stress using combinations of the basic invariants

$$\sigma_{eq} = \alpha \mu_1 I_1 + \beta \sqrt[2]{\mu_2 I_1^2 + \mu_3 I_2} + \gamma \sqrt[3]{\mu_4 I_1^3 + \mu_5 I_1 I_2 + \mu_6 I_3}$$

can be introduced. α, β, γ are curve fitting parameters and $\mu_1, \ldots, \mu_6$ denote material characteristics. Similar coefficients to the introduced α, β and γ are used for the characterization of different failure mechanisms by other authors [12.30, 12.38]. Below we assume $\alpha = \beta = \gamma = 1$. After some calculations we establish the creep law as [12.7]

$$\dot{\varepsilon}^{cr} = \dot{\varepsilon}^{cr}_{eq} \left[\mu_1 I + \frac{\mu_2 I_1 I + \mu_3 \sigma}{\sigma_2} + \frac{(\mu_4 I_1^2 + \frac{\mu_5}{3} I_2)I + \frac{2}{3}\mu_5 I_1 \sigma + \mu_6 \sigma \cdot \sigma}{\sigma_3^2} \right]$$

$$(12.6)$$

with $\dot{\varepsilon}^{cr}_{eq}$ following from tests. Suitable approximations are the power law, the sinh-law, the exp-law, etc. Equation (12.6) is based on the equivalence of the dissipation power for uniaxial and multi-axial cases.

Up to now the parameters μ_i in the non-classical creep law (12.6) are not specified and we have to identify them. There are different possibilities of identification [12.1]. For creep problem a suitable identification procedure was presented by the author and his co-workers [12.11, 12.13, 12.7, 12.5]. In all these cases the approximation of the uniaxial secondary creep was given in the form of NORTON's law. The question what kind of tests one has to perform is open and depends on the experimental equipment available for providing tests and the possibilities to find closed solutions for the creep mechanics problems similar to the tests.

12.5 Analysis of Thin-Walled Structures

In many creep applications the structural elements can be modelled as thin-walled structures that means one or two geometrical dimensions are much smaller than the other two or the third, respectively. Examples are beams, plates and shells. The governing equations for these models can be deduced starting from the thinness-hypothesis [12.16]. It is well-known that such theories have advantages (the set of field equations is much simpler), but also disadvantages (especially in the case of inelastic material behavior). Focussing our attention on creep problems we have to recognize at first that we have a time-dependent material behavior and the stresses, the strains and the displacements, for example, are changing with time. At second, since the damage behavior is not the same for tension and compression states the correct description of the stress state has a great influence on the accuracy of the solution. This is important because the reduction of the three-dimensional continuum mechanics equations to the two-dimensional or one-dimensional field equations is connected with a "loss of information" leading to some constraints [12.43] and/or an incomplete description of the stress state (neglecting of stress tensor components). For example, BERNOULLI's beam theory or KIRCHHOFF's plate theory are based on the assumption that the transverse shear and normal stresses can be ignored. For creep problems such assumptions can result in errors.

One problem among others arising by creep-damage analysis can be related to the quality of the finite element predictions. Structural analysis of pressure vessels, pipes, pipe bends, etc., can be performed using the mechanical models of plane stress (strain) states or the equations of the beam, plate or shell theory. In the case of plane stress (strain) problems numerous finite element simulations considering creep-damage effects have been made. Since the examples confirm the ability of finite element simulations to predict stress redistributions and failure times with accuracy enough for engineering applications, little effort has been made for the analysis of transversely loaded thin-walled structures.

If mechanical models of beams, thin shells or plates are used in finite element based creep analysis, the following question requires a special consideration: how sensitive are the long-term predictions of thin-walled structures with predominating bending stresses to the type of elements and to the mesh sizes. Even for steady state creep the deflection function of coordinates is a polynomial of an order significantly higher as that for the elastic solution and the order of the polynomials depends on the power in the NORTON creep law [12.19]. If the damage evolution is considered, a closed solution is not possible but using the RITZ method the approximation of unknowns necessary for a convergent solution can be established.

In [12.8] the results of finite element solutions using the ANSYS code are presented. Based on the numerical analysis only the following questions are discussed. First special solutions for beams and plates including damage effects based on the RITZ method are constructed. These solutions are used for the verification of the ANSYS user creep material subroutine which was modified taking into account the damage evolution. Based on various numerical tests the mesh sensitivity of long-term solutions in creep bending of beams and plates was demonstrated. The results for time-dependent deflections and stresses based on different plane stress, shell and solid elements available in the ANSYS code for the plasticity and creep analysis were compared. For the simplicity of numerical studies the discussion is limited to steady state loads and temperatures.

Two possibilities connected with different analytical and numerical effort are available for the analysis of thin-walled structures. The first one is the use of general three-dimensional equations and solid finite elements. The second one is, for instance, the application of shell equations and shell based elements usually of the REISSNER-MINDLIN type. The solid concept can be considered as a general and in many cases more accurate for structural analysis, particularly by studying damage effects. However, the numerical effort for the approximate solving of three-dimensional equations is not comparable with the effort applying theories based on dimension reduced equations, but the last one in some cases yields in over- or underestimations and the distributions of the mechanical field quantities are partly incorrect and far from the reality. One possibility to improve the results is the use of shear correction factors which are constant and material independent in the case of homogeneous elastic plates or beams. On the state of art of different approaches in establishing shear correction factors is reported [12.3, 12.4, 12.32].

From the creep-damage analysis of a pipe bend the following results were obtained [12.9]:

1. With the σ_I-dependency the stress state effect of the damage evolution is considered and a zero damage rate is assumed if all three principal stresses are negative [12.40]. The calculations which were based on the use of the SOLID 45 and the SHELL 43 elements have shown that the initial distributions of the first principal stress are approximately the same, whereas the final distributions differ significantly near the pipe bend edge. In addition, appreciable differences between the time variations corresponding to the type of elements are observable. In the case of the solid model the first principal stress increases during the transient stage and relaxes with the damage evolution, the shell model yields a slight increasing of the first principal stress during the transient stage only.

2. The critical damage state occurs on the inner surface of the pipe bend. The solid model yields the maximum damage at the edge of the pipe bend. The shell model predicts two other zones of maximum damage. The first zone is observable on some distance from the edge. The second zone appears at the middle of the pipe bend along the circumferential coordinate. The difference between the failure initiation time predicted by the shell and the solid models was approximately 20%. This difference

may be considered to be not significant since the phenomenological material model (12.3) allows to characterize the damage evolution in general only and does not reflect deformation or damage mechanisms.

However, due to obtained disagreement in the time variation of stresses and damage distributions the following case study was performed. Two factors were assumed to be the reason of the obtained differences. The first factor is how correctly the first order shear deformable shell theory represents the stress redistributions if the damage rate is influenced by the stress state. If different damage rates are induced by tensile and compressive stresses, the behavior of "compressive layers" across the shell thickness is dominantly controlled by the steady state creep rate without significant damage, whereas the "tensile layers" exhibit an increasing strain rate due to damage evolution. The non-symmetric strain distribution in the thickness direction is similar to effects well-known in the theory of laminates [12.6] and requires to examine the assumption of a straight normal which is the basis of the displacement based shell theory used in ANSYS. The second factor is the boundary layer stress redistribution which may differ applying the solid and shell models.

From the conclusion based on the pipe bend analysis a new shear correction factor was established [12.46, 12.10]. Let us consider a beam with a rectangular cross section. Assuming the beam as a plane stress problem the principle of virtual displacements yields

$$
\int_0^l \int_{A_c} (\sigma_x \delta\varepsilon_x + \tau_{xz}\delta\gamma_{xz} + \sigma_z\delta\varepsilon_z)dA_c dx = \int_0^l \bar{q}(x)\delta w(x, -h/2)dx \qquad (12.7)
$$

Here l denotes the beam length, A_c is the cross-sectional area, $\sigma_x(x, z), \sigma_z(x, z)$ are the normal stresses and $\tau_{xz}(x, z)$ is the shear stress. For the sake of simplicity we presume the absence of tractions on the edges $x = 0$ and $x = l$, and the geometrically linear theory. In addition to the displacement based approximations, a refined engineering beam theory can be established starting from the stress approximations. Applying the stress approximations, E. Reissner [12.59] derived the elastic plate equations by means of the variational equation for stresses and in [12.60] by the use of the mixed variational equation. The linear distribution of the normal stresses and the parabolic distribution of transverse shear stresses follow from the classical beam or plate theory in the case of elasticity. The creep solutions for beams and plates provide a strongly nonlinear through-thickness distribution of the normal stresses even for steady state creep [12.50]. It can be shown that the first order shear deformation beam equations can be derived without the assumption of the linear normal stress σ_x distribution [12.46]. The starting point is the assumption that the components of the creep strain tensor ε_{ij}^{cr} are known functions of the coordinates $x, \zeta = 2z/h$ for fixed time variable. The variation of the work of the internal forces W_i in (12.7) can be performed, and according to the strain-displacement equations can be transformed. For the further derivations the approximation of the transverse shear stress as a product of two functions

$$\tau_{xz} = R(x)\psi^{\bullet}(\zeta), \quad (\ldots)^{\bullet} \equiv \frac{d}{d\zeta} = \frac{h}{2}\frac{d}{dz}(\ldots) \tag{12.8}$$

was introduced. The relation between the transverse shear stress τ_{xz} and the shear force $Q(x)$ provides with the assumed transverse shear stress approximation (12.8)

$$R(x) = \frac{2Q(x)}{bh}\frac{1}{\psi(1)-\psi(-1)}$$

From the equilibrium condition and the boundary condition $\sigma_z(x,1) = 0$ we get

$$\sigma_z = \frac{Q'(x)}{b}\frac{\psi(1)-\psi(\zeta)}{\psi(1)-\psi(-1)} \tag{12.9}$$

With these approximations for the transverse shear and normal stresses some manipulations in the variational equation can be performed. The additive decomposition of the total strain into the elastic and the creep part and the HOOKE law

$$\gamma_{xz} = \gamma_{xz}^{el} + \gamma_{xz}^{cr} = \frac{\tau_{xz}}{G} + \gamma_{xz}^{cr} \tag{12.10}$$

with G as the shear modulus is assumed. If the linear through-thickness approximation

$$u(x,\zeta) = u_0(x) + \zeta\frac{h}{2}\varphi(x), \tag{12.11}$$

is used after some calculations the following ordinary differential equations are established $[d(\ldots)/dx \equiv (\ldots)']$

$$N' = 0, \quad M' - Q = 0, \quad Q' + \bar{q} = 0, \quad Q = Gbhk(\varphi + \tilde{w}' - \tilde{\gamma}^{cr}) \tag{12.12}$$

The first three equations are the classical equilibrium conditions of the beam. The last equation is the constitutive equation connecting the shear force and the average shear strain. If we express the rotation φ from this equation, we obtain according to the assumed linear approximation (12.11)

$$u(x,\zeta) = u_0(x) - \zeta\frac{h}{2}\tilde{w}'(x) + \zeta\frac{h}{2}\frac{Q(x)}{Gbhk} + \zeta\frac{h}{2}\tilde{\gamma}^{cr}(x)$$

The second term is the normal rotation in the sense of BERNOULLI's hypotheses, the third term denotes the influence of the shear force in the sense of the TIMOSHENKO theory and the last term is the contribution of the average creep shear strain. In the equations considered, the coefficient k and the average of the creep strain $\tilde{\gamma}^{cr}$ are unknown since the function $\psi(\zeta)$ is not specified. If a parabolic shear stress distribution function according to the solution of the elastic BERNOULLI beam is assumed one obtains the classical shear correction factor $k = 5/6$ for a homogeneous rectangular cross section.

As an example we discuss the classical steady state creep solution of a Bernoulli beam [12.50]. Presuming that the elastic strains are negligible and the creep strain rate depends on stress according to the power law (12.4) we can write

$$\dot{\varepsilon}_x \approx \dot{\varepsilon}_x^{cr} = a\sigma_x^n = -\dot{w}''\zeta\frac{h}{2}$$

with a, n denotes material constants. The stress σ_x can be expressed as

$$\sigma_x(x,\zeta) = \left(-\frac{\dot{w}''}{a}\right)^{1/n}|\zeta|^{(1/n)-1}\zeta\left(\frac{h}{2}\right)^{1/n} = \frac{M(x)}{\alpha bh^2}|\zeta|^{(1/n)-1}\zeta, \alpha = \frac{n}{2(2n+1)}$$

From this equation using the equilibrium condition one get the distribution function

$$\psi^\bullet(\zeta) = 1 - \zeta^2|\zeta|^{(1/n)-1}$$

The shear correction factor can be estimated by

$$\frac{1}{k_n} = \frac{2}{[\psi(1) - \psi(-1)]^2}\int_0^1 \psi^{\bullet^2}(\zeta)d\zeta = 2\frac{2n+1}{(3n+2)}$$

The case $n = 1$ corresponds to the shear correction factor of elastic beam with rectangular cross section. Since the value of n varies between 3 and 10 for metallic materials one can estimate $k_3 = 11/14$ for $n = 3$ and $k_{10} = 16/21$ for $n = 10$, respectively. It can be seen that k_n is not constant but influenced by the creep exponent even in the case of steady state creep without damage evolution. However, the differences between the above mentioned values is not significant (even for $n \to \infty$ we obtain $k_\infty = 3/4$). From this follows that if damage effects can be neglected and the standard first order shear deformable theories of beams, plates and shells can be used for creep analysis.

In [12.46] the function $\psi^\bullet$ and the shear correction factor have been estimated for the creep-damage problem of a clamped beam. The corresponding equations and the numerical procedure are presented in [12.8]. The non-symmetric behavior is a result of the damage law depending on the stress state. Figure 12.3a) presents the distribution of transverse shear forces. Near the beam edges where the maximum damage occurs the distribution of shear stresses τ_{xz} is non-symmetrical across the thickness direction. Figure 12.3b) shows the solution according to the beam equations presented in this section. The shear stress can be calculated as a product of the shear force and the distribution function $\psi^\bullet$ with a constant factor. Because for the assumed beam

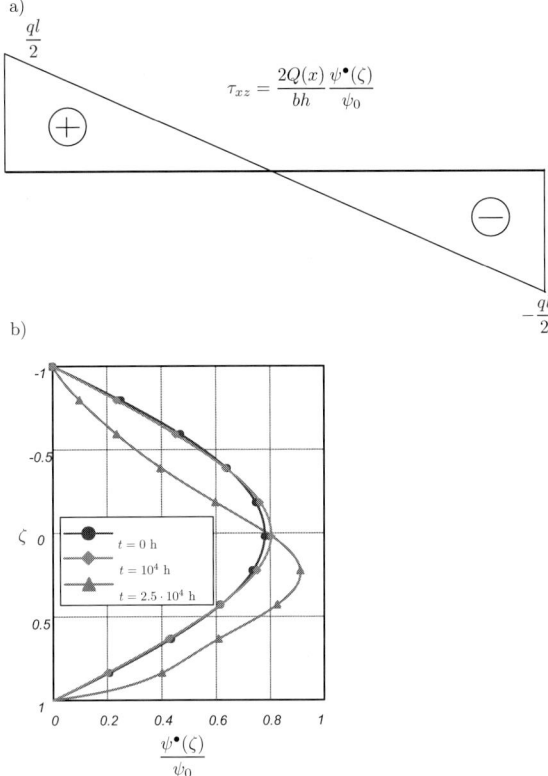

Fig. 12.3. Transverse shear stresses in a beam: a) shear force distribution according to the beam equations; b) function of the transverse shear stress distribution for different time steps

the shear force will be a constant value during the creep-damage process (statically determinate) the variation of the shear stress is determined by the time-dependence of $\psi^\bullet$. In the initial moment the shear stress distribution is symmetrical, after 25 000 h a non-symmetrical distribution can be obtained.

References

[12.1] H. Altenbach. Modelling of viscoelastic behaviour of plates. In M. Życzkowski, editor, *Creep in Structures*, pages 531 – 537. Springer, Berlin, Heidelberg, 1991.

[12.2] H. Altenbach. Creep-damage behaviour in plates and shells. *Mechanics of Time-Dependent Materials*, 3:103–123, 1999.

[12.3] H. Altenbach. An alternative determination of transverse shear stiffnesses for sandwich and laminated plates. *Int. J. Solids & Structures*, 37(25):3503 – 3520, 2000.

[12.4] H. Altenbach. On the determination of transverse shear stiffnesses of orthotropic plates. *ZAMP*, 51:629 – 649, 2000.

[12.5] H. Altenbach. Consideration of stress state influences in the material modelling of creep and damage. In S. Murakami and N. Ohno, editors, *IUTAM Symposium on Creep in Structures*, pages 141 – 150. Kluwer, Dordrecht, 2001.

[12.6] H. Altenbach, J. Altenbach, and R. Rikards. *Einführung in die Mechanik der Laminat- und Sandwichtragwerke*. Deutscher Verlag für Grundstoffindustrie, Stuttgart, 1996.

[12.7] H. Altenbach, J. Altenbach, and A. Zolochevsky. *Erweiterte Deformationsmodelle und Versagenskriterien der Werkstoffmechanik*. Deutscher Verlag für Grundstoffindustrie, Stuttgart, 1995.

[12.8] H. Altenbach, G. Kolarow, O. Morachkovsky, and K. Naumenko. On the accuracy of creep-damage predictions in thinwalled structures using the finite element method. *Comp. Mech.*, 25:87–98, 2000.

[12.9] H. Altenbach, V. Kushnevsky, and K. Naumenko. On the use of solid- and shell-type finite elements in creep-damage predictions of thinwalled structures. *Arch. Appl. Mech.*, 71:164 – 181, 2001.

[12.10] H. Altenbach and K. Naumenko. Shear correction factors in creep-damage analysis of beams, plates and shells. *JSME International Journal, Series A*, 45(1):77 – 83, 2002.

[12.11] H. Altenbach, P. Schieße, and A.A. Zolochevsky. Zum Kriechen isotroper Werkstoffe mit komplizierten Eigenschaften. *Rheol. Acta*, 30:388 – 399, 1991.

[12.12] H. Altenbach and J. Skrzypek, editors. *Creep and Damage in Materials and Structures*. CISM Courses and Lectures Notes No. 399. Springer, Wien, New York, 1999.

[12.13] H. Altenbach and A.A. Zolochevsky. Eine energetische Variante der Theorie des Kriechens und der Langzeitfestigkeit für isotrope Werkstoffe mit komplizierten Eigenschaften. *ZAMM*, 74(3):189 – 199, 1994.

[12.14] J. Altenbach and H. Altenbach. *Einführung in die Kontinuumsmechanik*. Teubner Studienbücher Mechanik. Teubner, Stuttgart, 1994.

[12.15] J. Altenbach, H. Altenbach, and K. Naumenko. Lebensdauerabschätzung dünnwandiger flächentragwerke auf der grundlage phänomenologischer materialmodelle für kriechen und schädigung. *Technische Mechanik*, 17(4):353–364, 1997.

[12.16] Y. Başar and W.B. Krätzig. *Mechanik der Flächentragwerke*. Vieweg, Braunschweig et al., 1985.

[12.17] A. Bertram. What is the general constitutive equation? In A. Cassius et al., editor, *Beiträge zur Mechanik: Festschrift zum 65. Geburtstag Rudolf Trostel*, pages 28 – 37. TU Berlin, Berlin, 1993.

[12.18] J. Betten. Anwendungen von Tensorfunktionen in der Kontinuumsmechanik anisotroper Materialien. *ZAMM*, 78(8):507 – 521, 1998.

[12.19] J.T. Boyle and J. Spence. *Stress Analysis for Creep*. Butterworth, London, 1983.

[12.20] B.J. Cane. Creep fracture of dispersion strengthened low alloy ferritic steels. *Acta Metall.*, 29:1581–1591, 1981.

[12.21] W.F. Chen and H. Zhang. *Structural Plasticity*. Springer, Berlin et al., 1991.

[12.22] R.M. Christensen. *Theory of Viscoelastiocity*. Academic Press, New York et al., 1971.

[12.23] T.J. Chung. *Continuum Mechanics*. Prentice-Hall, London et al., 1988.

[12.24] B.D. Coleman and M.E. Gurtin. Thermodynamics with internal state variables. *J. of Physical Chemics*, 47(2):597 – 613, 1967.

[12.25] B.F. Dyson and S. Osgerby. Modelling and analysis of creep deformation and fracture in a 1Cr0.5Mb ferritic steel. Technical Report 116, DMM(A), 1993.

[12.26] A.C. Eringen and G.A. Maugin. *Electrodynamics of Continua I - Foundations and Solid Media*. Springer, New York et al., 1989.

[12.27] P. Gummert. General constitutive equations for simple and non–simple materials. In H. Altenbach and J. Skrzypek, editors, *Creep and Damage in Materials and Structures*, CISM Courses and Lectures Notes No. 399, pages 1 – 43. Springer, Wien, New York, 1999.

[12.28] P. Haupt. Konzepte der Materialtheorie. *Techn. Mechanik*, 16(1):13 – 22, 1996.

[12.29] P. Haupt. *Continuum Mechanics and Theory of Materials*. Springer, Berlin, 2000.

[12.30] D.R. Hayhurst. Creep rupture under multiaxial states of stress. *J. Mech. Phys. Solids*, 20:381 – 390, 1972.

[12.31] D.R. Hayhurst. Materials data bases and mechanisms–based constitutive equations for use in design. In H. Altenbach and J. Skrzypek, editors, *Creep and Damage in Materials and Structures*, CISM Courses and Lectures Notes No. 399, pages 285 – 348. Springer, Wien, New York, 1999.

[12.32] J.R. Hutchinson. Shear coefficients for Timoshenko beam theory. *Trans. ASME. J. Appl. Mech.*, 68(1):87–92, 2001.

[12.33] L.M. Kachanov. O vremeni razrusheniya v usloviyakh polzuchesti (Time of the rupture process under creep conditions, in Russ.). *Izv. AN SSSR. Otd. Tekh. Nauk*, (8):26 – 31, 1958.

[12.34] L.M. Kachanov. *Introduction to Continuum Damage Mechanics*. Mechanics of Elastic Stability. Martinus Nijhoff, Dordrecht et al., 1986.

[12.35] Z.L. Kowalewski, D.R. Hayhurst, and B.F. Dyson. Mechanisms-based creep constitutive equations for an aluminium alloy. *J. Strain Anal.*, 29(4):309 – 316, 1994.

[12.36] D. Krajcinovic and J. Lemaitre, editors. *Continuum Damage Mechanics - Theory and Application*. CISM Courses and Lectures Notes No. 295. Springer, Wien, New York, 1987.

[12.37] A. Krawietz. *Materialtheorie. Mathematische Beschreibung des phänomenologischen thermomechanischen Verhalten*. Springer, Berlin et al., 1986.

[12.38] F.A. Leckie and D.R. Hayhurst. Constitutive equations for creep rupture. *Acta Metallurgica*, 25:1059 – 1070, 1977.

[12.39] J. Lemaitre. *A Course on Damage Mechanics*. Springer, Berlin et al., 1996.

[12.40] J. Lemaitre and J.-L. Chaboche. *Mechanics of Solid Materials*. Cambridge University Press, Cambridge, 1990.

[12.41] N.N. Malinin. *Raschet na polzuchest' konstrukcionnykh elementov (Creep calculations of structural elements, in Russ.)*. Mashinostroenie, Moskva, 1981.

[12.42] L.I. Manevitch, I.V. Andrianov, and V.G. Oshmyan. *Mechanics of Periodically Heterogeneous Structures*. Springer, Berlin et al., 2002.

[12.43] J. Meenen and H. Altenbach. A consistent deduction of von Kármán-type plate theories from threedimensional non-linear continuum mechanics. *Acta Mechanica*, 147:1 – 17, 2001.

[12.44] F.R.N. Nabarro and H.L. de Villiers. *The Physics of Creep. Creep and Creep–resistant Alloys*. Taylor & Francis, London, 1995.

[12.45] K. Naumenko. *Modellierung und Berechnung der Langzeitfestigkeit dünnwandiger Flächentragwerke unter Einbeziehung von Werkstoffkriechen und Schädigung*. Diss., Fakultät für Maschinenbau, Otto-von-Guericke-Universität, 1996.

[12.46] K. Naumenko. On the use of the first order shear deformation models of beams, plates and shell in creep lifetime estimations. *Technische Mechanik*, 20(3):215 – 226, 2000.

[12.47] W. Noll. A new mathematical theory of simple materials. *Archive for Rational Mechanics and Analysis*, 48:1 – 50, 1972.

[12.48] J.F. Nye. *Physical Properties of Crystals*. Oxford Science Publications, Oxford, 1992.

[12.49] F.K.G. Odqvist. *Mathematical Theory of Creep and Creep Rupture*. Clarendon, Oxford, 1974.

[12.50] F.K.G. Odqvist and J. Hult. *Kriechfestigkeit metallischer Werkstoffe*. Springer, Berlin u.a., 1962.

[12.51] P.R. Onck and E. van der Giessen. Growth of an initially sharp crack by boundary cavitation. *J. Mech. Phys. Solids*, 47:99 – 139, 1999.

[12.52] E. Sanchez Palancia. *Introduction aux méthodes asymptotiques et à l'homogénéisation*. Masson, Paris, 1992.

[12.53] V. Palmov. *Vibrations in Elasto-Plastic Bodies*. Springer, Berlin et al., 1998.

[12.54] P. Paul. Macroscopic criteria of plastic flow and brittle fracture. In H. Liebowitz, editor, *Fracture: An advanced treatise, Vol. II (Mathematical Fundamentals)*. Academic Press, New York, 1968.

[12.55] I.J. Perrin and D.R. Hayhurst. Creep constitutive equations for a 0.5Cr–0.5Mo–0.25V ferritic steel in the temperature range 600–675°C. *J. Strain Anal.*, 31(4):299 – 314, 1994.

[12.56] Yu. N. Rabotnov. *Creep Problems in Structural Members*. North-Holland, Amsterdam, 1969.

[12.57] Yu. N. Rabotnov. *Elements of Hereditary Solid Mechanics*. Mir, Moscow, 1977.

[12.58] M. Reiner. *Deformation and Flow. An Elementary Introduction to Rheology*. H.K. Lewis & Co., London, 3rd edition, 1969.

[12.59] E. Reissner. On the theory of bending of elastic plates. *J. Math. Phys.*, 23:184 – 191, 1944.

[12.60] E. Reissner. Variational theorem in elasticity. *J. Math. Phys.*, 29:90 – 95, 1950.

[12.61] H. Riedel. *Fracture at High Temperatures*. Materials Research and Engineering. Springer, Berlin et al., 1987.

[12.62] H. Rothert. Lineare konstitutive Gleichungen der viskoelastischen Cosseratfläche. *ZAMM*, 55:647 – 656, 1975.

[12.63] J.J. Skrzypek. *Plasticity and Creep*. CRC Press, Boca Raton et al., 1993.

[12.64] A. Służalec. *Introduction to Nonlinear Thermomechanics*. Springer, Berlin et al., 1992.

[12.65] P. Suquet. *Plasticité et homogénéisation*. Thèse de doctorat d'etat, Université Pierre et Marie Curie, Paris 6, 1982.

[12.66] E. van der Giessen, B.-N. Nguyen, and P.R. Onck. From a microstructural to a continuum model for creep fracture. In A. Benallal, editor, *Continuous Damage and Fracture*, pages 129 – 136. Elsevier, Paris et al., 2000.

[12.67] E. van der Giessen and V. Tvergaard. Development of final creep failure in polycrystalline aggregates. *Acta Metall. Mater.*, 42:959 – 973, 1994.

[12.68] M. Życzkowski. *Combined Loadings in the Theory of Plasticity*. PWN-Polish Scientific Publisher, Warszawa, 1981.

Part III

Modern Methods of Scientific Computing

13 Parallel Implementation Strategies
for Algorithms from Scientific Computing

T. Rauber[1] and G. Rünger[2]

[1] Universität Bayreuth, Fakultät für Mathematik und Physik,
rauber@uni-bayreuth.de
[2] Technische Universität Chemnitz, Fakultät für Informatik,
ruenger@informatik.tu-chemnitz.de

Abstract. Many applications from scientific computing are computationally intensive and can therefore benefit from a realization on a parallel or a distributed platform. The parallelization strategy and the resulting efficiency strongly depends on the characteristics of the target architecture (shared address space or distributed address space) and the programming model used for the implementation. For selected problems from scientific computing, we discuss parallelization strategies using message-passing programming for distributed address space.

13.1 Introduction

Simulations in natural sciences usually require large computing power in order to produce realistic results in acceptable execution time and today parallel computation is a standard way to achieve this. And although the performance of one-processor machines and PCs has increased tremendously in recent year there is still need for parallel computers due to the ever increasing demands of new complex and realistic simulation programs. Many regular algorithms can be realized as parallel programs in a straightforward way but especially when more complex algorithms are involved some more effort is needed to exploit the specifics of the algorithms and the parallel hardware platform to get fast and efficient programs.

Monolithic parallel machines with one processor per node and a homogeneous interconnection network did play an important role for quite a long time. Today fast interconnection networks like Myrinet, SCI (scalable coherent interface) or Gigabit-Ethernet enable to build cluster machines offering large computing power at a reasonable price [13.2, 13.12]. For the resulting distributed address space, the Message Passing Interface (MPI) allows efficient and portable programming and has developed into the de-facto standard [13.11]. There are other ways of programming those machines such as PVM [13.3] and HPF but their role is diminishing. Emerging from the server market, there are also parallel machines for which each processor has the same view to the shared address space (SMPs, symmetric multi-processors). OpenMP [13.6] has been proposed as a standard for programming these SMPs, but Pthreads [13.1] or Java threads are also popular. Using SMPs

T. Rauber and G. Rünger, Parallel Implementation Strategies for Algorithms from Scientific Computing, Lect. Notes Phys. **642**, 261–281 (2004)
http://www.springerlink.com/

as nodes of larger parallel machines results in an address space, for which the processors of one node share a common address space but the global address space is distributed among the nodes. All those types of computer systems are currently used for high performance computing. The TOP500 list gives statistical information about the current 500 most powerful computer systems (www.top500.org).

The possibilities to express a parallel program strongly depends on the parallel programming model provided by the parallel system consisting of a specific hardware platform and the software components provided. A parallel programming model can be described as a combination of the following aspects [13.9]: Which kind of parallelism can be exploited (instruction level, command level, procedure level, or parallel loops)? Is the parallelism specified explicitly with parallel commands or exploited implicitly? On which level is the parallelism specified (as independent tasks, threads, processes, or by data decomposition)? What kind of execution mode is used (e.g. asynchronous or not, SPMD, i.e. single program multiple data, MIMD, i.e. multiple instruction multiple data, or master-slave-model)? How do parallel program parts exchange information (by message passing or via a shared address space)? Is there a cost model to evaluate the performance of parallel programs? Many regular programs from scientific computing are implemented in a data parallel way. For a data parallel execution the program data like arrays are decomposed into parts and are assigned to the local memory of different processors which perform executions on their local data parts. This parallelization strategy usually leads to good load balance for regular programs, if each processor has about the same amount of operations to execute.

Cost models are important tools to evaluate the efforts of parallel programming since parallel programming in scientific computing is mainly used to reduce the execution time. The effect of execution time reduction is measured in term of speedup and efficiency. The speedup $S_p(n)$ expresses the relative performance gain of a parallel program compared to a sequential program according to $S_p(n) = \frac{T^*(n)}{T_p(n)}$, where n is the problem size, e.g. the size of a system, $T_p(n)$ is the execution time of the parallel program, and $T^*(n)$ is the execution time of an optimal sequential implementation of the same problem. Thus, the speedup for p processors is theoretically bounded by $S_p(n) \leq p$. The efficiency $E_p(n)$ of a parallel program measures the percentage of parallel computing power effectively used according to $E_p(n) = \frac{S_p(n)}{p} = \frac{T^*(n)}{p \cdot T_p(n)}$. The ideal speedup $S_p(n) = p$ corresponds to an efficiency $E_p(n) = 1$.

The execution times for the sequential and the parallel program version can come from measurements of an existing implementations. In this case speedup values are available after the entire implementation work is done. Using performance prediction models or cost models, the application programmer can get a fast a priori estimation of the effects of different implementation decisions and can thus compare their effect on the resulting performance before the actual implementation. There are many approaches for performance modeling, including statistical and queuing methods, analytical models as well as experimental approaches. Analytical models are suitable for data parallel programs [13.8] as well as for mixed task and data parallel programs [13.5], and they can also be used for large application programs [13.4].

A mixed task and data parallel implementation strategy might be suitable for a specific algorithmic behavior or innovative parallel machines for which a pure data parallel or SPMD implementation may lead to scalability problems. i.e. an increasing number of processors results in a decreasing speedup. In the following we give a short introduction to the SPMD programming with MPI. Based on this we briefly present a hybrid programming approach for implementation strategies with mixed parallelism.

13.2 A Short Introduction to MPI

The Message Passing Interface (MPI) is a widely used standard for writing message passing programs with a specified set of library operations. The interface mainly provides communication operations and the library routines can be used with C or Fortran (also with C++ in the extension MPI-2); www.mpi-forum.org contains the official MPI documents, errata, and archives of the MPI Forum.

There are different implementations of MPI on a wide range of computers including MPICH (www-unix.mcs.anl.gov/mpi/mpich) from the Argonne National Lab and the Mississippi State University and LAM-MPI (www.lam-mpi.org) originally developed at the Ohio Supercomputing Center. LAM (Local Area Multicomputer) is an MPI programming environment and development system for heterogeneous computers on a network. MPICH is a freely available complete implementation of the MPI specification, designed to be both portable and efficient. The "CH" in MPICH stands for "Chameleon", symbol of adaptability to one's environment and thus of portability. There are also many vendor specific implementations.

An MPI program consists of autonomous processes executing their own code in an MIMD style but in many programs an SPMD style is adopted. We give a short introduction to point-to-point communication between two processes and to a selected set of collective communication involving more than two processes. The basic point-to-point operations are send and receive operations. For a message transmission, the sending process calls

```
int MPI_Send(void *smessage, int count, MPI_Datatype datatype,
             int dest, int tag, MPI_Comm comm)
```

and the receiving process calls a corresponding

```
int MPI_Recv(void *rmessage, int count, MPI_Datatype datatype,
             int source, int tag, MPI_Comm comm, MPI_Status *status).
```

In MPI_Send, the argument smessage is a send buffer containing the message to be sent. The message contains count elements of type datatype,

```
#include <stdio.h>
#include <string.h>
#include "mpi.h"

int main (int argc, char *argv[]) {
    int my_rank, p, source, dest, tag=0;
    char msg [20];
    MPI_Status status;

    MPI_Init (&argc, &argv);
    MPI_Comm_rank (MPI_COMM_WORLD, &my_rank);
    MPI_Comm_size (MPI_COMM_WORLD, &p);
    if (my_rank == 0) {
        strcpy (msg, "Hello ");
        MPI_Send (msg, strlen(msg)+1, MPI_CHAR, 1, tag, MPI_COMM_WORLD);
    }
    if (my_rank == 1)
        MPI_Recv (msg, 20, MPI_CHAR, 0, tag, MPI_COMM_WORLD, &status);
    MPI_Finalize ();
}
```

Fig. 13.1. A first MPI program in which process 0 sends a message `msg` to process 1. `MPI_Comm_rank()` and `MPI_Comm_size()` provide the Id of the calling process in `my_rank` and the number of processors in the communicator `MPI_COMM_WORLD` in `p`.

which is one of the MPI specific data types (e.g. MPI_INT, MPI_FLOAT, MPI_DOUBLE); `dest` is the Id of the destination process, `tag` is a specific tag for the message, and `comm` denotes a communicator containing the sending and the receiving process. The receive operation has similar parameters. The argument `rmessage` is a buffer to store the received message; `count` is a buffer limit for the message size, and `source` is the Id of the sending process; `status` is needed to store information about the message actually received.

An MPI program requires a call of `MPI_Init()` at the beginning and a call of `MPI_Finalize()` at the end. In between MPI operations can be called in the context of a specific *communicator*. A communicator is a communication context consisting of all or a subset of the processors executing the MPI-program (also called group of processors) with a unique consecutive enumeration of Ids starting with 0. The standard communicator that contains all processes of a program is `MPI_COMM_WORLD`. Figure 13.1 shows a simple MPI program.

The total number p of processors executing the program is specified when starting an MPI program in the command line by giving the program name, the program arguments, and the number p of processors according to

```
mpirun -np 4 program_name program_arguments
```

Program version 1	Program version 2	Program version 3
if (my_rank == 0) {	**if** my_(rank == 0){	**if** my_(rank == 0){
MPI_Recv from P1	MPI_Send to P1	MPI_Send to P1
MPI_Send to P1	MPI_Recv from P1	MPI_Recv from P1
}	}	}
else if (my_rank == 1){	**else if** (my_rank == 1){	**else if** (my_rank == 1){
MPI_Recv from P0	MPI_Send to P0	MPI_Recv from P0
MPI_Send to P0	MPI_Recv from P0	MPI_Send to P1
}	}	}
deadlock	deadlock without system buffer	safe program

Fig. 13.2. Illustration of possible deadlock situations for process P0 with rank 0 and process P1 with rank 1 when using blocking MPI_Send() and MPI_Recv().

(here $p = 4$). This command may have slightly different syntax for different MPI realizations. Usually, each processor executes one MPI process.

MPI_Send() and MPI_Recv() are blocking and asynchronous operations. For a blocking MPI operation the return from the function call indicates that the user can re-use all resources, e.g. buffer arguments. There is also an MPI send operation in synchronous mode which requires the send operation to block until the corresponding receive operation has been issued [13.11].

The use of blocking send and receive operations may lead to a deadlock situation (i.e., the processes cyclically wait on each other) when used carelessly. We illustrate this by the code fragments in Fig. 13.2 in which a communication from process P0 to process P1 and from process P1 to process P0 is expressed. Program version 1 causes a deadlock since both receive operations wait for the corresponding send operation from the other process which cannot be executed since that process itself waits. The same effect results for program version 2 for MPI implementations which do not provide a system buffering of messages send by MPI_Send. A safe implementation of the data exchange is given by program version 3 [13.7].

Collective communication is defined as communication that involves a group of processes. We present some of the collective communication operations provided by MPI. A more detailed overview is given in [13.11, 13.9].

A broadcast operation MPI_Bcast is called by all processes of a communicator comm. The root process with Id **root** sends the message A0 in buffer **smessage** of size **count** and type **type** to all other processes in comm.

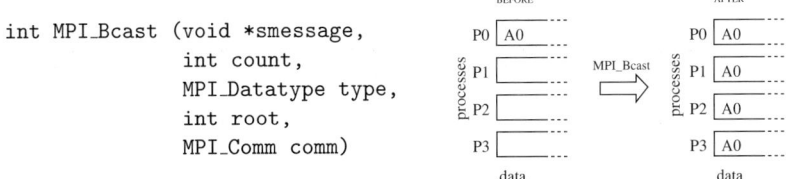

```
int MPI_Bcast (void *smessage,
               int count,
               MPI_Datatype type,
               int root,
               MPI_Comm comm)
```

The global reduction operation `MPI_Reduce` also uses a specific root process with Id `root`. All processes provide a buffer with `count` elements in `sendbuf`. `MPI_Reduce` combines those data using a reduction operation of type `MPI_Op`, like `MPI_MAX`, `MPI_MIN` or `MPI_SUM`, and returns the combined value in buffer `recvbuf` of the root process.

```
int MPI_Reduce (void *sendbuf,
                void *recvbuf,
                int count,
                MPI_Datatype type,
                MPI_Op op,
                int root,
                MPI_Comm comm)
```

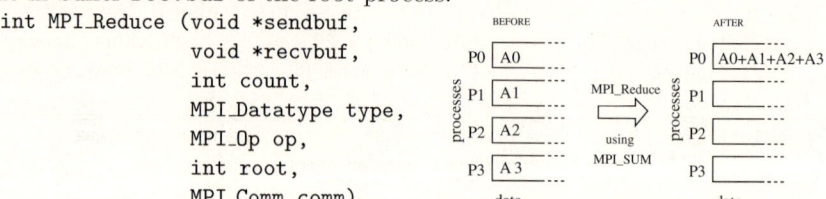

For an `MPI_Allgather` operation, each process contributes a block of data in `sendbuf` with `sendcount` elements of type `sendtype`. The result of the operation is that each process stores all data blocks in rank order in its target buffer `recvbuf`.

```
int MPI_Allgather (void *sendbuf,
                   int sendcount,
                   MPI_Datatype sendtype,
                   void *recvbuf,
                   int recvcount,
                   MPI_Datatype recvtype,
                   MPI_Comm comm)
```

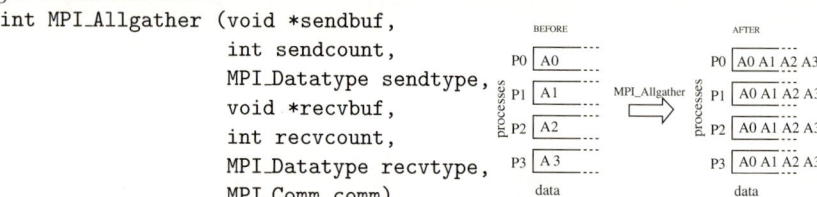

An `MPI_Allreduce` performs an `MPI_Reduce` operation followed by an `MPI_Bcast`, thus making the combined data block available to each processor.

```
int MPI_Allreduce (void *sendbuf,
                   void *recvbuf,
                   int count,
                   MPI_Datatype type,
                   MPI_Op op,
                   MPI_Comm comm).
```

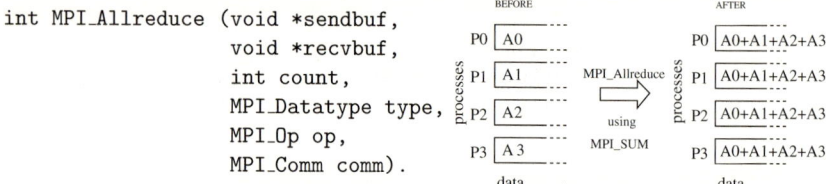

Collective communication operations are often used in algorithms from scientific computing, e.g. when the norm for a convergence criterion is computed collectively and information is send out to all processes. From the parallel programming point of view it is important to know how much time is needed to perform those collective communication operation.

13.3 Modeling the Execution Time of MPI Operations

The execution times of MPI communication operations can be modeled by runtime formulas that depend on various machine parameters, including the number of processors, the bandwidth of the interconnecting network, and the startup time for the corresponding operations. Some of the runtime functions are summarized in Table 13.1 [13.5]. The value b is the message size in bytes, p is the number of processors. The specific values of the parameters τ, t_c,

τ_1, and τ_2 depend on the execution platform and the MPI implementation used and can be determined based on measured execution times by curve fitting using the least squares method. Figure 13.3 compares measured and predicted execution times of an `MPI_Bcast` on a PC Cluster (CLiC).

Table 13.1. Runtime formulas for some MPI collective communication operations. `MPI_Reduce` can be described by the same runtime formula as `MPI_Bcast`, using different parameter values. Similarly, `MPI_Allreduce` uses the same formula as `MPI_Allgather`.

operation	runtime formula
point-to-point	$t_{s2s}(b) = \tau + t_c \cdot b$
`MPI_Bcast`	$t_{sb}(p,b) = \tau \cdot \log(p) + t_c \cdot \log(p) \cdot b$
`MPI_Allgather`	$t_{mb}(p,b) = \tau_1 + \tau_2 \cdot p + t_c \cdot p \cdot b$

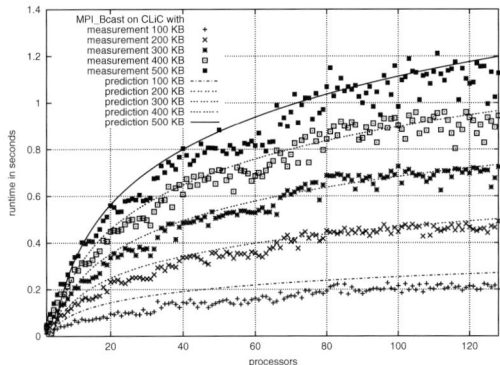

Fig. 13.3. Modeling the execution times of MPI_Bcast() on a PC cluster.

13.4 Example: Solving Systems of Linear Equations

As example we consider iterative solution methods for linear equation systems $Ax = b$ with matrix $A \in \mathbb{R}^{n \times n}$ and vector $b \in \mathbb{R}^n$, which compute a sequence of approximation vectors $\{x^{(k)}\}_{k=1,2,\ldots}$ that converge to the solution $x^* \in \mathbb{R}^n$ under certain conditions [13.13]. Iteration methods are based on a partitioning of A into $A = M - N$ with $M, N \in \mathbb{R}^{n \times n}$, where M is a non-singular matrix for which M^{-1} can be easily computed. The form $Mx = Nx + b$ of the linear equation system induces the iteration method

$$x^{(k+1)} = Cx^{(k)} + d, \quad k = 0, 1, \ldots \tag{13.1}$$

with iteration matrix $C := M^{-1}N$ and vector $d := M^{-1}b$.

13.4.1 Standard Iterative Methods

The Jacobi method is based on the specific partitioning $A = D - L - R$ where D is a diagonal matrix, L is the lower triangular and R is the upper triangular matrix of A. The iteration is based on $Dx^{(k+1)} = (L + R)x^{(k)} + b$ and uses the iteration matrix $C_G := D^{-1}(L + R)$. The component-wise formulation of the iteration process is

$$x_i^{(k+1)} = \frac{1}{a_{ii}} \left(b_i - \sum_{j=1,j\neq i}^{n} a_{ij} x_j^{(k)} \right) , \quad i = 1, \dots, n . \qquad (13.2)$$

The Gauß-Seidel method is based on the same partitioning $A = D - L - R$, but uses a different iteration method $(D - L)x^{(k+1)} = Rx^{(k)} + b$. The iteration matrix is $C_E := (D - L)^{-1}R$ and the iteration in component-wise form is

$$x_i^{(k+1)} = \frac{1}{a_{ii}} \left(b_i - \sum_{j=1}^{i-1} a_{ij} x_j^{(k+1)} - \sum_{j=i+1}^{n} a_{ij} x_j^{(k)} \right) , \quad i = 1, \dots, n . \qquad (13.3)$$

Over-relaxation methods like SOR are more efficient due to a better convergence behavior [13.13]. But the data dependencies which are important for a parallel implementation are the same so that it is sufficient to consider the Jacobi and he Gauß-Seidel method.

Figure 13.4 shows an SPMD implementation of the Jacobi method in which the components of $x^{(k+1)}$ are computed in parallel. The implementation assumes that the system size n is a multiple of the number p of processors. A row-oriented block distribution of the iteration matrix is used so that each processor stores n/p rows locally in array loc_A. In each iteration step, each processor computes n/p scalar products. The arrays x_old and x_new are used to store the previous and the current global iteration vector. Each processor stores the local components of the iteration vector in loc_x. At the end of each iteration step, MPI_Allgather() is used to collect the components for the next iteration step.

In the Gauß-Seidel method the computation of $x_i^{(k+1)}$ depends on the values $x_1^{(k+1)}, \dots, x_{i-1}^{(k+1)}$ computed in the same iteration step so that the components of the new iteration vector have to be computed one after another. A parallel execution is restricted to the parallelization of the scalar products by letting each processor compute a partial sum that is then accumulated to the new value. This corresponds to a column-wise distribution of the iteration matrix. Processor P_q, $q = 1, \dots p$ computes the partial sum

$$s_{qi} = \sum_{\substack{j=(q-1)\cdot n/p+1 \\ j<i}}^{q\cdot n/p} a_{ij} x_j^{(k+1)} + \sum_{\substack{j=(q-1)\cdot n/p+1 \\ j>i}}^{q\cdot n/p} a_{ij} x_j^{(k)}$$

and the partial sums s_{qi} are accumulated by an MPI_Reduce() operation. This results in the program in Fig. 13.5.

```
int Parallel_jacobi(int n, int p, int max_it, float tol)
{
 int i_loc, i_global, j, i;
 int n_loc, it_num;
 float x_temp1[GLOB_MAX], x_temp2[GLOB_MAX];
 float *x_old, *x_new, *temp;

 n_loc = n/p; /* local block size */
 MPI_Allgather(loc_b, n_loc, MPI_FLOAT, x_temp1, n_loc,
               MPI_FLOAT, MPI_COMM_WORLD);
 x_new = x_temp1;
 x_old = x_temp2;
 it_num = 0;
 do {
    it_num ++;
    temp = x_new; x_new = x_old; x_old = temp;
    for (i_loc = 0; i_loc < n_loc; i_loc++) {
       i_global = i_loc + me * n_loc;
       loc_x[i_loc] = loc_b[i_loc];
       for (j = 0; j < i_global; j++)
         loc_x[i_loc] = loc_x[i_loc] - loc_A[i_loc][j] * x_old[j];
       for (j = i_global+1 ; j < n; j++)
         loc_x[i_loc] = loc_x[i_loc] - loc_A[i_loc][j] * x_old[j];
       loc_x[i_loc] = loc_x[i_loc]/ loc_A[i_loc][i_global];
    }
    MPI_Allgather(loc_x, n_loc, MPI_FLOAT, x_new, n_loc,
                  MPI_FLOAT, MPI_COMM_WORLD);
 } while ((it_num < max_it) && (distance(x_old,x_new,n) >= tol));
 output(x_new,global_x);
 if (distance(x_old, x_new, n) < tol ) return 1;
 else return 0;
}
```

Fig. 13.4. SPMD realization of the Jacobi method using C and MPI. The use of variable names x_loc and A_loc indicates that each processor stores only a part of the global array x or A in its local memory. Accordingly, index variable i_loc denotes the local numbering of array parts.

13.4.2 Sparse Iteration Matrices

The data dependencies in (13.3) restrict the potential for a parallel execution of the Gauß-Seidel method such that only for very large systems a speedup can be obtained. But if A is a sparse matrix there are less dependencies and a larger potential for a parallel execution results. More precisely: if $a_{ij} = 0$, then the computation of $x_i^{(k+1)}$ does *not* depend on $x_j^{(k+1)}$ for $j < i$ and their computation can be done in parallel.

```
n_loc = n/p;
do {
    delta_x = 0.0;
    for (i = 0; i < n; i++) {
        s_k = 0.0;
        for (j = 0; j < n_loc; j++)
            if (j + me * n_loc != i)
                s_k = s_k + loc_A[i][j] * x[j];
        root = i/n_loc;
        i_local = i % n_loc;
        MPI_Reduce(&s_k, &x[i_loc], 1, MPI_FLOAT, MPI_SUM, root,
                    MPI_COMM_WORLD);
        if (my_rank == root) {
            x_new = (b[i_loc] - x[i_loc]) / loc_A[i][i_loc];
            delta_x = max(delta_x, abs(x[i_loc] - x_new));
            x[i_loc] = x_new;
        }
    }
    MPI_Allreduce(&delta_x, &global_delta, 1, MPI_FLOAT,
                    MPI_MAX, MPI_COMM_WORLD);
} while(global_delta > tol);
```

Fig. 13.5. Parallel realization of the Gauß-Seidel method. The `MPI_Reduce` operation provides the currently computed component only for the processor which needs this component since it belongs to the parts of its partial sum.

In the following, we consider a sparse matrix with a banded structure emerging, e.g., when discretizing the Poisson equation, see Fig. 13.6. According to the Gauß-Seidel method the computation of $x_i^{(k+1)}$ needs row i of the banded matrix and the computation of (13.3) is reduced to

$$x_i^{(k+1)} = \frac{1}{a_{ii}} \left(b_i - a_{i,i-\sqrt{n}} \cdot x_{i-\sqrt{n}}^{(k+1)} - a_{i,i-1} \cdot x_{i-1}^{(k+1)} - a_{i,i+1} \cdot x_{i+1}^{(k)} \right.$$
$$\left. - a_{i,i+\sqrt{n}} \cdot x_{i+\sqrt{n}}^{(k)} \right), \quad i = 1, \dots, n . \tag{13.4}$$

Only the components $x_{i-\sqrt{n}}^{(k+1)}$ and $x_{i-1}^{(k+1)}$ have to be computed before the computation of $x_i^{(k+1)}$ can be performed. The data dependencies of $x_i^{(k+1)}$, $i = 1, \dots, n$, to $x_j^{(k+1)}$, $j < i$ are shown in Fig. 13.7 for a rectangular grid. The computation of grid point $x_i^{(k+1)}$ depends on all grid points $x_j^{(k+1)}$ that are located to the left and above grid point x_i. Computations of components $x_i^{(k+1)}$ in the diagonals shown in the figure are independent from each other and can be computed in parallel. The resulting maximum degree of parallelism is $\sqrt{n}$. The computation of the values $x_i^{(k+1)}$ for $i = 1, \dots, n$ can be organized as a doubly nested loop such that the inner loop represents independent computations.

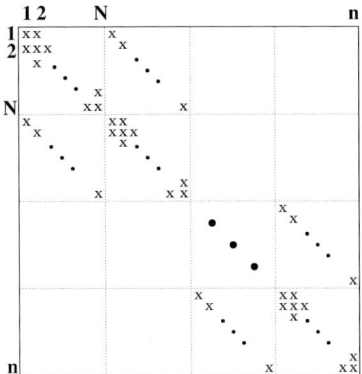

Fig. 13.6. Sparse matrix of size $n \times n$, $N = \sqrt{n}$, with banded structure.

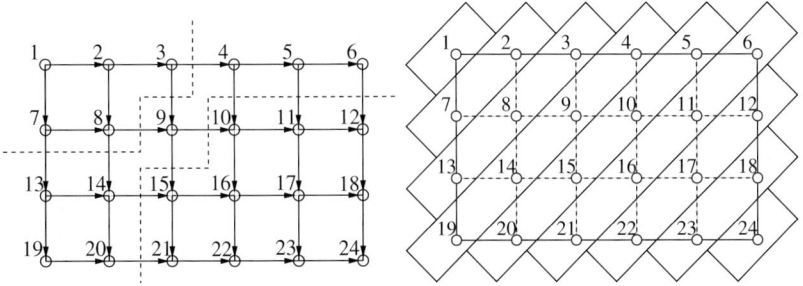

Fig. 13.7. Data dependencies for the Gauß-Seidel method or the SOR method for a rectangular grid of size 6×4 (left). The grid area the computation of which has to be finished before grid point 9 can be computed is indicated. The data dependencies induce computation phases (right) such that the computations within a computation phase are independent from each other.

For the special case of a square grid of size $\sqrt{n} \times \sqrt{n}$ with a row-wise numbering of the grid points, there are $2\sqrt{n} - 1$ diagonals. The first $\sqrt{n}$ diagonals contain l grid points for $l = 1, \ldots, \sqrt{n}$; the grid points in diagonal l are x_i with

$$i = l + j \cdot (\sqrt{n} - 1) \quad \text{for} \quad 0 \leq j < l .$$

The last $\sqrt{n} - 1$ diagonals contain $\sqrt{n} - l + 1$ grid points for $l = 2, \ldots, \sqrt{n}$; the grid points in diagonal l are x_i with

$$i = l \cdot \sqrt{n} + j \cdot (\sqrt{n} - 1) \quad \text{for} \quad 0 \leq j \leq \sqrt{n} - l .$$

A parallel realization on a parallel machine with distributed address space can be based on a row-cyclic distribution of the grid points among the processors. Figure 13.8 illustrates the corresponding parallel program. Two separate loops are used for computing the two sets of diagonals mentioned above. After the computation of a complete diagonal, the function `collect_elements()` is used to send the entries to the next processor.

```
sqn = sqrt(n);
do {
    for (l = 1; l <= sqn; l++) {
        for (j = me; j < l; j+=p) {
            i = l + j * (sqn-1) - 1; /* enumeration starts with 0 */
            x[i] = 0;
            if (i-sqn >= 0) x[i] = x[i] - a[i][i-sqn] * x[i-sqn];
            if (i > 0) x[i] = x[i] - a[i][i-1] * x[i-1];
            if (i+1 < n) x[i] = x[i] - a[i][i+1] * x[i+1];
            if (i+sqn < n) x[i] = x[i] - a[i][i+sqn] * x[i+sqn];
            x[i] = (x[i] + b[i]) / a[i][i];
        }
        collect_elements(x,l);
    }
    for (l = 2; l <= sqn; l++) {
        for (j = me -1 +1; (j <= sqn -1) && (j >= 0); j+=p) {
            i = l * sqn + j * (sqn-1) - 1;
            x[i] = 0;
            if (i-sqn >= 0) x[i] = x[i] - a[i][i-sqn] * x[i-sqn];
            if (i > 0) x[i] = x[i] - a[i][i-1] * x[i-1];
            if (i+1 < n) x[i] = x[i] - a[i][i+1] * x[i+1];
            if (i+sqn < n) x[i] = x[i] - a[i][i+sqn] * x[i+sqn];
            x[i] = (x[i] + b[i]) / a[i][i];
        }
        collect_elements(x,l);
    }
} while(convergence_test() < tol);
```

Fig. 13.8. Illustration of a sparse matrix computation using the Gauß-Seidel method.

13.4.3 Red-Black Ordering

A reordering of the equations can be used to increase the potential of parallelism. The *red-black ordering* partitions the grid points into red and black points such that red points only have black neighbors and vice versa. Red and black grid points are numbered separately in a row-wise way. For n grid points that are partitioned into n_R red and n_S black grid points, the red points are numbered $1, \dots, n_R$ and the black grid points $n_R+1, \dots, n_R+n_S$, see Fig. 13.9 (right). The entries of the solution vector are numbered accordingly. The resulting equation system for the Poisson problem is

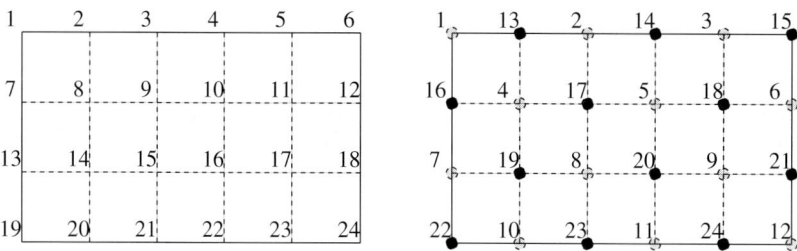

Fig. 13.9. Rectangular grid of size 6×4 with row-oriented numbering (left) and red-black ordering (right).

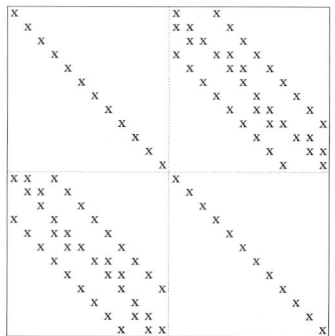

Fig. 13.10. Matrix of the linear equation system based on a five-point stencil corresponding to the red-black ordering in Fig. 13.9.

$$\hat{A} \cdot \hat{x} = \begin{pmatrix} D_R & F \\ E & D_S \end{pmatrix} \cdot \begin{pmatrix} \hat{x}_R \\ \hat{x}_S \end{pmatrix} = \begin{pmatrix} \hat{b}_1 \\ \hat{b}_2 \end{pmatrix}, \tag{13.5}$$

where $\hat{x}_R$ denotes the first n_R unknowns and $\hat{x}_S$ denotes the remaining n_S unknowns. The matrix $\hat{A}$ consists of four blocks $D_R \in \mathbb{R}^{n_R \times n_R}$, $D_S \in \mathbb{R}^{n_S \times n_S}$, $E \in \mathbb{R}^{n_S \times n_R}$ und $F \in \mathbb{R}^{n_R \times n_S}$. The sub-matrices D_R and D_S are diagonal matrices, the E and F are sparse banded matrices, see Fig. 13.10. Since D_R and D_S are diagonal matrices, there are no dependences between red and black grid points. Row i of matrix F shows the interactions between the red unknowns $\hat{x}_i$ ($i < n_R$) and the black unknowns $\hat{x}_j, j \in \{n_R + 1, ..., n_R + n_S\}$.

The solution of (13.5) with the Gauß-Seidel method is based on the partition $\hat{A} = \hat{D} - \hat{L} - \hat{R}$ with $\hat{D}, \hat{L}, \hat{R} \in \mathbb{R}^{n \times n}$,

$$\hat{D} = \begin{pmatrix} D_R & 0 \\ 0 & D_S \end{pmatrix}, \quad \hat{L} = \begin{pmatrix} 0 & 0 \\ -E & 0 \end{pmatrix}, \quad \hat{R} = \begin{pmatrix} 0 & -F \\ 0 & 0 \end{pmatrix},$$

where $\hat{D}$ is a diagonal matrix, $\hat{L}$ is a lower triangular matrix and $\hat{R}$ is an upper triangular matrix. One iteration step of the Gauß-Seidel method has the form

$$\begin{pmatrix} D_R & 0 \\ E & D_S \end{pmatrix} \cdot \begin{pmatrix} x_R^{(k+1)} \\ x_S^{(k+1)} \end{pmatrix} = \begin{pmatrix} b_1 \\ b_2 \end{pmatrix} - \begin{pmatrix} 0 & F \\ 0 & 0 \end{pmatrix} \cdot \begin{pmatrix} x_R^{(k)} \\ x_S^{(k)} \end{pmatrix} \tag{13.6}$$

for $k = 1, 2, \dots$. Iteration vector $x^{(k+1)}$ is partitioned into sub-vectors $x_R^{(k+1)}$ and $x_S^{(k+1)}$ for the red and black unknowns. Iteration (13.6) can be written as

$$D_R \cdot x_R^{(k+1)} = b_1 - F \cdot x_S^{(k)} \quad \text{for } k = 1, 2, \dots , \tag{13.7}$$

$$D_S \cdot x_S^{(k+1)} = b_2 - E \cdot x_R^{(k+1)} \quad \text{for } k = 1, 2, \dots . \tag{13.8}$$

which reflects the decoupling of the red and black sub-vectors $x_R^{(k+1)}$ and $x_S^{(k+1)}$. In (13.7), the new red iteration vector $x_R^{(k+1)}$ is computed depending only on $x_S^{(k)}$. In (13.8), the new red iteration vector $x_S^{(k+1)}$ is computed depending only on $x_B^{(k)}$. There is no dependence between the components of $x_R^{(k+1)}$ or $x_S^{(k+1)}$, respectively. Thus, the components of $x_R^{(k+1)}$ can be computed in parallel The same holds for $x_R^{(k+1)}$. The parallel implementation of the red-black method in Fig. 13.11 uses the component-wise form

$$\left(x_R^{(k+1)} \right)_i = \frac{1}{\hat{a}_{ii}} \left(\hat{b}_i - \sum_{j \in N(i)} \hat{a}_{ij} \cdot (x_S^{(k)})_j \right), \quad i = 1, \dots , n_S ,$$

$$\left(x_S^{(k+1)} \right)_i = \frac{1}{\hat{a}_{i+n_S, i+n_S}} \left(\hat{b}_{i+n_S} - \sum_{j \in N(i)} \hat{a}_{i+n_S, j} \cdot (x_R^{(k+1)})_j \right), \quad i = 1, \dots , n_R .$$

where $N(i)$ is the set of four indices of the neighboring grid points to grid point i.

13.5 Task and Data Parallel Execution

The data parallel execution of programs on parallel architectures with a distributed address space often leads to efficient programs. But depending on the communication operations used and the number of processors employed, scalability problems may occur. Often, the reason is the use collective communication operations, as their execution times exhibit a logarithmic or linear dependence on the number of processors participating in the operation [13.15].

```
local_nr = nr/p; local_ns = ns/p;
do {
    mestartr = me * local_nr;
    for  (i= mestartr; i < mestartr + local_nr; i++) {
        xr[i] = 0;
        for (j ∈ N(i))
            xr[i] = xr[i] + a[i][j] * xs[j] ;
        xr[i] = (xr[i]+b[i]) / a[i][i] ;
    }
    collect_elements(xr);
    mestarts = me * local_ns + nr;
    for (i= mestarts; i < mestarts + local_ns; i++) {
        xs[i] = 0;
        for (j ∈ N(i))
            xs[i] = xs[i] + a[i+nr][j] * xr[j];
        xs[i]= (xs[i] + b[i+nr]) / a[i+nr][i+nr];
    }
collect_elements(xs);
} while (convergence_test());
```

Fig. 13.11. Parallel realization of the Gauß-Seidel method with red-black ordering. The red and black grid points are stored in the arrays xr and xs.

In many cases, these problems can be avoided, if the application is partitioned into independent tasks that are executed concurrently to each other on disjoint subsets of the processors in SPMD style. This leads to a mixed execution scheme that is also referred to as mixed task and data parallel execution. In the following we give a short overview of the Tlib library that has been developed to support this style of programming. A more detailed description can be found in [13.10].

13.5.1 Overview of the Tlib Library

Tlib is a coordination library that has been built on top of MPI. The library supports the formulation of mixed task and data parallel programs with an arbitrary task structure. A program realized with the library consists of two parts: a collection of basic SPMD functions that are intended to be assigned as data parallel tasks to subsets of processors and a set of coordination functions describing the task structure. The basic SPMD functions can contain arbitrary MPI communication operations and a specific communicator that is generated by the library can be used to perform communication operations within the executing processor group. The coordination functions may contain calls of basic SPMD functions to activate data parallel executions as well as calls to library functions to partition the set of processors into subsets and to assign basic SPMD functions or coordination functions to processor groups for execution. The task structure can be nested arbitrarily which means that a coordination function can assign other coordination functions to subgroups for execution, which can then again split the corresponding subgroup and assign other coordination functions. It is also possible to specify recursive splittings of groups into subgroups and assignments of coordination functions to these subgroups, so recursive algorithms like divide-and-conquer algorithms or tree-based algorithms can be formulated quite naturally.

Interface Design. The interface of the library has been designed similarly to the Pthreads-interface, i.e., the functions to be executed concurrently are given as arguments to a specific library function together with an additional parameter of type `void *` containing the arguments of these functions. The programmer has to pack the function arguments in a single data structure.

Because of the specific way to pass functions as arguments to a library function, all basic SPMD tasks and all coordination functions that may be passed as an argument to a library function are required to have type

```
void *F (void * arg, MPI_Comm comm, T_Descr * pdescr)
```

where the MPI communicator `comm` can be used for group-internal communication and `pdescr` is a pointer to a group descriptor that contains information about the processor group onto which the function is mapped. The body of F may contain further calls to library functions to generate subgroups and initiate task parallel executions on these subgroups. F may also generate a recursive call of itself on a smaller subgroup, thus enabling the implementation of divide-and-conquer algorithms. As well as group-internal communication using the `comm` parameter, global communication is possible in the body of F by using a different communicator like `MPI_COMM_WORLD`.

The library provides functions for initialization, splitting of groups into two or more subgroups, assignment of tasks to processor groups, and getting information on the subgroup structure. The prototypes of the library functions are collected in `tlib.h` which has to be included in the program.

Initialization. To initialize a Tlib program, the library function `T_Init()` is provided. This function has to be executed before any other function of the library is called.

```
int T_Init( int argc, char * argv[], MPI_Comm comm, T_Descr * pdescr)
```

The parameter `pdescr` is a pointer to an object of type `T_Descr` which is filled by the call to `T_Init()` and is used by later calls of library functions. The data structures of type `T_Descr` are opaque objects that represent a communicator and the corresponding processor group.

Splitting operations. Before initiating a concurrent execution, appropriate processor groups and corresponding communicators have to be established.

By calling the following library function, two disjoint processor groups are generated:

```
int T_SplitGrp( T_Descr * pdescr, T_Descr * pdescr1, float p1, float p2)
```

where **pdescr** is a pointer to an existing group descriptor; **pdescr1** is a pointers to a group descriptor that is filled by the library for each processor calling the **T_SplitGrp()** function; **p1** and **p2** specify fractional values with $p1 + p2 \leq 1$. The effect of the call of **T_SplitGrp()** is the creation of two disjoint processor groups and their corresponding communicators such that the processor groups contain a fraction of **p1** and **p2** of the processors of the original processor group described by **pdescr**, respectively. The group descriptor **pdescr1** contains the descriptions of the new processor groups and their communicators after the return to the calling function.

A more general splitting operation is provided to split a processor group into more than two subgroups.

```
int T_SplitGrpParfor( int n, T_Descr * pdr, T_Descr * pdr1, float p[])
```

The first parameter **n** specifies the number of different groups in the partition; **pdr** is a pointer to an existing group descriptor; **pdr1** is a pointer to a group descriptor filled by the call of this function; **p** is an array of length **n** specifying fractional values with $\sum_{i=0}^{n-1} p[i] \leq 1$.

Concurrent Task Execution. After a splitting operation, the concurrent execution of two independent tasks can be initiated by the library function

```
int T_Par( void * (*f1)(void *, MPI_Comm, T_Descr *),
           void * parg1, void * pres1,
           void * (*f2)(void *, MPI_Comm, T_Descr *),
           void * parg2, void * pres2, T_Descr *pdescr)
```

where **f1** and **f2** are pointers to functions to be executed concurrently by the different processor groups induced by the descriptor **pdescr**; **parg1** and **parg2** contain pointers to the parameters for the function calls of **f1** and **f2**, respectively; **pres1** and **pres2** are pointers to the results to be computed by **f1** and **f2**, respectively; **pdescr** is a pointer to an object of type **T_Descr** that has previously been generated by a **T_SplitGrp()** operation.

The library function **T_Parfor()** allows the parallel activation of an arbitrary number of different functions according to a previously performed creation of concurrent subgroups. The function has the following form:

```
int T_Parfor( void *(*f[])(void *, MPI_Comm, T_Descr *),
              void * parg[], void * pres[], T_Descr * pdescr)
```

The parameter **f** is an array of length **n** with pointers to functions where **n** is the number of concurrent subgroups previously generated by a splitting operation like **T_SplitGrpParfor()** (the entry **f[i]** specifies the function to be executed by group **i** of the group partition); **parg** is an array of length **n** containing pointers to the arguments for the different function calls in f; pres is an array of length n containing pointers to results to be produced by the functions f[i]; pdescr is a pointer to a descriptor for a group decomposition into n disjoint subgroups. The number n of functions to be executed in parallel should correspond to the number of groups created in pdescr as well as to the number of arguments and results.

13.5.2 Example: Strassen Matrix Multiplication

To illustrate task-oriented recursive programming in the Tlib library we consider Strassen's algorithm [13.14] for the computation of the matrix multiplication $C = AB$ with $A, B, C \in \mathbb{R}^{n \times n}$. We denote the call to the algorithm by $C = strassen(A, B)$. Strassen's algorithm decomposes the matrices A, B into square blocks of size $n/2$ according to

$$\begin{pmatrix} C_{11} & C_{12} \\ C_{21} & C_{22} \end{pmatrix} = \begin{pmatrix} A_{11} & A_{12} \\ A_{21} & A_{22} \end{pmatrix} \begin{pmatrix} B_{11} & B_{12} \\ B_{21} & B_{22} \end{pmatrix} \tag{13.9}$$

and computes the submatrices $C_{11}, C_{12}, C_{21}, C_{22}$ separately according to

$$C_{11} = Q_1 + Q_4 - Q_5 + Q_7$$
$$C_{12} = Q_3 + Q_5$$
$$C_{21} = Q_2 + Q_4$$
$$C_{22} = Q_1 + Q_3 - Q_2 + Q_6$$

where the computation of the matrices $Q_1, \ldots, Q_7$ require 7 matrix multiplications for smaller matrices of size $n/2 \times n/2$. The seven matrix products can be computed by a conventional matrix multiplication or by a recursive application of the same procedure resulting in a divide-and-conquer algorithm with the following subproblems:

$$Q_1 = strassen(A_{11} + A_{22}, B_{11} + B_{22});$$
$$Q_2 = strassen(A_{21} + A_{22}, B_{11});$$
$$Q_3 = strassen(A_{11}, B_{12} - B_{22});$$
$$Q_4 = strassen(A_{22}, B_{21} - B_{11});$$
$$Q_5 = strassen(A_{11} + A_{12}, B_{22});$$
$$Q_6 = strassen(A_{21} - A_{11}, B_{11} + B_{12});$$
$$Q_7 = strassen(A_{12} - A_{22}, B_{21} + B_{22});$$

There are different possibilities to organize the algorithm into tasks. A straightforward implementation would create 7 tasks, one for each matrix multiplication of lower size. Here we use a task program with 4 tasks Task_C11(), Task_C12(), Task_C21(), Task_C22() where each tasks performs 2 matrix multiplications and the final assembly of one block of the

```
void * Strassen( void * arg, MPI_Comm comm, T_Descr *pdescr){
    T_Descr descr1;
    void * (* f[4])(), *pres[4], *parg[4];
    float per[4];

    per[0]=0.25; per[1]=0.25; per[2]=0.25; per[3]=0.25;
    T_Split_GrpParfor(4, pdescr, &descr1, per);
    f[0] = Task_C11; f[1] = Task_C12; f[2] = Task_C21; f[3] = Task_C22;
    parg[0] = parg[1] = parg[2] = parg[3] = arg;
    T_Parfor(f, parg, pres, &descr1);
    assemble_matrix (pres[0], pres[1], pres[2], pres[3]);
}

void * Task_C11(void * arg, MPI_Comm comm, T_Descr *pdescr){
    double **q1, **q4, **q5, **q7, **res;
    int i,j,n2;

    /* extract arguments from arg including matrix size n */
    n2 = n/2;
    q1 = Task_q1 (arg, comm, pdescr);
    q7 = Task_q7 (arg, comm, pdescr);
    /* allocate res, q4 and q5 as matrices of size n/2 times n/2 */
    /* receive q4 from group 2 and q5 from group 1 using father communicator */
    for (i=0; i < n2; i++)
      for (j=0; j < n2; j++)
        res[i][j] = q1[i][j] + q4[i][j] - q5[i][j] + q7[i][j];
    return res;
}

void * Task_q1(void * arg, MPI_Comm comm, T_Descr *pdescr){
    double **res, **q11, **q12, **q1;
    struct struct_MM_mul *mm, *arg1;

    *mm = (struct_MM_mul *) arg;
    n2 = (*mm).n /2;
    /* allocate q1, q11, q12 as matrices of size n2 times n2 */
    for (i=0; i < n2; i++)
      for (j=0; j < n2; j++) {
        q11[i][j] = (*mm).a[i][j] + (*mm).a[i+n2][j+n2]; /* A11+A22 */
        q12[i][j] = (*mm).b[i][j] + (*mm).b[i+n2][j+n2]; /* B11+B22 */
      }
    arg1 = (struct_MM_mul *) malloc (sizeof(struct_MM_mul));
    (*arg1).a = q11; (*arg1).b = q12; (*arg1).n = n2;
    q1 = Strassen(arg1, comm, pdescr);
    return q1;
}
```

Fig. 13.12. Realization of Strassen's multiplication using the Tlib library

final result C. For load balancing reasons and to save communication time, submatrix Q_1 is computed both by tasks `Task_C11()` and `Task_C22()`. The following execution scheme results:

Task C11	Task C12	Task C21	Task C22
on group 0	on group 1	on group 2	on group 3
compute Q_1	compute Q_3	compute Q_2	compute Q_1
compute Q_7	compute Q_5	compute Q_4	compute Q_6
receive Q_5	send Q_5	send Q_2	receive Q_2
receive Q_4	send Q_3	send Q_4	receive Q_3
compute C_{11}	compute C_{12}	compute C_{21}	compute C_{22}

Figure 13.12 sketches a part of the resulting parallel program using Tlib. Only `Task_C11()` is shown; the realization of `Task_C12()`, `Task_C21()`, and `Task_C22()` is similar. We only sketch `Task_q1()` to compute Q_1 in a recursive way. The data structures of type `struct_MM_mul` are used to store the two matrices to be multiplied along with their size. Function `assemble_matrix()` assembles the computed submatrices according to (13.9).

References

[13.1] D.R. Butenhof. *Programming with POSIX Threads*. Addison-Wesley, 1997.

[13.2] D.E. Culler, J.P. Singh, and A. Gupta. *Parallel Computer Architecture: A Hardware Software Approach*. Morgan Kaufmann, 1999.

[13.3] A. Geist, A. Beguelin, J. Dongarra, W. Jiang, R. Manchek, and V. Sunderam. *PVM Parallel Virtual Machine: A User's Guide and Tutorial for Networked Parallel Computing*. MIT Press, Camdridge, MA, 1996. www.netlib.org/pvm3/book/pvm_book.html.

[13.4] D. Kerbyson, H. Alme, A. Hoisie, F. Petrini, H. Wasserman, and M. Gittings. Predictive Performance and Scalability Modeling of a Large-Scale Application. In *Proc. of the ACM/IEEE Supercomputing*, Denver, USA, 2001.

[13.5] M. Kühnemann, T. Rauber, and G. Rünger. Performance Modelling for Task-Parallel Programs. In *Proc. of the Communication Networks and Distributed Systems Modeling and Simulation Conference (CNDS 2002)*, pages 148–154, San Antonio, USA, 2002.

[13.6] *OpenMP C and C++ Application Program Interface, Version 1.0*, October 1998.

[13.7] P.S. Pacheco. *Parallel programming with MPI*. Morgan Kaufmann Publ., 1997.

[13.8] T. Rauber and G. Rünger. PVM and MPI Communication Operations on the IBM SP2: Modeling and Comparison. In *Proc. 11th Symp. on High Performance Computing Systems (HPCS'97)*, pages 141–152, 1997.

[13.9] T. Rauber and G. Rünger. *Parallele und Verteilte Programmierung*. Springer Verlag, Heidelberg, 2000.

[13.10] T. Rauber and G. Rünger. Library Support for Hierarchical Multi-Processor Tasks. In *Proc. of the ACM/IEEE Supercomputing*, Baltimore, USA, 2002.

[13.11] M. Snir, S. Otto, S. Huss-Ledermann, D. Walker, and J. Dongarra. *MPI: The Complete Reference.* MIT Press, Camdridge, MA, 1996. www.netlib.org/utk/papers/mpi_book/mpi_book.html.

[13.12] Th. Sterling and J. Salmonand D.J. Becker. *How to Build a Beowulf: A Guide to the Implementation and Application of PC Clusters.* MIT Press, 1999.

[13.13] J. Stoer and R. Bulirsch. *Numerische Mathematik II.* Springer, 3 edition, 1990.

[13.14] V. Strassen. Gaussian Elimination is not Optimal. *Numerische Mathematik*, 13:354–356, 1969.

[13.15] Z. Xu and K. Hwang. Early Prediction of MPP Performance: SP2, T3D and Paragon Experiences. *Parallel Computing*, 22:917–942, 1996.

14 Multi-Grid Methods – An Introduction

G. Wittum

Universität Heidelberg, Interdisziplinäres Zentrum für Wissenschaftliches Rechnen (IWR), Im Neuenheimer Feld 368, 69120 Heidelberg, Germany

Abstract. The lecture will give an introductory overview on multi-grid methods. Multi-grid methods are fast solvers for algebraic equations derived from the discretiziation of pde. They have optimal complexity, are very flexible and can used for a wide variety of problems. The bad news is, however, that they often must be adapted to the problem.

After an introduction and the description of the main ideas will present guidelines for the construction of multi-grid methods and give some theoretical backup of those. Further we address the main multi-grid difficulties and workaround for a number of application problems.

14.1 Introduction

14.1.1 Historical Overview and Introduction to Multi-Grid Methods

One basic problem for the numerical solution of partial differential equations is solving large systems of linear equations, e.g.

$$Ax = b, \tag{14.1}$$

where A is a given n by n matrix.

$$A = \begin{pmatrix} a_{11} & \cdots & a_{1n} \\ \vdots & \ddots & \vdots \\ a_{n1} & \cdots & a_{nn} \end{pmatrix} \tag{14.2}$$

$x = (x_1, \ldots, x_n)^T$ is a vector of unknowns and $b = (b_1, \ldots, b_n)^T$ the corresponding right-hand side. A straightforward algorithm for solving (14.1) is the well-known Gauß algorithm which is usually applied as a lower-upper factorization, i.e. decomposing A as follows

$$A = LU \tag{14.3}$$

with

G. Wittum, Multi-Grid Methods – An Introduction, Lect. Notes Phys. **642**, 283–311 (2004)
http://www.springerlink.com/ © Springer-Verlag Berlin Heidelberg 2004

$$L = \begin{pmatrix} 1 & 0 & \cdots & 0 \\ * & \ddots & \ddots & \vdots \\ \vdots & \ddots & \ddots & 0 \\ * & \cdots & * & 1 \end{pmatrix}, \qquad U = \begin{pmatrix} * & \cdots & \cdots & * \\ 0 & \ddots & \ddots & \vdots \\ \vdots & \ddots & \ddots & \vdots \\ 0 & \cdots & 0 & * \end{pmatrix} \qquad (14.4)$$

where an asterisk stands for any number, and solving the system by forward and backward substitution

$$y = L^{-1}b, \qquad (14.5)$$
$$x = U^{-1}y,$$

which is easily accomplished.

This method has some severe drawbacks when applied to common discretizations of partial differential equations. First, the number of operations necessary to solve (14.1) is proportional to n^3, where n is the number of unknowns. Hence, computing-time increases rapidly with growing n. Second, the matrices arising from those discretizations are "sparse", i.e. there are only a few elements different from zero in each line. This feature drastically reduces the storage requirement for A. Unfortunately, LU-decomposition (14.2) produces "fill-in", additional non-zero entries, destroying the original sparsity structure. Thus, this direct method is not suited for larger systems. Already in 1823 C.F. Gauß realized this, having to solve systems of approx. 40 unknowns for the triangulation of Hannover. On this occasion he found the first iterative method solving system (14.1) which is still called the Gauß-Seidel method. An iterative method for problem (14.1) is given by splitting A into (cf. [14.1])

$$A = M - N \qquad (14.6)$$

M regular, "easily invertible",
and the following scheme: Let an arbitrary starting guess $x^{(0)}$ be given. Then the $i + 1$-st iterate is calculated from the i-th one via

$$x^{(i+1)} = x^{(i)} - M^{-1}(A\,x^{(i)} - b) \qquad (14.7)$$

Now we do not get the exact solution after one step as with the direct method but rather a sequence of iterates converging towards x. M^{-1} is considered an "approximate inverse" of A. There are many possibilities to choose M. So, the original Gauß-Seidel method uses

$$M = L + D, \qquad (14.8)$$

where

$$A = L + D + U,$$

L containing the strictly lower triangular part of A, U the strictly upper triangular and D the diagonal one.

Concerning the superiority of these "indirect" methods over the direct ones for this kind of system, we quote a few words from a letter of Gauß to Gerling (see [14.2]):

"Ich empefehle Ihnen diesen Modus zur Nachahmung. Schwerlich werden Sie je wieder direct eliminiren, wenigstens nicht, wenn Sie mehr als 2 Unbekannte haben. Das indirecte Verfahren lässt sich im Schlafe ausführen, oder man kann während desselben an andere Dinge denken."

In 1845 Jacobi introduced a second iterative method (see [14.3]) with the choice

$$M = D, \tag{14.9}$$

D from (14.8).

These two methods remained the basic tools of geodesists and engineers to solve large systems for about 100 years.

Then with the first computers the interest concentrated on these methods as it was possible now to treat larger and larger systems. So for Gauß $n = 40$ was large, as he had to do his calculations alone and by hand. Then the number of unknowns increased more and more as groups of scientists worked together. In 1890 Nagel had to solve a system with $n = 159$ for a triangulation of Saxony. Thereafter, about 1900, Boltz had a system of 670 unknowns. Imagine that they were solved by hand!

The use of large n revealed a severe drawback of such iterative methods. The larger n, the more steps are needed to obtain a certain accuracy, as can be seen applying the Gauß-Seidel method to the following model problem, the Poisson equation on a square

$$-\Delta u = f \quad \text{in } (0,1) \times (0,1) \tag{14.10}$$

with Dirichlet boundary conditions

$$u = u_R, \quad \text{on the boundary,} \tag{14.11}$$

which is the standard paradigm for such methods. To obtain a finite dimensional system of equations, we introduce an equidistant grid with mesh size h. Now we replace the differential operator by a central difference quotient in each grid point and the functions u and f by discrete "grid-functions" u_h, f_h defined in each grid-point. In the present case this discretization yields the well-known five-point formula

$$K_l = h_l^{-2} \begin{bmatrix} & -1 & \\ -1 & 4 & -1 \\ & -1 & \end{bmatrix} \tag{14.12}$$

which by numbering the grid points lexicographically corresponds to the matrix showing the typical sparsity-structure of such a system (missing entries are zero!).

Of course, the discretization process introduces errors and it is one main task of numerical analysis to obtain estimates for these "truncation errors". Apparently, the finer the grid becomes, the better the approximation will be. Hence it is a substantial requirement to find new methods which are asymptotically optimal, i.e. need not more work than absolutely neccessary to solve the discrete system.

The performance of the Gauß-Seidel method for this problem is shown in Table 14.1 and Fig. 14.1.

Table 14.1. Performance of a Gauß-Seidel method for problem (14.9, 14.10). The rate of convergence in the first ten steps of a classical iteration scheme (14.6), depending on the number of unknowns n.

Step	n = 49	n = 225	n= 961	n = 3969	n = 15876
1	0.56047	0.78939	0.89753	0.94947	0.97491
2	0.61741	0.85628	0.93860	0.97142	0.98619
3	0.61969	0.87089	0.94948	0.97728	0.98919
4	0.61942	0.87627	0.95505	0.98036	0.99077
5	0.61915	0.87840	0.95851	0.98234	0.99178
6	0.61903	0.87925	0.96085	0.98374	0.99251
7	0.61898	0.87960	0.96253	0.98479	0.99305
8	0.61895	0.87973	0.96377	0.98563	0.99349
9	0.61895	0.87978	0.96470	0.98630	0.99384
10	0.61894	0.87979	0.96541	0.98686	0.99414

Much research has been done on that topic. The first to accelerate the original iterative methods was Young, introducing the so-called *SOR* method in 1950, [14.4]. In 1952 the conjugate-gradient technique was invented by Stiefel (cf. [14.5]) which is still one of the most important techniques. However it does not usually provide asymptotically optimal efficiency.

In 1961 Fedorenko introduced the first multi-grid method (cf. [14.6]), a technique yielding optimal work estimates. Their actual efficiency was first realized by Brandt in 1973 (cf. [14.7]) and applied to a large class of problem. In 1976 Hackbusch developed this class of methods independently, see [14.8], and since then many main ideas as well as a comprehensive mathematical treatment are due to him.

The development of multi-grid started with a detailed analysis of classical iterative methods. From this analysis it is readily seen that standard iterations effect a "smoothing" of the error rather than a reduction. Smoothing means a strong reduction of high-frequency error components, while the "smooth", i.e. low-frequency, components are not touched substantially. Thus to obtain a fast method this smoothing has to be combined with a method which strongly damps the smooth components of the error, while leaving the high-frequency parts unchanged. To that end consider a smooth function in one dimension discretized on a "fine" grid, as shown in Fig. 14.2. In Fig. 14.3

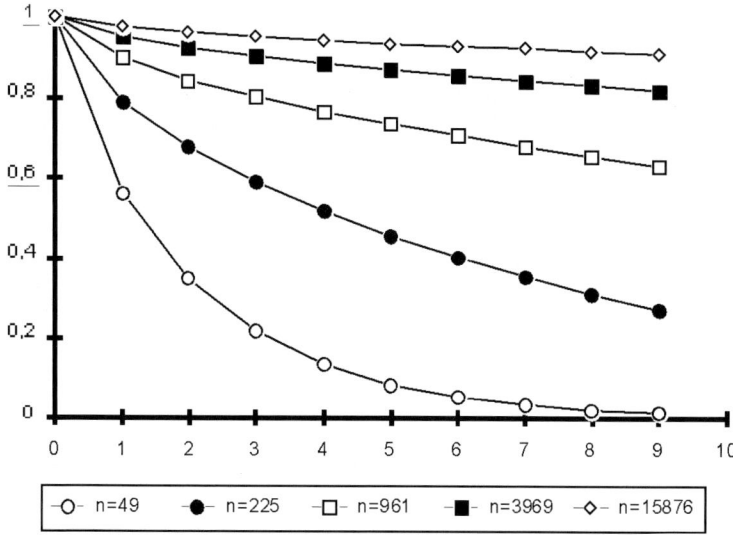

Fig. 14.1. Performance of a Gauß-Seidel method for problem (14.9,14.10). The norm of the residual after the first ten steps of a classical iteration scheme (14.6), depending on the number of unknowns n.

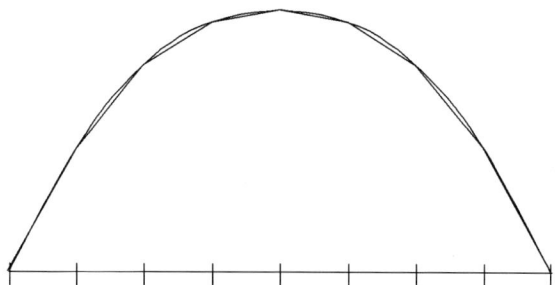

Fig. 14.2. Discretization of a smooth function on a given grid. The grid function is represented by the piecewise linear interpolation of the function values in the grid points.

the same function is depicted using a coarser discretization which is obtained leaving out each second point of the original grid. Apparently the coarser discretization still provides a reasonable representation. This is no longer true if the function under consideration is highly oscillating as demonstrated in 14.4.

That means, if the function is smooth, we may approximate it on a "coarser" grid, yielding a matrix of lower dimension, for which the classical iteration is faster. Thus, we found a way of improving the performance of

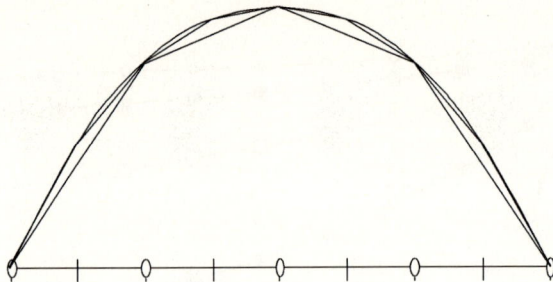

Fig. 14.3. The function from Fig. 14.2 restricted to a coarser grid. The coarse-grid points are distinguished by ovals.

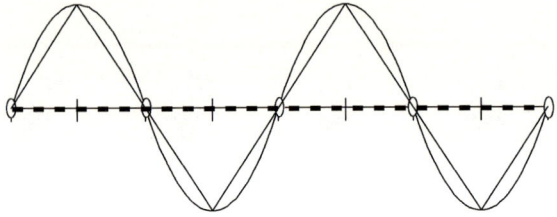

Fig. 14.4. Unsmooth function discretized on fine grid and restriction to a coarser one. The restricted function is zero, indicated by a bold dashed line. The coarse-grid points are distinguished by ovals.

classical methods in a simple way by the so-called multi-grid algorithm which is described as follows.

Let the partial differential equation

$$Ku = f \quad \text{in } \Omega \qquad u|_{\partial\Omega} = u_R \qquad (14.13)$$

be discretized on a hierarchy of grids Ω_l, $l = 1, \ldots, l_{\max}$, with decreasing stepsize yielding the discrete problem

$$K_l u_l = f_l \text{ in } \Omega_l \qquad u_l|_{\partial\Omega_l} = u_R|_{\Omega_l} \text{ on the boundary.} \qquad (14.14)$$

Let further a prolongation p and a restriction r be given, transforming a grid-function from coarse to fine or fine to coarse resp. as well as a smoother S which usually consists of a classical iterative scheme. Then the multi-grid algorithm reads as follows

algorithm 14.1. :

procedure mgm(l,u,f); **integer** l ;**array** u,f;
        if l = 1 **then** u := K_l^{-1} f **else**
        begin array d, v ; **integer** j ;
            u := S^{ν_1}(u,f);
            d := r*(K_l u - f);

```
        v := 0;
        for j := 1 step 1 until γ do mgm(l-1,v,d );
        u := u - p*v ;
        u := S^{ν_2}(u,f );
end;
```

The advantage of this method for the same example as used to compute the values from Table 14.1 can be seen from Tables 14.2 and 14.3 as well as Fig. 14.5.

Table 14.2. Performance of a multi-grid method for problem (14.9, 14.10).The norm of the residual (defect) after the first ten steps of a multi-grid method, depending on the number of unknowns n.

Step	n = 49	n = 225	n= 961	n = 3969	n = 16129
0	1.000E+00	1.000E+00	1.000E+00	1.000E+00	1.000E+00
1	8.045E-02	1.050E-01	1.222E-01	1.323E-01	1.377E-01
2	2.505E-03	5.262E-03	6.650E-03	7.159E-03	7.314E-03
3	7.971E-05	2.864E-04	3.557E-04	3.869E-04	4.079E-04
4	2.524E-06	1.557E-05	2.090E-05	2.343E-05	2.449E-05
5	7.903E-08	8.576E-07	1.271E-06	1.456E-06	1.530E-06
6	2.436E-09	4.795E-08	7.923E-08	9.121E-08	9.567E-08
7	7.381E-11	2.710E-09	5.015E-09	5.759E-09	5.986E-09
8	2.201E-12	1.543E-10	3.207E-10	3.659E-10	3.758E-10
9	6.462E-14	8.812E-12	2.067E-11	2.340E-11	2.370E-11
10	roundoff	5.042E-13	1.342E-12	1.506E-12	1.512E-12

Table 14.3. Performance of a multi-grid method for problem (14.9, 14.10). The rate of convergence after the first ten steps of a multi-grid method (1.1.5), depending on the number of unknowns n.

Step	n = 49	n = 225	n= 961	n = 3969	n = 16129
1	8.045E-02	1.050E-01	1.222E-01	1.323E-01	1.377E-01
2	3.114E-02	5.013E-02	5.444E-02	5.413E-02	5.311E-02
3	3.182E-02	5.443E-02	5.349E-02	5.404E-02	5.577E-02
4	3.167E-02	5.437E-02	5.876E-02	6.056E-02	6.004E-02
5	3.131E-02	5.508E-02	6.080E-02	6.217E-02	6.249E-02
6	3.082E-02	5.591E-02	6.234E-02	6.263E-02	6.252E-02
7	3.031E-02	5.653E-02	6.330E-02	6.314E-02	6.257E-02
8	2.982E-02	5.691E-02	6.395E-02	6.354E-02	6.278E-02
9	2.936E-02	5.713E-02	6.446E-02	6.394E-02	6.307E-02
10	roundoff	5.722E-02	6.490E-02	6.436E-02	6.377E-02
11		roundoff	roundoff	roundoff	roundoff

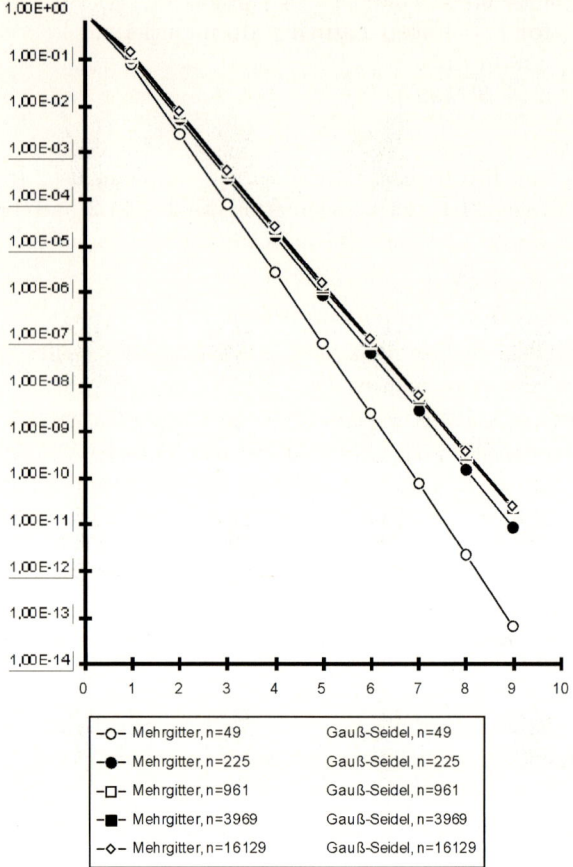

Fig. 14.5. Descent of defect in 10 iterations od smoother only and multi-grid.

Apparently, using multi-grid we get a convergence behaviour wich is independent of the number of unknowns and thus solve the problem in 10 staps instead of about 10000. A deteiled introduction to multi-grid technique and applications can be found in [14.10].

14.1.2 Additive Multigrid

Parallelization problems as well as thinking about adaptivity gave rise to the introduction of another multigrid variant. The additive multigrid algorithm is given by the following scheme.

algorithm 14.2. : additive multi-grid method

```
amgm(l,u,f)
integer l; grid function u[l], f[l], v;
{
        integer j;
        [l] := K_l u[l]- f[l]; v := u[l];
        for j:=l step -1 to 1 do f[j-1] := r_{j-1} f[j];
        for j:=1 step 1 to l do u[j] := S_j^ν(u[j],f[j]);
        u[0] := K_0^{-1} f[0];
        for j:=1 step 1 to l do u[j] :=u[j] + p_{j-1} u[j-1];
        u[l] := v - u[l];
}
```

The basic structure of the two multigrid variants can be seen from Fig. 14.6. The main difference between the so-called multiplicative multigrid algorithm and the additive one is that in the multiplicative method smoothing and re-stricition of the defect to the next coarser level are performed on one level after the other sequentially, while in the additive method smoothing on different levels can be performed in parallel. Restriction and prolongation, however, are sequential in the additive method too. Thus on a parallel machine the additive method also has a logarithmic complexity. Usually, the additive methods are applied as preconditioners, since acceleration methods like cg directly pick an optimal damping parameter, the multiplicative methods are used as solvers. According to [14.11], these methods can be formulated as additive Schwarz methods. It is easily proved that on a serial computer the multiplicative version mgm is twice as fast as the additive one. In parallelization, however, the additive method may play a role.

14.2 Convergence Theory

14.2.1 General Setting

For the following brief description of multi-grid convergence theory we use the setting from [14.10]. For a detailed introduction to multi-grid theory and algorithms see there.

Let K_l be the discretization of a differential operator K of order $2m$ on a hierarchy of grids Ω_l, $1 \le l \le l_{\max}$. Let further S_l be a smoother for the discrete equation

$$K_l u_l = b_l \tag{14.15}$$

and p and r denote the prolongation and restriction between the grids. Then the two-grid operator is given by

$$T_{2,l}(\nu,0) := (I_l - pK_{l-1}^{-1}rK_l)S_l^\nu \tag{14.16}$$

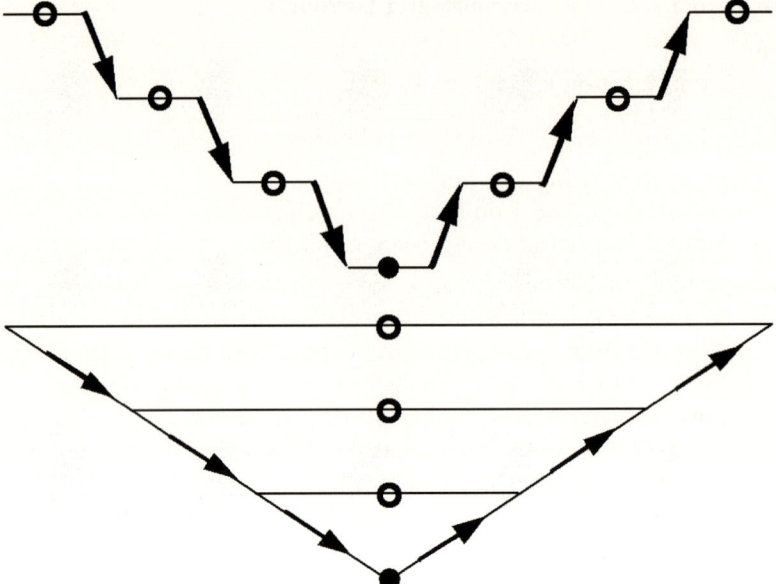

Fig. 14.6. Outline of multiplicate multigrid algorithm $mmgm$ (V-cycle) above and of the additive multigrid algorithm $amgm$ below. Symbols used:

○	smoothing	●	coarse-grid solution
↘	restriction	↗	prolongation

(see [14.10]). To establish monotonic convergence we have to estimate the norm of $T_{2,l}(\nu,0)$.

Among a lot of different approaches to prove multi-grid convergence, two of them turned out to be the most powerful ones. First, the splitting approach introduced by Hackbusch in [14.9], second the V-cycle proof, introduced in [14.12] and generalized to larger classes of problems in [14.14, 14.13, 14.10].

To estimate the norm of $T_{2,l}(\nu,0)$ the first approach uses the splitting

$$\|T_{2,l}(\nu,0)\| \le \|K_l^{-1} - pK_{l-1}^{-1}r\|\,\|K_l S_l^\nu\|. \tag{14.17}$$

The two factors are estimated separately by the

approximation property

$$\|K_l^{-1} - pK_{l-1}^{-1}r\| \le C\,h_l^{2m}, \tag{14.18}$$

and the

smoothing property

$$\|K_l\, S_l^\nu\| \le C\, h_l^{-2m} \eta(\nu) \quad \text{with } \eta(\nu) \,\to\, 0 \; for \; \nu \,\to\, \infty. \tag{14.19}$$

Together both properties yield the h-independent two-grid estimate

$$\|T_{2,l}(\nu,0)\| \le \xi < 1 \quad \forall\, l \ge 1. \tag{14.20}$$

For scalar elliptic problems as treated in the present paper usually the spectral norm is used for the above estimates. This is especially convenient since the spectral norm corresponds to the discrete L^2-norm, thus allowing the use of merely algebraic techniques when considering the smoothing property and the use of tools provided by the discrete Sobolev-spaces for the proof of the approximation property.

In a sequence of papers, Hackbusch established the approximation property for several discretizations of large classes of partial differential equations and the smoothing property for simple iterative schemes as damped Jacobi and Gauß-Seidel in linear and semi-iterative versions for these discretizations (see [14.10] and the references there). In [14.15] and [14.16] the smoothing property was proved for the simple 5-point ILU scheme applied to model problem (14.21), mainly aiming towards robustness.

The smoothing property of convergent iterations has been proved by the author for symmetric problems in [14.16]. This theory has been generalized by Reusken [14.17].

Due to splitting 14.17, this approach will yield overestimates for the norm of the two-grid operator and in particular it will not give sharp predictions for the number of smoothing steps, n, necessary to obtain convergence of the whole process (see [14.16]). Multi-grid convergence can be based on two-grid estimate (14.20), yielding convergence for a W-cycle (see [14.10]).

To obtain convergence results for the V-cycle, however, we need a different approach, requiring stronger assumptions and abandoning the separation between the algebraic part (smoothing property) and the differential one (approximation property). Thus the family of V-cycle proofs (see [14.12, 14.14, 14.13, 14.10]) is based on a relation between the coarse grid correction and the approximate inverse, M, in the smoother. There the smoother is required to have nonnegative eigenvalues.

14.2.2 The Smoothing Property

First we establish the smoothing property for convergent iterations. We investigate the smoothing property of a linear iterative solver for the equation

$$K\,u = f \tag{14.21}$$

constructed by splitting

$$K = M - N \qquad \text{M regular, easily invertible.} \qquad (14.22)$$

The iteration is given by

$$u_{\text{new}} = u_{\text{old}} - M^{-1}(K\, u_{\text{old}} - f) \qquad (14.23)$$

with the corresponding operator

$$S = I - M^{-1}K. \qquad (14.24)$$

Let A, B be $n \times n$-matrices. By

$$A \underset{(\geq)}{\geq} B \qquad (14.25)$$

we denote the positive (semi-) definiteness of the matrix $A - B$.

Theorem 1. *Let*

$$K = K^T > 0 \quad and \quad M = M^T > 0. \qquad (14.26)$$

Let further

$$\|M\| = C_M \|K\| \qquad (14.27)$$

with the spectral norm $\|\cdot\|$ and assume S to be convergent.
Then the damped iteration

$$u_{\text{new}} = u_{\text{old}} - \omega M^{-1}(K u_{\text{old}} - f) \qquad (14.28)$$

with $\omega < 1$ satisfies the smoothing property (14.18) in the spectral norm.

Proof.

$$\|KS^\nu\| = \|(M - N)(M^{-1}N)^\nu\| \qquad (14.29)$$
$$= \|M^{\frac{1}{2}}(I - M^{-\frac{1}{2}}NM^{-\frac{1}{2}})(M^{-\frac{1}{2}}NM^{-\frac{1}{2}})^\nu M^{\frac{1}{2}}\| \qquad (14.30)$$
$$\leq \|M\| \, \|(I - X)X^\nu)\| \quad \text{with } X = M^{-\frac{1}{2}}NM^{-\frac{1}{2}}. \qquad (14.31)$$

Since S is convergent, the eigenvalues of the damped iteration S_ω are contained in $(-\vartheta, 1)$ with $\vartheta = 2\omega - 1$. Thus X satisfies $-\vartheta I < X < I$. We have

$$\|(I - X)X^\nu)\| \leq \max\{\|(1 - \xi)\xi^\nu\| \; for \; (-1, 1)\} \qquad (14.32)$$
$$= \max\{\nu^\nu/(\nu + 1)^{\nu+1}, (1 - \vartheta)\vartheta^\nu\} \qquad (14.33)$$

Thus we obtain the assertion with
$$\eta(\nu) := \max\{\nu^\nu/(\nu + 1)^{\nu+1}, (1 - \vartheta)\vartheta^\nu\} = \mathcal{O}\left(\nu^{-1}\right).$$

$\square$

Thus for any convergent iteration the damped variant will be suited for smoothing. However, from the proof of Theorem 1 it can be seen that damping is essential. This is essential for practice too which is shown by the simple example of the Jacobi method for standard Laplace model problem. The iteration will not smooth the problem unless it is damped. Instead of damping one can use other methods shifting the spectrum of the smoother. In particular in the case of incomplete decompositions this can be achieved by modification as introduced in [14.15] or shifting as proposed in [14.18].

In the V-cycle convergence proof an additional condition for the smoother is involved. Nevertheless a properly damped convergent iteration will be sufficient as smoother as shown in [14.16].

The nonsymmetric case has been treated by Reusken, [14.17]. He established the same result as above in the maximum norm on the assumption that the damping parameter ω satisfies $0 < \omega < \frac{1}{2}$. Hackbusch in [14.19] and [14.20] generalized this result in the manner of the V-cycle proof from [14.16].

14.2.3 Approximation Property

The proof of the approximation property needs stronger assumptions on the differential problem. To avoid overloading the lecture with theory, we start discussing a formal splitting of the coarse-grid correction matrix . For a detailed discussion of the approximation property we refer to [14.10].

$$
\begin{aligned}
K_l^{-1} - pK_{l-1}^{-1}r &= (I - pr)K_l^{-1} + p(rK_l^{-1} - K_{l-1}^{-1}r) \\
&= (I - pr)K_l^{-1} + pK_{l-1}^{-1}((K_{l-1} - rK_lp)r + rK_lpr - rK_l)K_l^{-1} \\
&= \left\{ (I - pr) + pK_{l-1}^{-1}((K_{l-1} - rK_lp)r + rK_l(pr - I)) \right\} K_l^{-1}.
\end{aligned}
$$

Based on this splitting we formulate the following theorem.

Theorem 2. *Let*

$$
\|\!|\!| I - pr \|\!|\!|_{0 \leftarrow 2m} \leq C_l \, h_l^{2m} \tag{14.34}
$$

$$
\|\!|\!| K_{l-1} - rK_l p \|\!|\!|_{-2m \leftarrow 2m} \leq C_A \, h_l^{2m} \tag{14.35}
$$

$$
\|\!|\!| K_l^{-1} \|\!|\!| \leq C_i \ for \ \{(2m, 0), (0, -2m)\} \tag{14.36}
$$

$$
\|\!|\!| K_l \|\!|\!|_{-2m \leftarrow 0} \leq C_K, \tag{14.37}
$$

$$
\|\!|\!| p \|\!|\!|_{0 \leftarrow 0} \leq C_p \tag{14.38}
$$

and

$$
\|\!|\!| r \|\!|\!|_{s \leftarrow s} \leq C_r, \qquad with \ s \in \{-2m, 2m\}, \tag{14.39}
$$

where $\|\!|\!| A \|\!|\!|_{s \leftarrow t}$ denotes the norm of the operator A between discrete Sobolev spaces: $A : H^t \to H^s$.

Them the approximation property

$$\||K_l^{-1} - p\,K_{l-1}^{-1}\,r\||_{0\leftarrow 0} \leq C\,h_l^{2m} \qquad (14.40)$$

holds with

$$C = C_i(C_I + C_p C_i(C_A C_r + C_r C_K C_I)). \qquad (14.41)$$

Proof. Using formula 14.34 we estimate:

$$\||K_l^{-1} - p\,K_{l-1}^{-1} r\||_{0\leftarrow 0} \leq \{\||I - pr\||_{0\leftarrow 2m}$$
$$+\||p\||_{0\leftarrow 0}\,\||K_{l-1}^{-1}\||_{0\leftarrow -2m}$$
$$(\||K_{l-1} - r\,K_l\,p\||_{-2m\leftarrow 2m}\||r\||_{2m\leftarrow 2m}$$
$$+\||r\||_{-2m\leftarrow -2m}\,\||K_l\||_{-2m\leftarrow 0}\||p\,r - I\||_{0\leftarrow 2m}$$
$$\}\||K_l^{-1}\||_{2m\leftarrow 0}.$$

□

The above assumptions are usually fulfilled, provided that discretization, coarse-grid operator and grid transfers satisfy the following requirements.

- The prolongation is of the order of the differential operator.
- The discretization is of the order of the differential operator.
- The discrete problem is fully regular.

This holds e.g. for a discretization of Laplace's equation on a smooth domain. The approximation property not only gives multigrid convergence, but is also connected to the quality of the coarse-grid discretization.

From Theorem 2 we directly deduce the following rules for the construction of grid transfers and coarse-grid operators.

Note 1. :

- The order of the prolongation m_p and the one of the restriction m_r should satisfy

$$m_p + m_r > 2m \qquad (14.42)$$

m denoting the order of the differential operator.
- The restriction should be chosen as the adjoint of the prolongation with respect to the underlying scalar product:

$$r = p^*. \qquad (14.43)$$

- For finite element discretisations the restricition is canonically constructed representing the coarse-grid basis functions by fine grid ones. The prolongation is then defined by (14.43).

– In smooth cases it is usually sufficient to use the discretization of the differential operator on the respective grid as coarse-grid operator. From Theorem 2, however, it is apparent that the so-called Galerkin ansatz for the coarse grid operator

$$K_{l-1} = r K_l \, p \tag{14.44}$$

which is automatically satisfied in case of conforming finite elements with canonical grid transfers. Then assumption 14.35 is satisfied automatically and the bound for the approximation property only depends on the interpolation error and on the regularity.

In practice the multigrid convergence rate depends on the regularity of the problem. The proof above, however, requires too much. In case of no regularity the method works a bit more slowly, but still converges quickly. A proving technique which does not use assumptions on the regularity of the problem is due to Xu, [14.21], see also [14.22], [14.23].

14.3 Robustness

The treatment of singular perturbations is a severe problem in solving partial differential equations. The discretization of a singularly perturbed operator contains characteristic difficulties which may affect multi-grid convergence, as the discretization process is hidden in the hierarchy of grids Ω_l, $l = 1, \ldots, l_{max}$, used in multi-grid technique. To handle such problems robust multi-grid methods are required. In the present paper we introduce a robust multi-grid method based on a variant of incomplete LU-smoothing. This method is proven to be robust for the anisotropic model-problem and related ones.

14.3.1 Robustness for Anisotropic Problems

An operator $K = K(\varepsilon)$ depending on a parameter e is called singularly perturbed, if the limiting operator $K(0) = \lim_{\varepsilon \to 0} K(\varepsilon)$ is of a type other than $K(\varepsilon)$ for $\varepsilon > 0$. We are especially interested in operators of the form

$$K(\varepsilon) = \varepsilon \, K' + K'', \tag{14.45}$$

where $K(\varepsilon)$ is elliptic for $\varepsilon > 0$ and K'' is non-elliptic or elliptic of lower order.

A simple model for this situation is

$$K(\varepsilon)u = f \quad \text{in } \Omega, \qquad u = g \text{ on } \partial\Omega$$

with

$$K(\varepsilon) := -\varepsilon \frac{\partial^2}{\partial x^2} - \frac{\partial^2}{\partial y^2} \qquad (14.46)$$

in a domain $\Omega \subset \mathbb{R}^2$. For the moment let $\Omega = (0,1) \times (0,1)$ and (14.3.1) be discretized by

$$K_l(\varepsilon) := h_\lambda^{-2} \begin{pmatrix} & -1 & \\ -\varepsilon & 2(1+\varepsilon) & -\varepsilon \\ & -1 & \end{pmatrix} \qquad (14.47)$$

on a rectangular equidistant grid with stepsize h_l, ordering the grid-points lexicographically. Stencil (14.47) applies to grid points lying totally inside Ω. For points having a boundary point as neighbour, the corresponding entry of the stencil has to be skipped.

This problem is very simple, however, some typical difficulties arising from singular perturbations show up there and can be analyzed thoroughly. In addition, the insight obtained by the analysis of the model-problem gives an understanding of more complex situations.

Applying the multi-grid method with Gauß-Seidel smoothing   from Sect. 14.1.1 to problem (14.45/14.47) and varying ε we get the convergence rates shown in Fig. 14.3.1. Apparently, the method deteriorates for smaller ε. The dependence of κ on ε closely resembles the one of the smoother alone on h, as described in Sect. 14.1.1

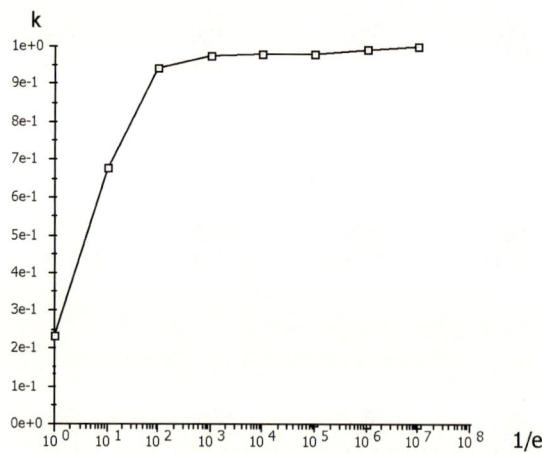

Fig. 14.7. Convergence rate κ of the multi-grid method from Sect. 14.1.1 applied to problem (14.45/14.47) averaged over 10 steps versus $1/\varepsilon$.

This is due to the fact that the averaging effect of the smoother in x-direction is scaled by ε and thus does not suffice to reduce high oscillations in

x-direction for small ε. This corresponds to the not appropriate discretization on the equidistant grid for small ε.

One possiblity to avoid this problem is to use a so-called semi-coarsening i.e. to coarsen the grids only in y-direction (see [14.24]. This process is quite natural, taking into account the different quality of the smoother in x- and y-direction. For more complicated practical problems where the anisotropy is not known in advance, however, it can hardly be used.

Another possibility to cope with these problems is to use a smoother which solves the limit problem for $\varepsilon \to q$, $q \in \{0, \infty\}$, exactly. Such a smoother is e.g. the alternating block GAUSS-SEIDEL method. Another and considerably cheaper possibility, however, is a five-point incomplete LU-decomposition (ILU) which is described in the following.

Defining one step of GAUSSIan elimination on K produces fill-in as shown in 14.8.

Now, the basic idea of ILU is to neglect fill-in and perform the factorization only on a prescribed sparsity pattern, in this case the pattern of K. Proceeding that way we obtain a lower and upper triangular matrix L and U with the same sparsity pattern as the corresponding parts of K. The product LU, however, is not the decompostion of K, but of another matrix K', which may be written as

$$LU = K' = K + N. \tag{14.48}$$

Thus we obtain

$$K = LU - N \tag{14.49}$$

which corresponds to (14.6) with $M = LU$ providing a new choice for the approximate inverse M. According to (14.7) this gives rise to the linear iterative method Let an arbitrary starting guess $x^{(0)}$ be given. Then the $i+1$-st iterate is calculated from the i-th one via

$$x^{(i+1)} := x^{(i)} - (LU)^{-1}(Kx^{(i)} - f). \tag{14.50}$$

Comparing the sparsity patters of LU and K we obtain the pattern of N

$$\begin{pmatrix} & * & \\ * & * & * \\ & * & \end{pmatrix} = \begin{pmatrix} & \square & \\ * & * & \square \\ & * & \end{pmatrix} \begin{pmatrix} & * & \\ \square & * & * \\ & \square & \end{pmatrix} - \begin{pmatrix} * & \square & \\ \square & \square & \square \\ & \square & * \end{pmatrix} \tag{14.51}$$

For large as well as for small values of ε the discretization matrix K approaches a matrix with three diagonals only the decomposition of which produces no fill-in. Thus our incomplete LU smoother solves the limit case exactly.

This is the substantial requirement a robust smoother has to satisfy:

$$\|\|N\|\| = \mathcal{O}(\varepsilon), \tag{14.52}$$

i.e. the iteration becomes exact in the limit case.

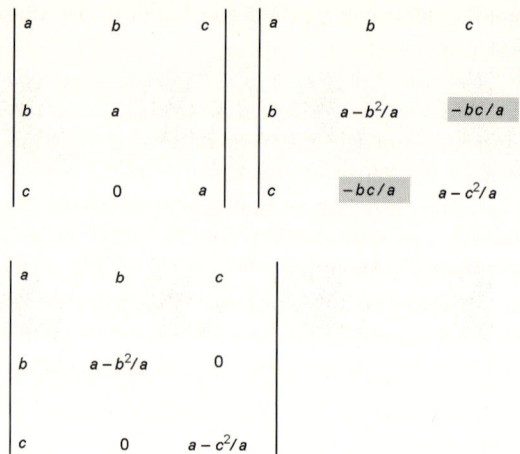

Fig. 14.8. Matrix, fully eliminated matrix and the corresponding incompletely eliminated one $(abc \neq 0)$. In the fully eliminated matrix fill-in is indicated by grey background

Replacing the symmetric GAUSS-SEIDEL smoother (SGS) in the multi-grid method in 14.3.1 by a 5-point ILU and applying this to the same test problem we obtain the results shown in 14.9. Apparently, this smoother improves the results by several orders of magnitude for small ε, which is due to the fact that ILU solves the limit systems exactly. Further, the convergence rate is bounded in $\kappa(\varepsilon)$. This means the method is robust for that problem

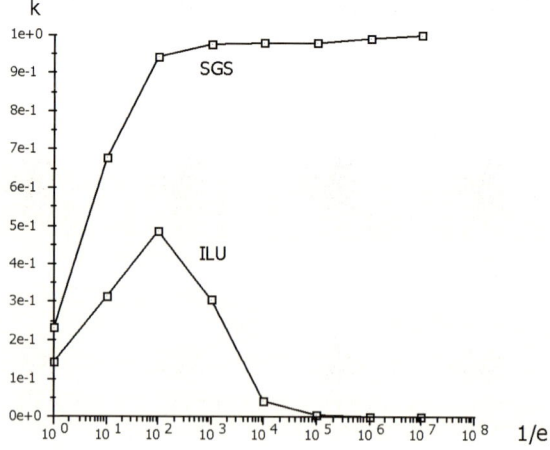

Fig. 14.9. Convergence rate κ of the multi-grid method from Sect. 14.1.1 applied to problem (14.45/14.47) averaged over 10 steps versus $1/\varepsilon$.

and fixed stepsize. In general we call a method with convergence rate $\kappa(\varepsilon)$ robust if and only if

$$\kappa(\varepsilon) \leq \kappa < 1 \quad \forall \varepsilon \geq 0. \tag{14.53}$$

This is still not possible with the simple incomplete decompostion mentioned above, since the peak shown in 14.3.1 tends to one for $(\varepsilon, h) \to (0,0)$. The reason is that the eigenvalues of the ILU smoother tend towards -1 for $(\varepsilon, h) \to (0,0)$.

To obtain robustness we have to ensure two conditions:

– The smoother has to solve the limit case exactly,
– The spectrum of the smoother has to be bounded away from -1 uniformly in ε.

To obtain the latter we should not use damping, since this spoils the first property. Here modification is preferable in connection with incomplete decompositions. Modification means that within the incomplete decomposition stage we add the modulus of neglected elements multiplied by some parameter β to the diagonal entry.

A modified incomplete decomposition of the sparse matrix $A = (a_{ij})_{i,j=1,\ldots,n}$ using the pattern $P \subset \{(i,j) : 1 \leq i,j \leq n\}$ is given by:

algorithm 14.3. : Modified ILU

```
void ilu_mod(A,P,β,n)
    matrix A; pattern P; real β; integer n;
    { integer i,j,k; real pivot, corr;
        for (i = 0, i < n; i++)
        { for (j = i+1, j<n; j++)
            if (a_ji != 0)
            { pivot = a_ji/a_ii;
                for (k = i+1; k<n; k++)
                if (a_ik != 0)
                { corr = pivot* a_ik ;
                    if ((j,k) ∈ P)
                        a_jk = -corr;
                    else
                        a_jj = a_jj + β |corr|;
                }
            }
        }
    }.
```

As proved in [14.15], for $\beta > 0$ this method yields an incomplete decomposition satisfying (14.52) and providing a bound $-\vartheta > -1$ on the eigenvalues of the smoother uniformly in ε as required in Theorem 1. In many practical computations this parameterβ has shown itself to be very important to obtain good efficiency.

14.3.2 Robustness for Convection-Diffusion Problems

Now we consider the convection-diffusion problem

$$-\varepsilon \Delta u + c\nabla u = f \ in\Omega \qquad u = u_R \ \text{on} \ \partial\Omega_l, \qquad (14.54)$$

with a convection field c and $\varepsilon > 0$.

We discretize the diffusive terms by some standard method e.g. finite volumes or finite elements, the convective terms are discretized by a simple first-order upwind scheme. In the limit case $\varepsilon \to 0$, pure convection will be left. Following Chap. 14.3.1 we have to provide a smoother which solves the limit problem exactly. Such a smoother can be constructed easily by renumbering the unknowns in convection direction. The algorithm called **downwind numbering** which is described in what follows, has been introduced in [14.25].

Let us switch off the diffusion for the time being. Due to the upwind scheme a given node depends only on its upwind neighbours. If we can find a global ordering of the unknowns in a way that the stiffness matrix has non-zero entries only in the lower triangle, then of course we will be able to solve the system of equations in one step even by a Gauß-Seidel method. Unfortunately in most of the relevant cases there are vortices in the flow and therefore cyclic dependencies. But nevertheless we will obtain good results if we introduce arbitrary cuts through the vortices by removing just enough of the "cyclic" nodes to get rid of the cylic dependencies. We start the numbering at the inlet going in layers downstream but taking only nodes depending on the already numbered ones (those nodes will form the beginning of our new list). In a similar way we go upstream from the outlet (those nodes will make up the end of our new list). Finally we are left with nodes with cyclic dependencies. We cut it, appending those nodes to the beginning of our list. Then steps one to three are repeated until every node is processed.

After this rough description we introduce the following algorithm that does the job (see [14.25], [14.26]):

algorithm 14.4. : Downwind Numbering
  while (some nodes are not numbered)

{
    /* find FIRST set */
    do {
        Find all nodes having at most such UPWIND neighbour nodes that
        are already numbered.

        Number them starting with the least number not used yet.
        } while (no further nodes are found).

    /* find LAST set */
    do {

Find all nodes having at most such DOWNWIND neighhour
nodes that are already numbered

Number them starting with the greatest number not used yet.

} while (no further nodes are found).

/* find CUT set (only cyclic dependencies are left) */

Cut one vortex transverse to the streamlines.

Number the nodes on this cut starting with the smallest number not
used yet.

}

Example 1. :
We consider the downwind-numbering algorithm for the backward facing step
flow field cf. Fig. 14.10.

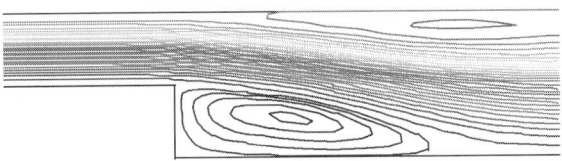

Fig. 14.10. Flow field over a backward facing step ($Re = 400$).

We consider the downwind-numbering algorithm for the backward facing
step flow fieldStages of the downwind-numbering algorithm: cf. Fig. 14.11.

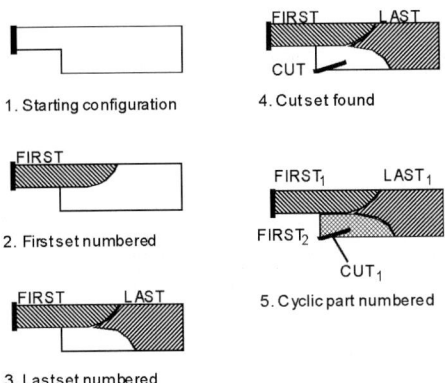

Fig. 14.11. Stages of the downwind numbering algorithm.

The corresponding stiffness matrix has the following pattern (Fig. 14.12):

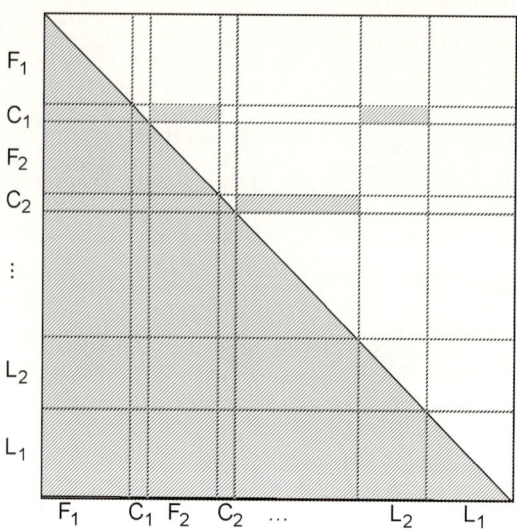

Fig. 14.12. Sparsity pattern of the corresponding stiffness matrix.

Example 2. :
Let us turn back to the full equations. Based on downwind numbering we perform an ILU decomposition and use this as a smoother. Numerical results are discussed in the following examples cf. Fig. 14.13(a)–(c).

Example 3. :
We consider a problem which is convection-dominated only in part of the region. There we really need multi-grid to treat the diffusion dominated parts. Using the smoother as a solver we will not get reasonable convergence. For simplicity let Ω be the unit square. To test this effect we choose c as follows

$$c = \left(1 - \sin[\alpha]\right) \left[2\left\{x + \frac{1}{4}\right\} - 1\right] 2\cos[\alpha] \left[y - \frac{1}{4}\right]\right)^4 (\cos[\alpha], \sin[\alpha])^T$$

$$(14.55)$$

where α is the angle of attack. The boundary conditions are: $u = 0$ on $\{(x,y) : x = 0, \ 0 \le y \le 1\} \cup \{(x,y) : 0 \le x \le 1, \ y = 1\} \cup \{(x,y) : x = 1, \ 0 \le y \le 1\} \cup \{(x,y) : 0 \le x < 0.5, \ y = 0\}$ and $u = 1$ on $\{(x,y) : 0.5 \le x \le 1, \ y = 0\}$. The jump in the boundary condition is propagated in direction a. We have $div\ c = 0$ and c varies strongly on Ω such that the problem is convection-dominated in one part of the region and diffusion-dominated in another part of it.

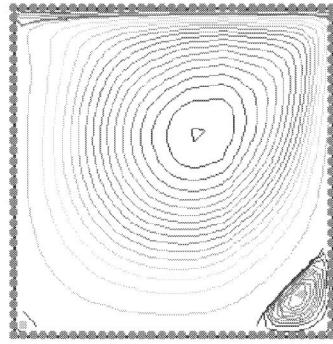

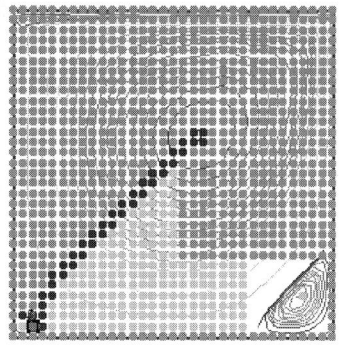

(a) FIRST$_1$ set (marked nodes), LAST$_1$ is empty

(b)   CUT$_1$ (black),   FIRST$_2$ (medium gray), LAST$_2$ (light gray)

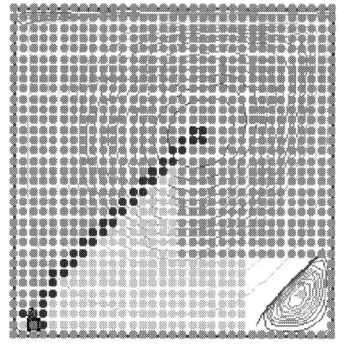

(c)   CUT$_2$ (black),   FIRST$_3$ (gray), LAST$_2$ is empty

Fig. 14.13. Driven cavity flow.

As discretization we use a finite volume scheme with first order upwinding for the convective terms on a triangular grid. The grid is refined adaptively using a gradient refinement criterion. As a smoother we took a Gauß-Seidel scheme with downwind numbering using Algorithm 14.4 in a $(1, 1, V)$-cycle local multi-grid. It is important to note that the smoother itself is not an exact solver. Thus we should see the benefit of multi-grid in the diffusion-dominated part and of the robust smoother in the convection-dominated one. This is confirmed by the results given in Fig. 14.14. There we show the residual convergence rate averaged over 10 steps for problem 14.54 on adaptively refined unstructured grids versus $1/\varepsilon$.

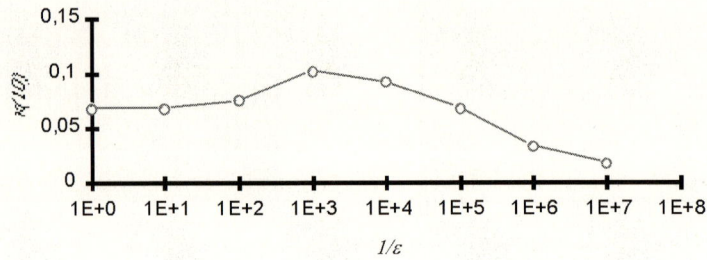

Fig. 14.14. Robustness of a $(1, 1, V)$-mmgm with ILU-smoother and downwind numbering. The method used 8 locally refined grids to discretize problem 14.54 with over 10.000 unknowns on level 8. The convergence rate $\kappa(10)$ is averaged over 10 steps and refers to finest grid

Unfortunately there is no theory for the robustness of this smoother. This is due to the nonsymmetry of the problem. Obtaining sharp estimates is very difficult there.

14.4 Treatment of Systems of PDE

Now we consider a system of partial differential equations denoted by

$$K = \begin{pmatrix} K_{11} & K_{12} \\ K_{21} & K_{22} \end{pmatrix} \tag{14.56}$$

Typically the blocks represent differential operators with different order. For example the Stokes operator reads

$$K_{\text{Stokes}} = \begin{pmatrix} -\Delta & \nabla \\ \nabla^T & 0 \end{pmatrix} \tag{14.57}$$

The construction of a smoother for such a saddle-point problem is not obvious, since all the standard ones described above need non-zero diagonal entries. To cope with this problem we use transforming smoothers, introduced in [14.27].

We construct an auxiliary matrix $\bar{K}$ which is supposed to be regular such that the product $K \bar{K}$ can be used for the construction of a splitting

$$K \bar{K} = M - N \tag{14.58}$$

The corresponding iterative scheme to solve $Kx = b$ is then given by

$$x_{\text{new}} = x_{\text{old}} - \bar{K}M^{-1}(K\,x_{\text{old}} - b). \tag{14.59}$$

In the case of Stokes equations this can be obtained choosing

$$\bar{K}_{\text{Stokes}} = \begin{pmatrix} I & D\nabla \\ 0 & I \end{pmatrix} \tag{14.60}$$

where D is an approximation to Δ^{-1}. Then the product system reads

$$K\bar{K}_{\text{Stokes}} = \begin{pmatrix} -\Delta & \nabla - \Delta D\nabla \\ \nabla^T & \nabla^T D\nabla \end{pmatrix} \tag{14.61}$$

For $D = \Delta^{-1}$ this is lower block triangular. In practice, however, we use an approximation but still neglect the entry in the upper left corner. Choosing

$$D = diag\,(\Delta)^{-1} \tag{14.62}$$

we obtain the SIMPLE method by Patankar and Spalding, [14.28]. Other variants of this widely used family of methods can be obtained by choosing more sophisticated approximations for D.

In the multigrid context we use this scheme as a smoother, combining it with an incomplete decomposition for the scalar equations. This can be shown to satisfy the smoothing property (see [14.29]) and yields robust convergence as shown in the following example.

Example 4. :
We compute the flow through a pipe and the one across a backward facing step at $Re = 100$. Then we plug the corresponding velocity fields u^0 into the Navier-Stokes equations:

$$-\Delta u + Re(\lambda u^0 \nabla)u + \nabla p = 0 \qquad \nabla^T u = 0 \tag{14.63}$$

Then we use the downwind numbering algorithm from Chap. 14.3.2 and apply a SIMPLE scheme with an ILU smoother for the scalar subproblems. We obtain the following convergence rates shown in Fig. 14.15. Apparently the resulting multigrid method is quite robust.

14.5 Adaptive Multigrid

Applying multi-grid methods to problems on locally refined grids one has to think about how to associate grid points with levels in the multi-grid hierarchy. Consider the hierarchy of grids Ω_l, $l = 0, \ldots, l_{\max}$. Early multi-grid approaches smooth all points in Ω_l. This may cause a non-optimal amount of work and memory of $\mathcal{O}(n \log n)$ per multi-grid step. This problem was the

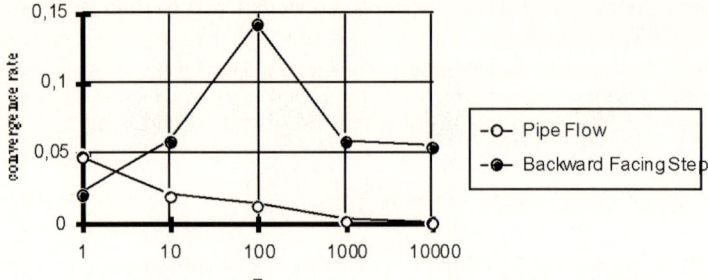

Fig. 14.15. The convergence rate of a multigrid method with downwind numbered SIMPLE-*ILU* smoother for problem (14.63).

starting point for Yserentant, [14.30], and Bank-Dupont-Yserentant, [14.31], to develop the method of hierarchical bases (HB) and the hierarchical basis multi-grid method (HBMG). These were the first multi-grid methods with optimal amount of work per step for locally refined grids. This is due to the fact that on level lonely the unknowns belonging to points in Ω_l/Ω_{l-1} are treated by the smoother. However, the convergence rate deteriorates with $\log n$. For the first time this problem was solved by the introduction of the additive method by Bramble, Pasciak and Xu, [14.32], (BPX). There on level l the smoother treats all the points in Ω_l/Ω_{l-1} and their direct neighbours, i.e. in all points within the refined region. Table 14.4 gives an overview of the multi-grid methods used for the treament of locally refined grids and classifies the variant we call "local multi-grid". The methods mentioned above differ in the smoothing pattern, i. e. the choice of grid points treated by the smoother. The methods in the first two lines are of optimal complexity for such problems. The amount of work for one step is proportional to the number of unknowns on the finest grid. However, only the methods in the second line, BPX and local multi-grid converge independently of h for scalar elliptic problems. The basic advantage of the multiplicative methods is that they do not need cg-acceleration and thus can be directly applied to unsymmetric

Table 14.4. Multi-grid methods for locally refined grids.

smoothing pattern	basic structure	
	additive	multiplicative
(1): new points only	hierarchical basis (HB) Yserentant, 1984, [14.30]	HBMG, Bank & Dupont Yserentant, 1987, [14.31]
(2): refined region only	BPX, Bramble & Pasciak & Xu, 1989, [14.32]	local multi-grid [14.33], [14.34], [14.35]
(3): all points	parallel multi-grid Greenbaum, 1986, [14.36]	standard multi-grid

problems; further they show a better convergence rate and on a serial computer the additive process does not have any advantage. The local multi-grid scheme is the natural generalisation of the classical multi-grid method to locally refined grids, since in case of global refinement, it is identical with the standard classical multi-grid method.

The local multi-grid was first analyzed in 1991 by Bramble, Pasciak, Wang and Xu, [14.33]. They considered it a multiplicative variant of their so-called BPX-method. However, they did not think about robustness. Further predecessors of this method have existed for a couple of years, [14.34]. Without knowledge of these two of the authors developed this method as a variant of standard multi-grid based on the idea of robustness (cf. [14.35]). The main advantage of this approach is that the application to unsymmetric and non-linear problems is straightforward (cf. [14.35]).

A simple analysis of the hierarchical basis methods (HB, HBMG) shows that the smoothing pattern is too poor to allow robust smoothing.

Remark 1. :
The hierarchical basis method and the hierarchical basis multigrid method do not allow robust smoothing for a convection-diffusion equation. The smoothing pattern used in these methods does not allow the smoother to be an exact solver for the limit case. This holds for uniformly as well as for locally refined grids.

Based on this observation, we extended the smoothing pattern, adding all neighbours of points in Ω_l/Ω_{l-1}. This yields the local multigrid method, allowing the smoother to solve the limit case exactly, provided the grid refinement is appropriate. This is confirmed by numerical evidence in practical computations. A more detailed treatment of this topic can be found in [14.35].

References

[14.1] Varga, R.S.: Matrix iterative analysis. Prentice Hall, 1962

[14.2] Gauß, C. F.: letter to Gerling from 25. of december 1823

[14.3] Jacobi, C.G.J.: über eine neue Auflösungsart der bei der Methods der kleinsten Quadrate vorkommenden linearen Gleichungen. Astronom. Nachrichten 32 (1845), 297-306.

[14.4] Young, D.: Iterative methods for solving patrial difference equations of elliptic type. Thesis Harvard University (1950)

[14.5] Stiefel, E.: über einige Methoden der Relaxationsrechnung. Z Angew. Math. Phys. 3 (1952) 1-33.

[14.6] Fedorenko, R.P.: Ein Relaxationsverfahren zur Lösung elliptischer Differentialgleichungen. (russ.) UdSSR Comput Math Phys 1,5 (1961), 1092-1096

[14.7] Brandt, A.: Multi-level adaptive technique for fast numerical solution to boundary value problems. Proc. 3rd Internat. Conf. on Numerical Methods in Fluid Mechanics, Paris, 1972. Lecture Notes in Physics 18, Springer, Heidelberg, (1973), 83-89.

[14.8] Hackbusch, W.: Ein iteratives Verfahren zur schnellen Auflösung elliptischer Randwertprobleme. Report 76-12, Universität Köln (1976).

[14.9] Hackbusch, W.: Survey of convergence proofs for multi-grid iterations. In: Frehse-Pallaschke-Trottenberg: Special topics of applied mathematics, Proceedings, Bonn, Oct. 1979. North-Holland, Amsterdam 1980, 151-164

[14.10] Hackbusch, W.: Multi-grid methods and applications. Springer, Berlin, Heidelberg (1985).

[14.11] Xu, J: Iterative methods by space-decomposition and subspace correction. SIAM Review, 34(4), 581-613, 1992.

[14.12] Braess, D.: The convergence rate of a multigrid method with Gau -Seidel relaxation for the Poisson equation. In: Hackbusch-Trottenberg: Multigrid Methods. Proceedings, K ln-Porz,1981.Lecture Notes in Mathematics, Bd. 960, Springer, Heidelberg,1982

[14.13] Bank, R.E., Dupont, T.: An optimal order process for solving elliptic finite element equations. Math. Comp. 36, 35-51, 1981

[14.14] Braess, D., Hackbusch, W.: A new convergence proof for the multigrid method including the V-cycle. SIAM J Numer. Anal. 20 (1983) 967-975

[14.15] Wittum, G.: On the robustness of ILU-smoothing. SIAM J. Sci. Stat. Comput., 10, 699-717, (1989)

[14.16] Wittum, G.: Linear iterations as smoothers in multi-grid methods. Impact of Computing in Science and Engineering, 1, 180-215 (1989).

[14.17] Reusken, A.: The smoothing property for regular splitting. In: W. Hackbusch and G. Wittum, editors, Incomplete Decompositions (ILU) - Algorithms, Theory, and Applications, p. 130-138, Vieweg, Braunschweig, 1992.

[14.18] Wittum, G.: Spektralverschobene Iterationen. ICA-Bericht N95/4, Universität Stuttgart, Mai 1995.

[14.19] Hackbusch, W.: A note on Reusken's Lemma. To appear in Computing, 1994.

[14.20] Ecker, A., Zulehner, W.: On the smoothing property for the non-symmetric case. Institutsbericht Nr. 489, Institut f r Mathematik, Johannes Kepler Universität Linz, April 1995.

[14.21] Xu, J: Iterative methods by space-decomposition and subspace correction. SIAM Review, 34(4), 581-613, 1992.

[14.22] Bramble, J.H., Pasciak, J.E., Xu, J.: Parallel multilevel preconditioners. Math. Comput., 55, 1-22 (1990).

[14.23] Yserentant, H.: Old and new convergence proofs for multigrid methods. Acta Numerica, 1993.

[14.24] Stüben, K., Trottenberg, U.: Multigrid methods: fundamental algorithms, model problem analysis and applications. In: Hackbusch,W., Trottenberg,U.(eds.): Multi-grid methods. Proceedings, Lecture Notes in Math. 960, Springer, Berlin (1982).

[14.25] Bey, J., Wittum, G.: Downwind Numbering: A Robust Multigrid Method for the Convection-Diffusion Equation on Locally Refined Grids. In: Hackbusch, W., Wittum, G. (eds.): Adaptive Methods - Algorithms, theory and Applications, NNFM 46, Vieweg, Braunschweig, 1994

[14.26] Reichert, H., Wittum, G.: Robust multigrid methods for the incompressible Navier-Stokes equations. ICA-Bericht N94/2, Universität Stuttgart, November 1994.

[14.27] Wittum, G.: Multi-grid methods for Stokes and Navier-Stokes equations. Transforming smoothers - algorithms and numerical results. Numerische Mathematik, 54 , 543-563(1989)

[14.28] Patankar, S.V., Spalding, D.B.: A calculation procedure for heat and mass transfer in three-dimensional parabolic flows. Int. J. Heat Mass Transfer, 15 (1972), 1787-1806

[14.29] Wittum, G.: On the convergence of multi-grid methods with transforming smoothers. Theory with applications to the Navier-Stokes equations. Numer. Math. 57:15-38 (1990)

[14.30] Yserentant, H.: Über die Aufspaltung von Finite-Element-R umen in Teilräume verschiedener Verfeinerungsstufen. Habilitationsschrift, RWTH Aachen, 1984

[14.31] Bank, R.E., Dupont, T., Yserentant, H.: The hierarchical basis multigrid method. Numer. Math. 52, 427-458 (1988)

[14.32] Bramble, J.H., Pasciak, J.E., Xu, J.: Parallel multilevel preconditioners. Math. Comput., 55, 1-22 (1990).

[14.33] Bramble, J.H., Pasciak, J.E., Wang, J., Xu, J.: Convergence estimates for multigrid algorithms without regularity assumptions. Math. Comp. 57, 23-45 (1991)

[14.34] Rivara, M. C.: Design and data structure of a fully adaptive multigrid finite element software. ACM Trans. on Math. Software, 10, 242-264 (1984)

[14.35] Bastian, P., Wittum, G.: On robust and adaptive multigrid methods. In: Hemker, P., Wesseling, P. (eds.): Multigrid methods, Birkh user, Basel, 1994

[14.36] Greenbaum, A.: A multigrid method for multiprocessors. Appl. Math. Comp. 19, 75-88 (1986)

Index

β-cyclodextrin-binaphtyl   187
11-cis-retinal chromophore   191, 192

ab initio molecular dynamics simulations   190
absorption spectrum   88, 92
adaptive multigrid   307
adatoms
– Co on Cu(111)   170
– motion   172
additive multi-grid method   290
ADE-FDTD algorithm   96
AL clusters   201
AMBER   179
annihilation operator   25
APW   40
ASW   40
atomic relaxation   166, 167, 173
atomistic simulation   226
Aufbau principle   10
auxiliary difference equation   93
auxiliary differential equation   89

B matrix   209
Bakers scheme   209
band gap
– photonic   56
band structure   17
– aluminium   18
– calculation   22
– Cu   116
– electronic   102, 115
– MgO   48
– photonic   56–58, 71, 102, 120
basis functions   109
– symmetry adapted   118
beamsplitter   69, 71, 72
bend geometries   67

Bernoulli beam   252
Bernoulli's hypotheses   251
Bessel functions   69
biological water   180
biomolecules   179
Birch-Murnaghan equation   218
Bloch condition   80
Bloch equations
– optical   76
Bloch functions   58, 60, 68, 113, 115
Bloch theorem   40, 112, 113
Bloch-Floquet theorem   57, 58
blocking operations   265
BLYP gradient-corrected exchange-correlation   181
Born-Oppenheimer approximation   180
Born-Oppenheimer surface   133, 180
bosons   24
bovine rhodopsin   189
Bragg resonance   56
Brillouin zone   58, 114, 118, 120
– fcc   115
– square lattice   121
broadcast   265
bulk modulus   218

Car-Parrinello MD simulations   181
CASTEP   207, 213, 218
catalyst   208
catalytic reactions   211
cavity
– frequencies   62
– localized modes   62
character   107
character projection operator   121
character table   108
– C_{4v}, C_{2v}, C_{1h} 121

– O_h 117
characters
– orthogonality theorems   108
CHARMM package   179, 192
chemical potential   28
chemical reaction   193
chemical vapor deposition   193
chromophore of rhodopsin   179
class   106
classical molecular dynamics   131
classical pairwise additive potential
    see CPAP
clusters   130
– Co_9 169
– CuO_n 135
– H_2O_n 185
– mixed Co-Cu   166
coarse grid operator   297
coarse-grid correction matrix   295
collective communication   265
COMPASS   226
COMPASS force-field   230
compatibility relations   117, 118, 120
computational costs   90
computer simulation   75
conjugate element   106
conjugate gradient refinement   213
conjugate-gradient technique   286
constitutive
– axioms   241
– equations   240, 241
– relations   242
continuum
– damage mechanics   245
– mechanics   240
convergence properties   64
convergence theory   291
core
– orbitals   15, 42
– states   12
Courant condition   80
CPA   20, 196
CPAP   184
creation operator   25
creep   238
– behavier   242, 243
– damage   246, 249
– effects   238
– microstructural   238

– modelling   238
– multi-axial model   239
– non-classical behavier   246
– potential   247
CuO_n clusters   135
cyclodextrin
– interacting with bulk water   180
cylindrical harmonics   68

damage variable   246
damped harmonic oscillator equation
    87
damping   295
deadlock   265
deductive approach   241
defect clusters   64
defect structures in photonic crystals
    59
delocalized coordinate method   209
delocalized coordinate optimizer   210
density
– of ground state   7
– polarisation   76
– representation   14
– spectral   28
density functional theory   *see* DFT
density-functional based tight-binding
    approach   179
DFT   7, 34, 160, 218
DFTB   181, 183, 186, 187, 190, 192
dielectric constant   57, 76, 82, 90
dielectric function   61
Diels-Alder reaction   216
diffusion coefficient   181
diffusivity   172
dimers   164
dipolar interaction hamiltonian   87
direct sum   107
discover molecular dynamics   230
disordered network-forming materials
    146
disordered networks   130
dispersion relation   65
– for straight waveguides   62
– photonic   58
distributed address space   274
$DMol^3$ 207, 209, 210, 213, 215
DOS
– aluminium   17

– photonic   58
double layer islands   172, 173
double Legendre transform   9
downwind numbering   302
– algorithm   302, 303
DPD   229
Dyson equation   32, 33, 160

EAM potential   196, 198, 202
effective potential   35, 42
eigenfrequencies   57
eigenmodes   57, 68
eigenvalue problem   39, 44
electron gas
– 2D   169
energy
– exchange-correlation   34
– excitation   33
– magnetic anisotropy   166, 168
– of ground state   7
– selfenergy   11
– total   34
equation of motion   26, 31, 177
evanescent guided modes   65, 66
exchange-correlation energy   34
excitation energy   33
external potential   7

fabricational tolerances   72, 73
FDTD   75, 76, 84, 98
– method   66, 67
fermi surface
– aluminium   18
fermions   24
field distribution   70–72
field operator   25
finite element discretization   59
finite element method   66, 67, 231
finite photonic crystals   68
finite-differnce time-domain   see
   FDTD
first-principles molecular dynamics
   131, 133
five-point formula   285
Flory-Huggins interaction parameter
   229
Flory-Huggins theory   230
force field   226
force-field expression   226

FPLAPW   40
FPLO   11, 15, 19
frequencies of localized cavity modes
   62
Friedel oscillations   170
full elastic constant tensor   218
full-potential combined basis method
   40
full-potential local-orbital minimum-
   basis method   40
fullerenes
– Si-doped   132

G-protein-coupled receptors   190
Galerkin ansatz   297
Gauß-Seidel method   268–270, 273,
   284, 286, 293, 298, 299
Gauß algorithm   283
Gaussian basis method   49
Gaussian thermostat   178
generalized gradient approximation
   134, 218
generators   106
genetic function approximation   225
geometry optimization   207, 208
GGA   10, 218
Glu113   190
Glu181   192
Green's function   23, 26, 28, 30, 31, 33,
   35, 69
– causal   29
– single particle   11
– structural   161
Green's function method   160
Green's tensor   68
ground state
– density   7
– energy   7
group
– abelian   106
– class   106
– conjugate element   106
– generators   106
– homomorphism   106
– isomorphism   106
– matrix representation   106
– pointgroup   112, 114
– spacegroup   112
– subgroup   106

group theory   103, 105
– computer algebra package   104
group velocity   56, 58, 59
growth process   179
guided modes   65
– evanescent   65, 66
– propagating   65, 66
GW approximation   47

H_2 193
Hall-Petch effect   198
– reverse   198
hamiltonian matrix   13, 44
Hankel functions   69
heat bath mass   178
Heaviside-Lorentz form   76
Heaviside-Lorentz units   89
Hedin equation   32, 47
Hellmann-Feynman theorem   133
heterofullerenes
– Si-doped   140
high spin ferromagnetic state   164
Hohenberg   8
HOMO   143
homoepitaxial growth   172
homoepitaxy   172
homomorphism   106
Hookean tensor   236
Hookes law   236, 244, 251
hopping diffusion   173
hopping rate   172
HPF   261
HSF   165
hydrogen bonded network   184, 192

ILU   299–301
– modified algorithm   301
– scheme   293
imperfect nanostructures   163
interaction energies   170
interaction parameter   161
intrinsic reaction coordinate   215
inversion density   88, 97, 98
islands   166
– double layer   172, 173
isomers   141
isomorphism   106

Jacobi method   268

Janak's theorem   10

kinetic Monte Carlo   179
KKR   40, 160, 167, 196
Kohn   8
Kohn anomalies   197
Kohn-Sham
– equations   10, 12, 35, 39, 42, 103, 133
– orbitals   9, 12
– parameterization   9
Korringa-Kohn-Rostoker method   see
    KKR

Lagrangian   133
LAM-MPI   263
LAPW   49
lattice
– harmonics   119
– periodicity   111
– reciprocal   114
– vector   112
LCAO   41
LDA   10, 35
LDOS   68, 169
Legendre transform   9
Lindemann's criteria of melting   185
linear and quadratic synchronous
    transit (LST/QST) method   213
liquid $SiSe_2$ 132
LMTO   40, 50
local coupling strength   68
local density approximation   161
local density of states   see LDOS, 169
local orbital representation   12
local strain field   172
local valence basis   43
local-density approximation   see LDA
localized basis functions   60
localized cavity modes   62
Lorentz line shape   87
low spin ferromagnetic state   164
LSDA   10
LSF   165
LU decomposition   284, 299
LUMO   143
Lys296   189, 190

macroscopic mismatch   171, 172
magnetic anisotropy energy   166, 168

magnetic orbital moments   168
magnetism of Rh nanostructures   163
magnetization
– noncollinear   169
– orientation   167
many-body
– hamiltonian   24
– potentials   161
– wave function   24
martensite   200
martensitic transformation   198
material behavior models   244
Materials Studio®   224
Mathematica
– group theory package   104
matrix eigenvalue problem   58
matrix element theorem   115
matrix multiplication
– Strassen   278
matrix representation   106
maximally localized Wannier functions
   60
Maxwell curl equations   76, 78
Maxwell equations   75, 89, 110, 111
mechanical behavior   235
melting temperature   186
MesoDyn   228, 230, 231
mesoscale   171
mesoscale methods   227
mesoscale simulation   223
mesoscopic misfit   171
mesoscopic mismatch   172, 173
mesoscopic relaxation   172
mesoscopic strain   172
Message Passing Interface   261
metallic nanostructures   159
metamagnetic states   164
micro-mechanical model   238
microdisk   96
microdisk laser   95, 98
microdisk resonator   95, 98
minimum energy path   215
mismatch
– macroscopic   171, 172
– mesoscopic   172, 173
– size dependent   172
mixed configurations   166
modeling   57
modelling of materials   238

models
– material behavier   244
– micro-mechanical   238
molecular dynamics   160, 177
– classical   131
Monte Carlo   179
– kinetic   179
Moore's law   75
MPI   261
– PI_Bcast   265
– blocking operations   265
– broadcast   265
– deadlock   265
– introduction   263
– MPI_Allgather   266
– MPI_Comm_rank   264
– MPI_Comm_size   264
– MPI_COMM_WORLD   264
– MPI_Finalize   264
– MPI_Init   264
– MPI_Recv   263
– MPI_Reduce   266
– MPI_Send   263
– Tlib library   275, 278, 279
MPICH   263
Mulliken population analysis   145
multi-grid algorithm   288
multi-grid methods   58, 282, 286
– adaptive   307
– additive   290
– convergence   291, 297
– historical overview   283
– robustness   297, 302
multiplication table   106
multiplicity of magnetic states   165
multipole expansion   72
multipole expansion technique   68
multipole method   68
multiscale modeling   230
Myrinet   261

nanomagnetism   159
nanostructures
– imperfect   163
– metallic   159
Navier-Stokes equations   307
Newton element   243
Newtonian equations of motion   177
noncollinear magnetization   169

nonlinear coupling   87
norm-conserving Troullier-Martins
   pseudopotential   181
Norton's law   245, 247
Nosé thermostat   192
Nosé-Hoover thermostat   135, 178
NPT ensemble   178
nudged elastic band method   215
numbering
– lexicographically   285
Nyquist criterion   80

one-electron wave function   42
OpenMP   261
operator
– annihilation   25
– creation   25
– field   25
– self energy   31
– time-ordering   27
– Wick's time ordering   26
optical Bloch equations   98
optimized bend design   69
optimized potential   163
OPW   41–43
orbital moment   169
ordering
– red-black   272
orthogonality theorems for characters
   108
overlap matrix   13, 44
oxide clusters   132

pair correlation function   149
Palmyra-GridMorph   231
paralleliziation   261
Parinello-Rahman scheme   178
partial neutron structure factor   151
PAW   40
perfect matching layers   80
phase diagram   199
– Fe-Ni   196
– Ni-Mn-Ga   196
phase shift   87
phase shift spectra   92
phonon dispersion relation   197
photonic band gap   56
photonic bandstructure   56, 57, 71
photonic crystals   55, 111

– circuits   66
– defect structures   59
– finite   68
– optical properties   57
photonic dispersion relation   58
photonic nano-materials   75
photonics   55
plane wave methods   49, 58
plane waves
– symmetry adapted   118
plane-wave basis set   134
pointgroup   106, 112, 114
Poisson equation   270, 285
Poisson problem   272
polarisation density   76
population analysis   145
potential
– chemical   28
– effective   35, 42
– external   7
– many-body   161, 163
– optimized   163
– representation   14
– TIP3P   185
– TIP4P   185
potential energy surface   177, 195
projection operator   109, 123
– character   121
propagating guided modes   65, 66
protein folding   179
protein-solvent interaction   180
pruning algorithm   209
pseudopotential
– norm-conserving Troullier-Martins
   181
pseudopotential methods   41
Pthreads   261, 276
PVM   261

QM/MM technique   179, 190
quantitative structure activity
   relationships   225
quantitative structure property
   relationships   225
quantum interference   160, 169
quantum state   8
quasi-particle   30

Rabi oscillations   93, 98

radial distribution function   181, 199
radial Schrödinger equation   43
random phase approximation   228
RDF   182, 183, 188
real space approach   58
reciprocal lattice   114
red-black ordering   272, 273, 275
relaxation
– atomic   166, 167
– mesoscopic   172
representation
– irreducible   107, 117, 118
– reducible   107, 120
representation theory   106
resonance frequency   87
retina   190
reverse Hall-Petch effect   198
rheological modelling   243, 245
rhodopsin   179, 189
– with internal water   180
Ritz method   248
robustness   297, 302
– anisotropic problems   297

scanning tunnelling microscope   170
Schrödinger equation   110, 111, 118
SCI   261
second moment tight-binding approxi-
    mation   161
second quantisation   25
secular problem   13
Seitz-operator   112
self-energy   11, 32, 34, 47
self-energy operator   31
semi-coarsening   299
shear
– force   251
– stress   250, 252
Si-doped fullerenes   132
Si-doped heterofullerenes   140
SiCl₄   193, 195
single particle Green's function   11
smoothing   286, 293, 297, 298
SMP   261
solid state reactions   179
solvent   188
SOR method   268, 271, 286
spacegroup   112
spacetime method   50

sparse iteration matrices   269
sparse matrix   271
spectral density   28
spectral function   29, 30
speedup   262
spherical harmonics
– symmetry adapted   119
spin magnetic moment   168
spin moment   167
spin-orbit coupling   167
SPMD   263, 268, 276
square   120
– symmetry   105
St. Venant element   243
state
– antiferromagnetic   164, 165
– ferromagnetic   165
– high spin ferromagnetic   164
– low spin ferromagnetic   164
step edges   173
STM   170
Stokes equations   307
Stokes operator   306
strain   171
strain relaxations   171
strain relief   173
strain variation   242
Strassen matrix multiplication   278
stress   171, 236
– function   242
– history   242
– shear   250, 252
structural Green's function   161
structural transformation   198
structure optimization   187
subgroup   106
super-cell calculation   62
surface alloying   166
surface-state electrons   169
surface-state fermi wavelength   170
susceptibility   87
symmetric multi-processors   261
symmetry
– adapted
– – basis functions   118
– – plane waves   118
– – spherical harmonics   119
symmetry group   105

t matrix   161
tetramers   164
The Flory-Huggins interaction
    parameter   230
thin-walled structures   248, 249
Thr94   190
tight-binding approximation   161
tight-binding representation   40
time correlation function   181
time-ordering operator   26, 27
TIP3P potentials   185
TIP4P potentials   185
Tlib library   275
TM-polarized radiation   57
total energy   34
total neutron structure factor   146,
    148
transition state   207
– confirmation algorithm   214
– searching   213
– theory   193
transmission spectrum   67
trimers   164, 165
two level atomic system   87

U-PML boundary condition   81

V-cycle   292, 293, 295
valence basis orbitals   15
valence orbitals   42

van Hove singularities   59
variational methods   39
variational principle   8–10
VASP   181, 190, 193, 197
vertex function   47
visual pigment   189
von Mieses   239, 246

W-cycle   293
Wannier functions   60, 65, 66, 71
– maximally localized   60
– optimized   60
– photonic   60
water   180, 181, 184
– biological   180
– clusters$(H_2O)_n$ 183
wave equation   57, 61, 62
wave function
– one electron   42
waveguides   64
Wavelength Division Multiplexing   66
whispering gallery mode   95–97
Wick's time-ordering operator   26
WIEN97   19

Yee   76
Yee algorithm   78
Yee cube   77, 79
Young's modulus   236